Technical Report Writing Today

TENTH EDITION

Technical Report Writing Today

Daniel G. Riordan
Emeritus Professor of English
University of Wisconsin–Stout

WADSWORTH
CENGAGE Learning®

Australia • Brazil • Japan • Korea • Mexico • Singapore • Spain • United Kingdom • United States

**Technical Report Writing Today,
Tenth Edition**
Daniel G. Riordan

Publisher: Michael Rosenberg

Development Editor: Megan Garvey

Assistant Editor: Erin Bosco

Editorial Assistant: Rebecca Donahue

Media Editor: Janine Tangney

Brand Manager: Lydia Lestar

Market Development Manager: Erin Parkins

Senior Marketing Communications Manager:
Linda Yip

Rights Acquisitions Specialist: Jessica Elias

Manufacturing Planner: Betsy Donaghey

Art and Design Direction, Production
Management, and Composition:
PreMediaGlobal

Cover Image: © Nuno Silva/iStockphoto
(royalty-free)

For product information and technology assistance, contact us at
Cengage Learning Customer & Sales Support, 1-800-354-9706

For permission to use material from this text or product,
submit all requests online at **www.cengage.com/permissions**.
Further permissions questions can be emailed to
permissionrequest@cengage.com.

Library of Congress Control Number: 2012954818

ISBN-13: 978-1-133-60738-0

ISBN-10: 1-133-60738-1

Wadsworth
20 Channel Center Street
Boston, MA 02210
USA

Cengage Learning is a leading provider of customized learning solutions with office locations around the globe, including Singapore, the United Kingdom, Australia, Mexico, Brazil and Japan. Locate your local office at **international.cengage.com/region**

Cengage Learning products are represented in Canada by Nelson Education, Ltd.

For your course and learning solutions, visit **www.cengage.com**.

Purchase any of our products at your local college store or at our preferred online store **www.cengagebrain.com**.

Instructors: Please visit **login.cengage.com** and log in to access instructor-specific resources.

Printed in the United States of America
1 2 3 4 5 6 7 16 15 14 13 12

For Mary, with love

Brief Contents

Contents

Preface

This edition of *Technical Report Writing Today* continues my love affair with teaching technical communication. My key idea is accessibility. I want the book to make accessible both the act of writing and current changes in the ways professionals must communicate. The changes and additions that I have introduced into this edition reflect that goal.

Before I detail those for you, let me tell you a bit about my teaching and, thus, my suggestions for using this book, whether you are a student or instructor. When I first began to teach in the 1970s, I lectured and showed examples. Students took notes and handed in papers that I graded and returned. In other words, I did what everyone did.

But as the years passed, I came to believe that I needed to alter the way I approached writing, and confidence about writing. I turned my classes into labs. Rather than lecture, I assigned chapters, and then worked on creating papers in class. I used fewer and fewer exercises and instead assigned the particular paper on day one and told students to begin to work on it. I circulated. I commented. I taught students to ask, "How do I handle this?" And when they finished the paper, they handed it in to me and then also handed in a "Learning Report." Those learning reports became the key to my teaching. My goal was to create ongoing self-reflection so that the course became, so to speak, one big assignment in developing awareness of how each individual wrote and in creating confidence that the student could handle any new situation. In other words I tried to set students on the trail to expertise, which grows by practice and reflection. I think my way worked. I encourage you to try it. I think that this edition provides a way to do that.

Organization

This text is organized so that instructors have maximum flexibility in creating a course based on it.

1. As has been true since I started writing this text, I have chapters on theory in the first half of the book, and applications in the second half. I have included both professional and student examples in order to illustrate the variety of ways in which a paper can be created so that it makes its topic accessible to its audience.

2. I have maintained the way in which most chapters are organized, drawing attention to the professional process of considering audience, organization, usual form, and visual aids—the four essentials in professional writing.

3. I have deleted and added material to keep the topics up to date, changing some chapters; I will detail that for you later.

4. I have kept exercises in all the chapters, retained all the Planning and Evaluating Worksheets, and all the assignments. A key assignment that I have in every chapter is the requirement to write a Learning Report. If you have not used that strategy, I introduce it in Chapter 5. I hope you will review it and use it.

5. I have seen a number of syllabuses that instructors have created in order to use the text. I am amazed at the creativity I find. You can work through the book chapter by chapter or skip around, often combining a theory and practice chapter as you assign a particular paper.

Special Pedagogical Aids and High-Interest Features

Learning reports.   As I have detailed above, I have continued the use of "Learning Reports," self-reflections on what the student did and learned while creating the assigned paper. My final exam is that students write a paper explaining what they learned in the class and what they will take forward into new communication situations.

Grants for non-profit organizations.   Because I am especially interested that my writing classes not be simply "figure out what he wants and hand that in" (which actually is pretty good reading of audience), I have added a new section on writing grants for non-profit organizations. I urge you to use this material as a way to have your students perform community service and also to write for an audience who will make an action decision on the writing, an action that the writers want to occur in their favor.

Social media.   This revision includes sections spread throughout the text on social media. While personal use of social media—texting, tweeting, facebooking—are now common in student life, the professional use of the same applications is not understood. Just because you can post quick notes about what you are doing tonight does not mean that students will know how to handle those applications when they arrive in the work force. I have consulted a number of people who manage social media outlets in order to provide your students with clear advice on what proper usage is for companies they might work for.

Slide presentations.   I have also changed the focus of the oral presentation chapter, focusing on slide presentations. My change is simple: stop using text, start using visuals. Speak to the visual, don't read the text to the audience. Communication regularly includes orally presenting the material, so skill in doing so is an important professional necessity.

Technical communication style.   Every time I talk to professional managers, they tell me not to worry about the forms or genres—they can teach those. They want clear writing. I have trimmed the style chapter so that it focuses on fewer issues, just the most essential ones, and I have added discussions of pronoun usage and comma splices in the Handbook. I have also added a long list of sentences I revised as I wrote this manuscript. I want students to see that revision occurs as one writes, not just later when searching for errors. I hope that you will return often to student sentence awareness, helping them learn for themselves how to craft their sentences. The need for clarity is especially important in e-mail where vague, loose sentences create the need for a lot of back and forth until clarity is achieved.

Old favorites retained.   Other pedagogical features have remained the same. The exercises, assignments, and worksheets are all there. I especially like the worksheets because they give the students the expectations before they begin to create the project.

Class as a lab.   While I do not have sections on "Class as a Lab" in the text, I urge you to consider this approach. Assign a number of tasks at the beginning of the class time, then circulate to discuss issues with students or groups of students. The class is a lot noisier but the results in student confidence are impressive.

New to This Edition

In keeping with my goals of creating an accessible, up-to-date text, I have made numerous changes. I have added a number of new sections, and, regrettably due to lack of space, deleted some old sections. Here they are in priority order.

Chapter 14.   I have added a new section on Writing Grants for Non-Profit Organizations. My reviewers suggested this section, and I am delighted to add it. This section is a dramatic refocusing of the old "External Proposals" section. Non-profit organizations write grants regularly. Many of our students will be involved in such writing either because they work for a non-profit, or because as a community member they join a board that requires such work in order to facilitate the daily running of the organization. In addition this topic allows students a wonderful opportunity for a community service project. Whether you are in a major urban or a rural area, you will be able to find a non-profit organization that will be delighted to have your students help them with this important task. Creating a grant proposal will also place your students into the world where their writing "counts," not for a grade but to make a difference in other people's lives. I have had the good fortune to find grant writers who were willing to share their successful professional examples, which demonstrate that successful papers take many forms, not just a "text book formula" form to achieve their goals.

Chapter 11. There is an entirely new section on Social Media. Since the last edition of this book, communication has been dramatically changed by social media—Facebook, Twitter, and many other applications. This edition provides you with a way to include social media in your class and teaching strategy. With the aid of several people who "do" social media for a living, I have provided discussions of the way in which these media are used professionally. Students need to understand the model that is developing for social media use—at the time of this writing an interaction between websites, Facebook sites, Twitter sites and YouTube. I have included a number of examples that show how these sites interact with one another. Creating assignments for social media usage is difficult. You have to have students create and use the media, but you also have to evaluate it. I have tried to give you enough information so that you can confidently wade into this new stream in our teaching. One engineering manager that I interviewed before beginning the revision told me that he was hoping to hire "someone young" in the near future who could take up the task of creating and using social media for his company. He felt that he and his colleagues were both too busy and a bit "out of it" to take up the task. I hope that you can prepare your students for the opportunities that exist in this new field.

Chapter 10. A third area in which I have updated and refocused material is a new section regarding e-mail. People don't write memos any more. They send e-mails with attachments. All professionals I know are inundated with e-mail (so much so that in my last position I took to sending paper mail to people I wanted a response from). The new focus suggests ways to use the elements of the message (e.g., subject line) to create accessible messages.

Chapter 16. As almost all speeches from committee meetings to keynote addresses use PowerPoint (which has become a generic term for any digital slide presentation regardless of the application used to create it), a new section revolving around slide presentations has been added. You have heard, no doubt, "Death by PowerPoint," the cliché that indicates the ubiquity and dullness of PowerPoint presentations. I have focused on the rising call for less text and more visuals. The text is for the speaker to deliver, the visual is for the audience to consider as the points are made. I urge you to require your student presenters to use this strategy. Not only will it make the reports you have to listen to more pleasant, you will be preparing them to be a welcome breath of fresh air in the presentation world that they will enter after their university work.

Chapter 4. Yet another topic that I am happy to refocus is the various places in which I discuss technical communication style, in Chapter 4. Every time

I interview professional managers about what they need in a new hire, they emphasize clear writing. While clear writing includes many topics, I have chosen to focus on creating sentences. I have deleted some of the advice so that Chapter 4 is more succinct focusing on important issues—I believe that skill in using active voice and parallelism are the two most important issues for any professional writer to possess.

Appendix A.   New sections on Pronouns, Comma Splices, and Revised Sentences have been incorporated. Many writers, young and old, can't "see" the mistakes they make, so I have introduced new features in Appendix A: text boxes on recognizing misuse of pronouns and recognizing comma splices, and lengthy set of rewritten sentences. Pronouns are a bugaboo for many writers ("they," "I," "you" and "we" are used, as you know, interchangeably by many writers for whom the interchanging is invisible). I hope that my sections help you make students aware of the issues that attend these problems. I know that they were supposed to learn all this somewhere else, long ago. But if they didn't, there they are in your class. I hope that you find a way to use the strategies I present. The revised sentences are ones that I rewrote as I composed the manuscript. I want students to see that revision is an ongoing practice, not just something you do at the end looking for obvious grammar mistakes. Most of my revised sentences are more concise and clearer. I urge you to make your students collect their own samples of sentence revision, maybe even hand them in regularly or make them somehow part of a class Facebook site.

Some Other Topics Have Been Revised

Chapter 5—Researching.   Completely updated to include new examples of online searching and current rules for MLA (7th ed.) and APA (6th ed.) citations.

Chapter 13—Professional feasibility examples.   Such examples are difficult to find. My local engineering firm, Cedar Corp., was generous enough to supply me with short ones that they used in a recent project.

Chapter 12—Extended explanation of executive summary.   This section of reports has become a major necessity in a world awash in information. It is what people read. The new section focuses attention on creating executive summaries that convey quickly and clearly the contents of the attached report.

Other additions in the text.   New examples have been added in many chapters, including new professional résumés (Chapter 17) and new IMRD examples (Chapter 10).

The Expansion of Globalization sections have been revised to include comments from European students and professional writers.

All web citations in the Works Cited sections in all chapters have been updated.

To make room for the new sections I reluctantly concluded to leave some old friends behind.

Some Sections That Have Been Removed

Old Chapter 8 (Summarizing) and 9 (Defining) are gone. In new Chapter 12 I focus on executive summaries, which is a skill graduates often will need. I have included operational definitions in Chapter 10 with Informal Reports.

The Letters chapter (old Chapter 19) is gone. I merged the section dealing with the elements of a letter into the chapter on Job Application materials (new Chapter 17).

Ancillaries

Save time and streamline your course preparation with the Instructor's Resource Manual, available upon request from the publisher. This useful guide for instructors includes a wealth of resources such as sample syllabi, chapter notes, teaching suggestions, assessment sheets, and sample documents. Instructors will also have access to a downloadable version of the Instructor's Resource Manual, as well as chapter-specific PowerPoint lecture slides, on the protected instructor's companion website.

The student companion website is a rich study tool that includes such resources as chapter overviews, student samples, relevant web links, and a step-by-step guide to developing a website.

To access these ancillaries or to learn more, go to www.cengagebrain.com.

Acknowledgments

As always a book of this complexity depends upon the good offices of many people. I am delighted and honored to acknowledge Jane Henderson, Dr. Joe Hagaman, Bill Wikrent, Dr. Steve Nold, Tracy Babler, Amy Jomantas, Tim Riordan, Clare Riordan, Jane Riordan, Mike Riordan, Nathan Riordan, Shana Goldman, April Riordan, Paul Woodie, Mary H. Riordan, Russ Kviniemi, Simon Riordan, Dr. Quan Zhou, Dr. Daisy Pignetti, Ryan Peterson, Marion Lang, Paul Vliem, Laurie Boetcher, Tim Laughlin, Dr. Jill Klefstad, Jake Riordan, Jon Hove, Heidi Leeson, Dr. Kat Lui, Richard Jahnke, Pam Fricke, Laura Kreger, Laurel Verhagen, and Dr. Alan Block.

I am especially indebted to colleagues who took on the chore of revising/reworking some of my chapters.

Dr. Matt Livesey, University of Wisconsin-Stout, Chapter 1
Dr. Andrea Deacon, University of Wisconsin-Stout, Chapters 2 and 3
Dr. Julie Watts, University of Wisconsin-Stout, Chapters 6 and 7
Dr. Paul Anheir, University of Wisconsin-Stout, Chapter 13
Ms. Elizabeth Barone, University of Wisconsin-Stout, Chapter 5 and
  Appendix B

Ms. Heidi Decker-Mauer, University of Wisconsin-Stout, who performed extensive editing on Chapters 1, 2, 3, 10, and 11 and who is responsible for guiding me into the world of Social Media while writing the basic text on that subject.

Dr. Birthe Mouston, Aarhus University, Aarhus, Denmark, for revising many of the Global focus sections from a European point of view.

I would also like to thank the many reviewers who carefully reviewed the 9th edition so I could prepare for this 10th one:

Jerry DeNuccio, Graceland University
Kerry Duncan, Mesabi Community College
Laura Wilson, University of Cincinnati
Janet L. Reed, Crowder College
Elizabeth Lohman, Tidewater Community College
Beth Leslie, Southeastern University
Cheryl Cardoza, Truckee Meadows Community College

I thank my editors who gave me a free hand to create the book as I felt best, but who have provided that guiding hand, that kept me from wandering off into the wilderness of tangents so likely to occur in this type of project. In particular I thank Michael Rosenberg, Publisher for Humanities; Megan Garvey, Associate Development Editor for Humanities; Erin Bosco, Assistant Editor for Humanities; Rebecca Donahue, Editorial Assistant for Humanities; and Preetha Sreekanth, Manager, Project Management, PreMediaGlobal.

If the book has brilliances, most of the credit goes to my collaborators who have shared so much with me. The errors are mine, for which I take full responsibility.

Dan Riordan
Menomonie, Wisconsin

SECTION

1

Technical Communication Basics

Definition of Technical Communication

CHAPTER CONTENTS

CHAPTER 1 IN A NUTSHELL

Here are the basics for getting started in technical communication:

Focus on your audience. Your audience needs to get work done. You help them. To help them, you must stay aware that your goal is to enable them to act.

Think of audiences as members of your community who expect that whatever happens will happen in a certain way and will include certain factors—your message is expected to include certain sections covering specific topics. When you act as members of the community expect other members to act, your message will be accepted more easily.

Use design strategies. Presenting your message effectively helps your audience grasp your message.

▶ Use the top-down strategy (tell them what you will say, then say it).

▶ Use headings (like headlines in newspapers).

▶ Provide navigation to guide users to the content they need.

▶ Use chunks (short paragraphs).

▶ Establish a consistent visual logic through your formatting choices.

▶ Use a plain, unambiguous style that lets readers easily grasp details and relationships.

These strategies are your repertoire. Master them.

Assume responsibility. Because readers act after they read your document, you must present a trustworthy message. In other words, readers are not just receptacles for you to pour knowledge into by a clever and consistent presentation. They are stakeholders who themselves must act responsibly, based on your writing. Responsible treatment of stakeholders means that, among other things, you will use language and visuals with precision and hold yourself responsible for how well your audience understands your message.

Think globally. Much technical communication is distributed to audiences around the world. To communicate effectively, you must learn to *localize*. *Radical localization* requires a significant commitment to take into account the audience's broad-based cultural beliefs, while *general localization* involves tailoring the details of your document to locally expected methods of description—for instance, designating the date as day/month/year, or weights in kilograms.

Welcome! Technical communication is a large and important field of study and professional activity. Universities worldwide offer courses and programs in technical communication. Professionals either are technical communicators or produce technical communication documents as part of their jobs. The goal of this book is to make you an effective, confident technical communicator. This chapter introduces you to the basic concepts you need to know in order to communicate effectively. All the rest of the ideas in the book are based on three concepts: technical communication is audience-centered, presentational, and responsible.

This chapter introduces the field with two major sections: A General Definition of Technical Communication and Major Traits of Technical Communication.

A General Definition of Technical Communication

What Is Technical Communication?

Technical communication is "writing that aims to get work done, to change people by changing the way they do things" (Killingsworth and Gilbertson, *Signs* 232). Authors use this kind of writing "to empower readers by preparing them for and moving them toward effective action" (Killingsworth and Gilbertson, *Signs* 222). This is a brief definition; later in this chapter, you will learn more about the implications of empowering readers.

What Counts as Technical Communication?

Technical communication is an extremely broad field. It encompasses a wide range of skills and writing types. The Society for Technical Communication, an international professional organization, says that technical communication is any item of communication that includes one or more of these characteristics (STC, "Defining"):

▶ Communicating *about technical or specialized topics*, such as computer applications, medical procedures, or environmental regulations.
▶ Communicating *by using technology*, such as Web pages, help files, or social media sites.
▶ Providing *instructions about how to do something*, regardless of how technical the task is or even if technology is used to create or distribute that communication.

STC offers a certification to become a professional technical communicator. In such a capacity, a communicator is able to do all of the following:

▶ **User, Task, and Experience Analysis**—Define the users of the information and analyze the tasks that the information must support.

▶ **Information Design**—Plan information deliverables to support task requirements. Specify and design the organization, presentation, distribution, and archival for each deliverable.

▶ **Process Management**—Plan the deliverables schedule and monitor the process of fulfillment.

▶ **Information Development**—Author content in conformance with the design plan, through an iterative process of creation, review, and revision.

▶ **Information Production**—Assemble developed content into required deliverables that conform to all design, compliance, and production guidelines. Publish, deliver, and archive (STC, "Certification").

Technical communicators apply these skill areas to deliver diverse information products, including technical reports, articles, books, periodicals, tutorials and training, training materials, brochures, posters, websites, quick start guides, context-sensitive help, organizational manuals, quick reference, reference documents, user guides, and interactive knowledge bases (based in part on STC, "General"). Further, the content they produce may be drawn upon to meet other needs of the company or organization, such as sales and marketing, product development, and regulatory compliance.

Broadly considered, technical communication is a part of almost everyone's life on a regular basis.

Who Creates Technical Communication?

Two different types of writers create technical communication—technical communication professionals and those professionals who write as part of their jobs.

Professional technical communicators are hired to write the content that companies need to explain their products or services, often to help customers and technicians interact efficiently with the product or service. For instance, technical communicators work with software engineers to understand their software and then write guides and tutorials that users need. Whatever is needed to make information available to help people with their work, technical communicators produce.

Technical communicators are also those professionals who write about issues in their specific field or workplace. Sometimes these experts write for other experts. For instance, an engineer might write a progress report explaining to a division manager the actions and issues with a current project; a dietitian could write a proposal to fund a new low-fat breakfast program at a hospital; a packaging engineer may offer a solution for an inefficient method of filling and boxing jars of perfume. Sometimes these experts write to help nonexperts with technical material. Dieticians, for instance, often write brochures or Web content explaining the components of a healthy diet to hospital patients. Engineers write reports for nontechnical users, perhaps a county board, explaining an issue that has arisen in a bridge project.

Both groups and their activities center on the basic definition of technical writing given by Killingsworth and Gilbertson. The goal is to empower readers who depend on the information for success.

How Important Is Technical Communication?

Communication duties are a critical part of most jobs. Survey after survey has revealed that every week people spend the equivalent of one or more days communicating. In one survey ("How do they"), engineers reported that they spend 34 percent of their time writing on the job. In addition they report that in their writing, they collaborate up to 30 percent of their time. E-mail takes up to 38 percent of their time. Bob Collins, a corporate manager, puts it this way: "The most critical skill required in today's business world is the ability to communicate, both verbally and in writing. Effective communication has a direct impact on one's potential within an organization." Holly Jeske, an assistant technical designer for a department store chain, says "communication is my job." Her comments demonstrate the importance and complexity of everyday, on-the-job writing:

> I have to say that I depend a lot on my computer and e-mail for communicating with our overseas offices. I send and receive a lot of e-mails daily. A huge part of my job depends on writing and communicating in that way. I don't get the chance to hop on a plane every time there is a fit issue so that I can verbally communicate with them or even call them on the phone. . . . If I were never able to communicate through writing what I want the factory to change about a garment, I probably never would be moving from my current position. Communication is my job and pretty much anyone's job, . . . e-mail is a huge part of the corporate world.

Major Traits of Technical Communication

Technical Communication Is Audience Centered

Let's return now to the implications of our brief definition of technical communication—"writing that aims to get work done" and writing "to empower readers." What do those phrases imply? Technical communicators create documents that aim to help readers act effectively in the situations in which they find themselves. Janice Redish, an expert in communication design, explains that "a document . . . works for its users" in order to help them

> Find what they need
> Understand what they find
> Use what they understand appropriately (163).

In order to create a document in which readers can find, understand, and use content appropriately, writers need to understand how writing affects readers and the various ways in which readers approach written content. *Audience centered*, in this larger explanation, means that technical communication

- Has definite purposes
- Enables readers to act
- Enhances relationships
- Occurs within a community
- Is appropriate
- Is interactive

Technical Communication Has Definite Purposes

Technical writers enable their readers to act in three ways: by informing, by instructing, and by persuading (Killingsworth and Gilbertson, "How Can"). Most writers use technical writing to inform. To carry out job responsibilities, people must supply or receive information constantly. They need to know or explain the scheduled time for a meeting, the division's projected profits, the physical description of a new machine, the steps in a process, or the results of an experiment.

Writers instruct when they give readers directions for using equipment and for performing duties. Writing enables consumers to use their new purchase, whether it is a garden tool or a laptop computer. Writing tells medical personnel exactly what to do when a patient has a heart attack.

Finally, with cogent reasons writers persuade readers to follow a particular course of action. One writer, for example, persuades readers to accept site A, not site B, for a factory. Another writer describes a bottleneck problem in a production process in order to persuade readers to implement a particular solution.

Technical Communication Enables Readers to Act

According to Killingsworth and Gilbertson, it is helpful to view technical writing as "writing that authors use to empower readers by preparing them for and moving them toward effective action" (*Signs* 221–222). "Effective action" means that readers act in a way that satisfies their needs. Their needs include anything that they must know or do to carry out a practical activity. This key aspect of technical writing underlies all the advice in this book.

Figure 1.1 (p. 7) illustrates this concept in a common situation. The reader has a need to fulfill a task that she must do. She must assemble a workstation. A writer, as part of his job, wrote the instructions for assembling the workstation. The reader uses the instructions to achieve effective action—she successfully assembles the workstation. This situation is a model, or paradigm, for all technical writing. In all kinds of situations—from announcing a college computer lab's open hours to detailing the environmental impact of a proposed shopping mall—technical writers produce documents that enable effective action. The writing enables the reader to act, to satisfy a need in a situation.

Figure 1.1   Writing Makes Action Possible

Technical Communication Enhances Relationships

The starting point for creators of documents is the realization that their documents enhance relationships (Schriver, "Foreword"). Audiences don't exist in a vacuum. They exist in situations. Those situations mean that they have relationships with many people. Writing, and all communication, enhances those relationships. Audiences read because documents help them relate to someone else.

This may strike you as a strange way to think about writing. Many beginners tend to see the goals of writing as "being clear" or "having correct spelling and grammar," both of which are fine and necessary goals. But the modern conception of writing asks you to consider the issues related to those goals later. First, you need to understand the relationship issue. Let's take a personal example. Suppose a father has to assemble a tricycle for a birthday present. To assemble it, he first opens the box it came in, reads the instructions included, collects the correct tools, and then puts the parts together. Perhaps he visits the manufacturer's website to view an assembly tutorial. He is able to assemble the trike because you produced clear instructional content, identifying the parts and presenting the steps so that at the end the father has completed a functional toy ready for a child to ride.

If you think about the example for a moment, you can see that the father is using your instructions to enhance his relationship with his child. His goal in this situation is not just to turn a pile of parts into a working machine. It is to give a present to another person, someone with whom he has an ongoing relationship. This present will enhance that relationship, and the content you produced is a helpful factor to that end.

Now let's take a business example. Your department is in the process of upgrading its computer network. Your job is to investigate various vendors and models in order to suggest which brand to buy. When you finish your investigation and produce a report, the equipment is purchased and the network upgraded. Here, too, if you think about it, the report is about enhancing relationships. The goal is not just to get the cheapest, best equipment, but to facilitate the effectiveness of the work flow between people. If the system is effective, the people can interact more easily with one another, thus enhancing their relationships. Your report is not just about selecting a supplier; ultimately, it is about the relationships people have with one another in the department.

In both examples, you can see the same dynamic at work. Documents enhance relationships. Documents function to make the interaction of people better, more effective, more comfortable. Documents then empower people in a rather unexpected way—not only is the tricycle assembled, the child rides it, and the gift is exciting. Not only is the network upgraded efficiently, the office workers can cooperate in effective, satisfactory ways as they exchange and analyze their data.

Technical Communication Occurs Within a Community

Action occurs within a *community*, a loosely or closely connected group of people with a common interest. The key point for a writer to remember is that belonging to a community affects the way a person acts and expects other members to act (Allen; Selzer). Think about it this way: When people join a community, they learn how to act. For instance, at a new job people watch to see how everyone dresses and then dress similarly. If a man shows up at work on his first day in a three-piece suit and everyone else is in sport shirts and jeans, he will quickly change his clothing choices. But more than clothing choices, people learn how to communicate. In high school that might mean picking and using certain slang phrases, but on the job it means understanding how to present your material so that readers get the information that they need in the form that they expect it. This concept means that readers expect writing—all communication, actually—to flow in a certain way, taking into account various factors that range from how a document should look to what tone it projects. Effective writers use these factors, or *community values*, to produce effective documents.

If you conduct research into customer satisfaction to present to the sales force, they expect to know the method and results of your research. However, if you come to the meeting to report and you sing your report as if you were in a 1950s musical, you would not be presenting it in the form they expect. If you arrived with a perfectly formatted presentation, just like everyone else presents, and filled the entire report with lengthy details of all the personal concerns that made it hard for you to get the report finished, you would not be presenting the information that the sales managers wanted. The result very likely would be that no one would remember the contents of your report, only

Figure 1.2   Writing Occurs Within a Community

that you were off base; you were not following the community's values. If you sang your reports three times over a few months, you would likely be fired (Figure 1.2).

One researcher (Schriver, *Dynamics*) found that one group failed to produce an effective brochure that delivered an antidrug message because the visual aid used in the brochure offended the teens' sense of what was the correct way to send the message. Rather than focus on the message, the teens focused on the image and, interestingly, on the writer. Their conclusion was that, like the singer in previous the example, the writer was off base and thus had little or no credibility. Other brochures on the same topic were rejected again and again because the writer had failed to find the "community connection" with the teenage audience (171–185).

In other words, community values affect the way you write. The writing you do is deeply affected by your awareness of what members of your community need and expect. They need certain facts; they expect a certain format. They cannot know how to act on the facts you discover until you give the facts to them in the e-mail. Technical communication is based on this sense of community. "We write in order to help someone else act" (Killingsworth and Gilbertson, *Signs* 6).

Technical Communication Is Appropriate

Because communication takes place within a community, it must be *appropriate*, which can have two meanings in communication: the material needed in the situation is present (Schriver, "Foreword"), or the material is socially acceptable (Sless).

The first meaning implies that the wording must be more than clear and well structured. Suppose, for instance, that a reader consults a user manual to discover how to connect a videogame system to a wireless home network. If that topic is not covered in the manual, or if the manual explains networking

Internet and Interactivity

As mentioned below, technical communication has always been interactive. However, with the aid of fast-evolving technology, this interaction is no longer limited to writers creating and readers acting on a static text. Instead, both parties (and more) can become involved in a conversation, can collaborate, and can participate on a scale limited only by their access to the technology.

Easy-to-use software programs enable technical writers to create dynamic electronic texts that are adaptable and allow two-way communication. Many companies are using blogs with interactive comment sections, social media platforms that customers can respond to, product-specific interactive Internet forums, and knowledge-sharing wikis to communicate not only to—but also with—their readers. The interactivity of these platforms means that customers contribute comments, questions, and content—all of which would have been unthinkable in the past. This subject will be covered in more detail in Chapter 11.

but does not deal with the particular steps needed to connect to the reader's particular network—in other words, if the reader can't find the instructions that she or he needs—then the manual is useless, or inappropriate. Writers must learn to conceptualize the reader's needs in several situations and create the sections that help her or him to act.

The second meaning deals with what can be called *social appropriateness*, or accurately representing the relationships in the situation. In this meaning word choice is often very important. Consider this sentence written in an e-mail: "I felt that there was needless repetition in what you wrote." That sentence can easily be read as a reprimand, indicating that the writer of the e-mail was unhappy with the reader, making the text a scolding. Not wishing to convey that impressing, the e-mail author rewrote the sentence: "I felt that the text would be easier to read if the sections were combined." That sentence takes the focus away from the scolder/scoldee relationship and turns it appropriately to the issue of combining texts for clarity.

Social appropriateness also has ethical and global dimensions, which are discussed later in this chapter. The ethical dimension arises because writing affects relationships and empowers action. The global dimension arises because readers may be members of communities based in other cultures than your own. Writers, aware of the role of writing to empower action, must learn to take into consideration the sometimes radically different needs of these other cultures.

Technical Communication Is Interactive

The key to all community exchanges is that they are interactive. Readers read the words in the document, but they also apply what they know or believe from past experiences. As the words and the experiences interact, the reader

Figure 1.3 Communication Is Interactive

in effect recreates the report so that it means something special to him or her, and that something may not be exactly what the writer intended.

Figure 1.3 shows how this interaction works. The writer presents a report that tries to enable the reader to act. The writer has already completed researching the subject. Acting on an awareness of community values, the writer chooses a form (an e-mail), assembles the data into a coherent report (visualized here as "1," suppose the fact is that the company Web page has received fewer hits in each of the past three months), and interprets those facts (perhaps you suggest that the color combination of the site is unattractive). The reader interacts with the e-mail, using the document's words and format and her past experiences to make it meaningful to her. With her personal meaning, the reader may take a different course of action from the one that the writer may have intended. For instance (visualized here as "2"), because of reading a previous report (knowledge from a prior experience), the reader knows the Facebook page has had 200 "likes" this month and Twitter following has grown by 300 (visualized here as "2"). She also knows from attending a recent conference that websites must be actively integrated with Facebook and Twitter feeds (visualized here as "2"). With the knowledge gained from these sources, she can conclude that color scheme is probably not the problem, but the interaction of the three ways of sending out the company information. The statements in the report also tell her that the IT department needs to be informed because there is an obvious communicatio problem that must be resolved. The report is more than a report on a problem. Because the report is read interactively, the reader constructs meaning (visualized here as "1+2+3") that tells her how to act in a situation that the writer in this case did not know about.

This interactive sense of writing and reading means that the document is like a blueprint from which the reader recreates the message (Green). The reader relates to certain words and presentation techniques from a framework of expectations and experiences and makes a new message (Rude; Schriver, *Dynamics*). Communication does not occur until the reader recreates the message.

Technical Communication Is Designed

Technical writers use design to help their readers both find information and understand it. Design has two ingredients—the appearance on the page and the structure of the content. Technical communicators design both the appearance *and* the content of both print and electronic documents.

Design the Appearance

Designing the appearance means creating a page that helps readers locate information and see the relationship among various pieces of information. Figure 1.4 (p. 13) illustrates the use of basic design strategies. You can tell immediately by the design that the message has two main divisions, that the first division has two subdivisions, and that the text in the second division is supported by a visual aid. Technical writers use this kind of design to make the message easy to grasp (Cunningham; Hartley). The basic theory is that a reader can comprehend the message if he or she can quickly grasp the overall structure and find the parts (Rude; Southard). The basic design items that writers use are

- Headings
- Chunks
- Visual aids
- Hyperlinks

Headings. Headings, or heads, are words or phrases that name the contents of the section that follows. Heads are top-down devices. They tell the reader what will be treated in the next section. In Figure 1.4 the boldfaced heads clearly announce the topics of their respective units. They also indicate where the units begin and end. As a result, the readers always have a "map" of the message, and, in online documents, can navigate easily to the information they need.

Chunks. A chunk is any block of text. The basic idea is to use a series of short blocks rather than one long block. Readers find shorter chunks easier to grasp.

Visual Aids. Visual aids—graphs, tables, and other media—appear regularly in technical writing. In Figure 1.4, the visual aid reinforces the message in the text, giving an example that would be impossibly long, and ineffective, as a piece of writing. Writers commonly use visual aids to present collections of numerical data (tables), patterns or trends in data (graphs), and examples of action (a short video clip showing how to connect to a computer network). Documents that explain experiments or projects almost always include tables or graphs. Manuals and sets of instructions rely heavily on drawings and photographs, and those delivered in digital format may include audio or video clips. Feasibility reports often include maps of sites. More discussion of visual aids appears in Chapter 7.

Hyperlinks. Specific to Web documents, hyperlinks are words embedded in the document that help the reader navigate to more information about a

Top

List two major
sections of the
document

Primary
subdivision

Secondary
subdivisions

Primary
subdivision

Technical writing is the practical writing that people do on their jobs. The goal of technical writing is to help people get work done. This report explains two key characteristics of technical writing: audience centered and designed.

Audience Centered.   Writing is audience centered when it focuses on helping the audience. To help the audience, the writer must help the reader act and must remember community values.

> **Help Act.**   Writing helps readers get a job done or increase their knowledge so that they can apply it another time in their job.
>
> **Community Values.**   Everyone who belongs to any organization agrees with or lives by some of that organization's values. The writer must be sure not to offend those values.

Designed.   Technical writing appears in a more designed mode than many other types of writing. Design strategies help readers grasp messages quickly. Three key strategies are the top-down approach, the use of heads, and the use of chunks. Figure 1 illustrates the two methods. The first sentence is the top, or main, idea. The boldfaced words are the heads, which announce topics, and the x's represent the chunks or ideas.

There are two methods: heads and chunks.

Heads

xxxxxxxxxxxxxxxxxxxxxxxxxx
xxxxxxxxxxxxxxxxxxxxxxxx
xxxxxxxxxxxxxxxxxxxxxxxxx

Chunks

xxxxxxxxxxxxxxxxxxxxxxx
xxxxxxxxxxxxxxxxxxxx
xxxxxxxxxxxxxxxxxxxxx

Figure 1.
Two Design Strategies

Remember, to be a good technical writer, always put your audience first and always design your material.

Figure 1.4   Sample Page

particular subject. Hyperlinks are denoted by a different-colored text, and are typically underlined. Clicking on a hyperlink within a document will take you to a different page—or a different location on the same page—that contains more information about the highlighted word or phrase.

Design the Content

Designing the content means selecting the sequence of the material and presenting it in ways that help the reader grasp it. Two common methods to use are

▶ Arrange the material top-down
▶ Establish a consistent visual logic

Arrange the Material Top-Down.   *Top-down* means putting the main idea first. Putting the main idea first establishes the context and the outline of the discussion. In Figure 1.4 (p. 13), the entire introduction is the top because it announces the purpose of the document. In addition, the list at the end of the introduction sets up the organization of the rest of the document. When the reader finishes the first paragraph, she or he has a clear expectation of what will happen in the rest of the message. With this expectation established, the reader can grasp the writer's point quickly. Top-down messaging is especially important in Web documents, because Web readers have a tendency to scan information on a page. If they don't see what they're looking for right away, they will often navigate away from your page and find something else that gives them more quickly the information they seek.

Establish a Consistent Visual Logic.   A consistent visual logic means that each element of format is presented the same as other similar elements. Notice in Figure 1.4 that the heads that indicate the primary subdivisions ("Audience Centered" and "Designed") look the same: boldfaced, the first letter of each word capitalized, and placed at the left margin. Notice that the heads that indicate the secondary subdivisions ("Help Act" and "Community Values") also look like each other, but differ from the primary heads because they appear indented a half inch. Notice the position of the visual aid, placed at the left margin, and the caption of the visual aid, italicized and in a smaller print size. If there were another visual aid, it would be treated the same way. The key to this strategy is *consistency*. Readers quickly grasp that a certain "look" has a particular significance. Consistent treatment of the look helps the reader to grasp your meaning. In fact, many companies and organizations have an official style guide that may require certain standards of formatting.

Technical Communication Is Responsible

Earlier sections focused on the audience and the text, but this section focuses on you, the writer. It is not enough just to help people act and to design your work to that end. Because readers count on you to be their guide, you must do what you can to fulfill their trust that you will tell them what—and all—they need to know.

In other words, technical writing is an ethical endeavor (Griffin). The key principle here is to take responsibility for your writing (Mathes). In short, technical communicators must act ethically as they create and present documents. "[C]ommunication has always been fundamentally about people interacting with other people, and ethical communication has always been about our responsibilities in relation to others" (Dombrowski, "The Evolving Face of Ethics" 317).

You take responsibility because your readers, your employer, and society—also called "stakeholders"—rightfully expect to find in your document all the information necessary to achieve their goals, from assembling a tricycle to opening a factory (Harcourt). According to one expert, "Ethically it is the technical writer's responsibility to [ensure] that the facts of the matter are truly represented by the choice of words" (Shimberg 60).

In the text of your documents, then, you must tell the truth and you must do all you can to ensure that your audience understands your message. To help you with these important concepts, this chapter includes a definition of ethics and strategies to use for ethical presentation.

Definition of Ethics

Ethics deals with the question, What is the right thing to do? Philosophers since Plato have written extensively on the topic. It is a concern in daily life, in political life, in corporate life. Instances of its importance appear daily in our decisions about how to act and in news stories probing public actions. Ethics is a matter of judging both private and communal action. Individuals are expected to do the right thing, for their own personal integrity and for the well-being of their communities.

The issue, of course, is that the answer to the question, What is the right thing to do? is problematic. It is not always clear what to do or what value to base the decision on. Philosophers' answers to that dilemma have not always been consistent, but in relation to communication several common threads have emerged.

One major thread is that the communicator must be a good person who cares for the audience. Communicators must tell the full truth as convincingly as possible, because truth will lead to the good of the audience. A second thread is that communicators must do what is right, regardless of the cost to themselves. A third thread is that communicators must act for the greatest good for the greatest number of people (Dombrowski 16–18, 45–62). Of course, there are many ethical standards and writers on ethics, but it is commonly held that one must act not for self-gain but for the good of the community, or for the stakeholders in the situation.

Ethical Situations

The situations in which a person would have to make ethical decisions, and consequences from those decisions, vary dramatically. For instance, there are "this could cost me my job" situations, or *whistle-blowing*, a practice protected in the United States by federal law. In these situations, the employee becomes aware that the company is doing something illegal or that could cause great harm, perhaps because OSHA, FDA, or EPA standards are not being followed. For instance, before the terrible *Challenger* disaster, one employee had written a very clear report outlining serious problems concerning the O-rings. This report was subsequently used legally as the smoking gun to prove negligence on the part of those in charge. The writer subsequently lost his job, fought back, and was reinstated under the law, only to leave the company because of difficulties posed by remaining employed (Dombrowski 132–140).

This kind of decision—and action—is incredibly intense, requiring more than just a sense of what is the right thing to do. It requires courage to accept the negative consequences on self, and family, that losing employment entails. Each person must ask himself or herself how to respond in a situation like this, but the ethical advice is clear—you should blow the whistle.

Much more common, however, are the everyday issues of communication. People rely on documents to act. These actions influence their well-being at all levels of their lives, from personal health, to financial indebtedness, to accepting arguments for public policy. As a result, each document must be designed ethically.

Two examples from an ethics survey will give you a sense of the kind of daily decision that can be judged unethical. Dragga ("Is This Ethical") interviewed several hundred technical communicators and asked them to evaluate these two issues, among others:

> You have been asked to design materials that will be used to recruit new employees. You decide to include photographs of the company's employees and its facilities. Your company has no disabled employees. You ask one of the employees to sit in a wheelchair for one of the photographs. Is this ethical?
>
> You are preparing materials for potential investors, including a 5-year profile of your company's sales figures. Your sales have steadily decreased every year for five years. You design a line graph to display your sales figures. You clearly label each year and the corresponding annual sales. In order to de-emphasize the decreasing sales, you reverse the chronology on the horizontal axis, from 1989, 1990, 1991, 1992, 1993 to 1993, 1992, 1991, 1990, 1989. This way the year with the lowest sales (1993) occurs first and the year with the highest sales (1989) occurs last. Thus the data line rises from left to right and gives the viewer a positive initial impression of your company. Is this ethical? (256–257)

Of the respondents, 85.6 percent found the first case and 71.8 percent found the second case "mostly" or "completely" unethical (260). Dragga found that the basic principle that the practitioners used was "The greater the likelihood of deception and the greater the injury to the reader as a consequence of that deception, the more unethical is the design of the document" (262–263).

If technical communication is ethical, how does one find out what is ethical behavior in technical communication situations? Dragga ("A Question") pursued this question in 2009. He found out that experienced technical communicators suggest three actions: talk to your colleagues, trust your intuition, and talk to your boss (167–169). When Dragga asked, "Whom should you talk to if you are asked to perform something you feel is unethical?" he found that communicators advised talking to your boss, to whomever asked you to perform the activity, and to colleagues.

Ethical considerations are integral parts of every project. In order to be a responsible member of the community, every communicator must investigate and find the principles—and courage—upon which to act ethically.

Codes of Ethical Conduct

Many companies and most professional associations—Johnson & Johnson and the American Marketing Association, for instance—publish codes of conduct for their employees or practitioners. These codes provide guidelines for ethical action. They include a variety of topics, but several are typically addressed: fundamental honesty, adherence to the law, health and safety practices, avoidance of conflicts of interest, fairness in selling and marketing practices, and protection of the environment (Business Roundtable).

In its Statement of Ethical Principles, the STC lists six broad areas of ethical standards: legality, honesty, confidentiality, quality, fairness, and professionalism. Your work for a particular employer will often also be guided by a corporate code of conduct, with which you should familiarize yourself.

Here is the STC code of ethics:

As technical communicators, we observe the following ethical principles in our professional activities.

Legality We observe the laws and regulations governing our profession. We meet the terms of contracts we undertake. We ensure that all terms are consistent with laws and regulations locally and globally, as applicable, and with STC ethical principles.

Honesty We seek to promote the public good in our activities. To the best of our ability, we provide truthful and accurate communications. We also dedicate ourselves to conciseness, clarity, coherence, and creativity, striving to meet the needs of those who use our products and services. We alert our clients and employers when we believe that material is ambiguous. Before using another person's work, we obtain permission. We attribute authorship of material and ideas only to those who make an original and substantive contribution. We do not perform work outside our job scope during hours compensated by clients or employers, except with their permission; nor do we use their facilities, equipment, or supplies without their approval. When we advertise our services, we do so truthfully.

Confidentiality We respect the confidentiality of our clients, employers, and professional organizations. We disclose business-sensitive information only with their consent or when legally required to do so. We obtain releases from clients and employers before including any business-sensitive materials in our portfolios or commercial demonstrations or before using such materials for another client or employer.

Quality We endeavor to produce excellence in our communication products. We negotiate realistic agreements with clients and employers on schedules, budgets, and deliverables during project planning. Then we strive to fulfill our obligations in a timely, responsible manner.

Fairness We respect cultural variety and other aspects of diversity in our clients, employers, development teams, and audiences. We serve the business interests of our clients and employers as long as they are consistent with the public good. Whenever possible, we avoid conflicts of interest in fulfilling our professional responsibilities and activities. If we discern a conflict of interest, we disclose it to those concerned and obtain their approval before proceeding.

Professionalism We evaluate communication products and services constructively and tactfully, and seek definitive assessments of our own professional performance. We advance technical communication through our integrity and excellence in performing each task we undertake. Additionally, we assist other persons in our profession through mentoring, networking, and instruction. We also pursue professional self-improvement, especially through courses and conferences.

Reprinted with permission from The Society for Technical Communication, Arlington, Virginia, USA. http://www.stc.org/

To find out more about codes of ethics, browse through "Codes of Ethics Collection." Center for the Study of Codes of Ethics in the Professions at IIT. http://ethics.iit.edu/ecodes/ethics-area/8

Throughout this book, you will learn strategies for the clear presentation of language, format, and visual aids. Use these communication devices responsibly to ensure that your writing tells the audience everything they have a right to know. The audience trusts you because you are an expert. Be worthy of that trust.

Technical Communication Is Global

Today, business is international, and so too are writing and communication. As a result, people must now deal with the many languages and cultures throughout the world on a regular basis. For instance, since the passage of the North American Free Trade Act (NAFTA), many manuals for products sold in North America routinely appear in three languages—English, French, and Spanish. Workers, even at relatively small firms, indicate that they must e-mail colleagues across the globe. Websites, easily accessible to anyone in the world with a network connection, must now be understandable to people who speak many different languages and are members of many different cultures.

All of these factors mean that you as a technical communicator must understand the strategies of effective international communication.

While the goal of all communication is to use words and forms that enable the receiver to grasp your meaning (Beamer, "Learning"), in intercultural communication you need to give special consideration to cultural factors and to strategies for adapting communication for a variety of audiences.

The basic strategy for adapting writing and communication to other cultures is *localization*. Nancy Hoft defines localization as "The process of creating or adapting an information product for use in a specific target country or a specific target market" (12). According to Hoft, there are two levels to localization: radical and general.

Radical Localization

Radical localization deals with those areas that affect the way users think, feel, and act (Hoft 13). These areas include rules of etiquette; attitudes toward time and distance; the rate and intensity of speech; the role of symbols; and local systems of economics, religion, and society—even the way people go about solving problems (60–77). In order to perform radical localization you must be able to look at social behavior from another culture's point of view, so that you can understand the thinking patterns of the other person's culture, the role of the individual in the other person's culture, and that culture's view of direct and indirect messages (Beamer, "Teaching"; Martin and Chancey).

Another Point of View.   Your ability to look at the meaning of behavior from a point of view other than your own is crucial to good communication. Failure to appreciate an alternative point of view results in culture bias. When a person exhibits culture bias, he or she sends a "community" message, indicating that the recipient is not part of the sender's community and that furthermore the sender doesn't care. This subtext to any message makes communication much more difficult. In order to eliminate culture bias, you need to investigate what is important to the members of the other culture (Hoft). The associations commonly made by one culture about some objects, symbols, words, ideas, and the other areas mentioned earlier are not the same as those made by another culture for the same items—and remember, the differences do not indicate that one group is superior to the other.

For example, in China the color red is associated with joy and festivity; in the West red can mean stop, financial loss, or revolution (*Basics*). In the United States, *janitor* usually means a person who maintains a building, and is often associated with sweeping floors. But in Australia that same job is called a *caretaker*—a word that in the United States usually means someone who maintains the health of another person (Gatenby and McLaren). To take another example (Hoft 74, 94), conceptions of authority may differ—the French often prefer to come to conclusions after appeals to authority, but many Scandinavians prefer more individual exploration. Levels of personal acquaintance differ in business relationships in other cultures. In the United States, people

Globalization and Cultural Awareness

Our world is shrinking incredibly. Almost every document has the potential to end up in the hands or on the screen of a person from another country and culture. The risk of miscommunication increases exponentially, and therefore in a technical communication context we face huge demands on our intellect, cultural understanding, and capability of being business-savvy.

English language and culture are not necessarily linked, because we see Scottish, Irish, English, American, Indian, and other local cultures where English is used as a language. English-study scholars talk about the *Englishes* of the world as a response to the combination of English and culture. Sometimes English is a common second language: a so-called lingua franca, a shared language used by two parties who each has a second language as his or her native language. Today, the most obvious examples of lingua franca English are found in international business, science, technology, and aviation. The multiple uses of English mean that the challenges of writing in English are huge.

In addition the varieties of English each express a culture, such as Irish or Indian. English used for one culture might make little sense when read by readers of another English culture. As companies are becoming international or partnering with multinational organizations, technical communicators need to become culturally informed. To use inappropriate language and be misunderstood may have enormous negative consequences. Likewise, lack of cultural adeptness may be interpreted as arrogance, elitism, or plain ignorance.

A good mental picture of the intricacies of culture is Hall's iceberg model. Spotting an iceberg in the sea, one sees only one-eighth of the ice; the main bulk is invisible under sea level. The visible parts of culture are what we say and what we do. That is the top one-eighth. The invisible parts are our norms and beliefs, assumptions, and values, as shown in Figure 1.5.

The English of one culture is built on a different iceberg from the English of another culture. The bottom of the Irish iceberg is different than the bottom of the Indian. It therefore takes some explanation to understand what "foreign" ways of behaving really mean. To understand the meaning of this excerpt of a newspaper article takes a lot of background knowledge (or awareness of the lower iceberg): "They frankly dislike the RSC's actors, them with their long hair and beards and sandals and roistering habits." The setting must be understood: England, Stratford-upon-Avon, Shakespeare, and the RSC people (Royal Shakespeare Company). In other words, not understanding the context and the deeper layer of knowledge and situation would make us fail to dive below sea level and decode the embedded cultural information (Swan).

Communicators within the same cultural circle need to understand context (or the bottom of the iceberg), too, but they need to express that context less explicitly. They know more about the layers underneath, and from the way people say, articulate, or gesticulate something or their body language in general, they know the codes for interpreting the sub–sea layers.

To decode other cultures, communicators need to do their desk research by studying the do's and don't's of the cultures, either through books and cultural guides, or by learning the basics of the language. But the field research, by learning through actively participating, by listening, by asking is just as paramount to

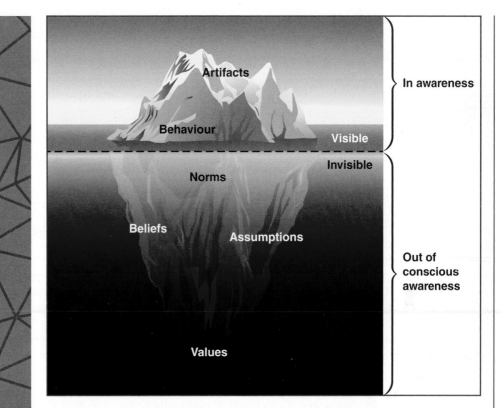

Figure 1.5 Hall's Iceberg Model of Cultural Awareness
Source: Based on Hall, E.T. and Mildred Reed Hall. Understanding Cultural Differences. Intercultural Press, Yarmouth. 1990.

become acquainted with other customs and needs. Developing a good cultural sensibility is not only good business, but it is the ethical thing to do.

The ethics of understanding other cultures, however, must always take its starting point in the understanding of the culture where a person or a company is based. Only then can a person start realizing, observing, and understanding other cultures.

During the 1980s and 1990s the idea developed that cultures globally would soon be unified and that all national and regional differences could be ignored. Coca-Cola followed this trend at first, but in 2000, it realized the fallacy of this as expressed by Coca-Cola CEO Donald Daft (Levitt)

The world was demanding greater flexibility, responsiveness and local sensitivity, while we were further centralising decision-making and standardising our practices, moving further away from our traditional multi-local approach. We were operating as a big, slow, insulated, sometimes even insensitive "global" company; and we were doing it in a new era when nimbleness, speed, transparency and local sensitivity had become absolutely essential to success.

(Continued)

Many corporations now have corporate social responsibility (CSR) guidelines as to how employees can address people in documents, as well as how the company at an overall level should request suppliers to obey minimum rules as regards work policies, environment, and so on, and the United Nations Global Compact lays down benchmark CSR guidelines. It also publishes results about how different companies obey CSR guidelines (Rasche and Kell; UN Global Compact).

Expanding awareness of the words and images used is the key to expressing yourself so that readers, irrespective of cultural assumptions and beliefs, will understand you.

often conduct business, including very large sales, with people whom they hardly know, but in many countries people prefer to achieve some kind of personal relationship before entering into any significant business arrangement with them.

The best method for gaining familiarity with another culture is to interact with members of that culture, whether those interactions occur in person or online. In order to communicate effectively, you must spend some time considering cultural differences and make changes in your documents accordingly.

Thinking Patterns.   Much of U.S. thought focuses on cause–effect patterns and problem solving—identifying the causes of perceived effects and suggesting methods to alter the causes. In other cultures, however, a more common thought pattern is "web thinking." In Chinese tradition, for instance, everything exists not alone but in a relationship to many other things, so that every item is seen as part of an ever-larger web, but the web is as important as the individual fact. These thinking patterns become part of the way people structure sentences. In American English, one says "I go to lunch every day," but in Chinese, one says "Every day to lunch I go." The first sentence emphasizes the individual, and the second emphasizes the web or context (Beamer, "Teaching").

Role of the Individual.   The individual is often perceived differently in a group dominated by web thinking, and web and group ideas can greatly affect the tone and form of communication. In the United States, long influenced by a tradition of individualism, many people feel that if they can just get their message through to the right person, action will follow. In other cultures, representatives of a group do not expect that same kind of personal autonomy or ease of identification from their readers.

Role of Direct and Indirect Messages.   In the United States we teach that the direct method is best: State the main point right away and then support it with facts. In some other cultures, that approach is unusual, even shocking. Although in the United States a writer would simply state in an e-mail that he or she needs a meeting, in a web culture, like China's, that request would come near the end of the message, only after a context for the meeting had been established (Beamer, "Teaching").

General Localization

General localization deals with items, usually details of daily life, that change from country to country, for instance, date format, currency, and units of measurement. Much of the literature that contains advice for writing in a global context deals with these concerns. Expert writers change these details when preparing documents for another country. These concerns fall into two broad areas: culture-specific references and style.

Culture-Specific References.   Culture-specific items are those that we use every day to orient ourselves. These items are often so ingrained that they are "invisible" to people in the culture—they are just the way things are done. The most common (based on Bacah; Hoft; Potsus; Yunker) of these are the following:

Time formats. Countries configure the calendar date differently; some use month/day/year, others use day/month/year. However, a common practice is to present dates in numeral format, for instance, 01/03/12. Depending on the common configuration, these numbers could mean January 3, 2012; March 1, 2012; or even March 12, 2001. Be careful to use the appropriate configuration.

Weights and measurements. The United States is one of the few countries that does not use the metric system. Most of the world travels in kilometers, measures in grams and liters, and is hot or cold in degrees Celsius. While it is easy to interpret those weights and measurements you are familiar with, if you are not, the numbers can be very difficult to translate into common experience. Change miles to kilometers, Fahrenheit to Celsius. Americans know it takes about an hour to go 60 miles, but in Europe it would be better to say 96 kilometers. In the United States 95°F is hot, but in France the same temperature is 35°C. Switching between systems is difficult, and you can help readers by performing the switch for them.

Currency. Try to express values in the country's money. Americans know that $8.50 is not a lot of money, but in Japan that figure is many hundreds of yen. (For help on the Web see, for instance, "XE—Universal Currency Converter" at <www.xe.com/ucc/>)

Number formatting. In English, the comma divides a number into thousands, then millions, and so on. The decimal point divides the number into tenths or less—1,234,567.89. But in other countries, the same numbers use different punctuation. In Germany, that number is 1.234.567,89.

Telephone numbers and addresses. In the United States, telephone numbers are grouped in threes and fours—715-444-9906, but in other countries they are often grouped by twos—33 (0)1 23 34 76 99. In the United States, it is common practice to address an envelope with the name at the top and list in descending order the street address and city. In some countries, Russia, for instance, the address list is reversed; the country is placed on the first line and the name of the person on the bottom line.

Page size. In the United States, the standard paper size is 8.5 × 11 inches; most documents are designed with these basic dimensions in mind. In many other parts of the world, however, the basic size is called A4 (8.25 × 11.66 inches). The difference in size can cause difficulties in copying material.

Style.   Style items are the subjects of many articles on globalization. The goals of managing style are to make English easier to understand and to make it easier to translate. Many of the style tips are simply calls for good, clear, unambiguous writing. Here are a few common style items (based on Hoft 214–236; Locke; Potsus) to consider:

Avoid using slang and idioms. Most of these are simply impossible to translate: He is a brick. She hit a home run with that presentation.

Avoid using humor. When a joke fails to get a laugh, the lame excuse is often, "You had to be there." Much humor is so culture dependent that what is hilarious to people of one culture is nearly incomprehensible to people from another culture. Humor often just does not work except in very small communities. Good writers generally avoid humor in their writing for other cultures.

Avoid puns, metaphors, and similes. Metaphors and similes compare items to indicate worth or appearance. These devices are helpful, but only if the reader gets the point of the comparison. Puns are plays on words, often used in ads. But puns are virtually untranslatable. Use these devices only if you are sure the reader would understand them.

Use glossaries. If you must use jargon or other specialized language, be sure to include a glossary of definitions.

Don't omit little words: a, an, the, of, these. Often, they are omitted to save space and to get to the point, but their absence may obscure the exact nature of the phrase. Compare "Click down arrow to bring up menu" with "Click on the down arrow to bring up the menu."

Include relative pronouns. The relative pronouns are *who, whom, whose, which,* and *that. That* is often the problem. A sentence like "A fire alarm losing power will beep" can be changed to "A fire alarm that is losing power will beep," or "The switches found defective were replaced" can become "The switches that were found defective were replaced" or "Maintenance personnel replaced the defective switches."

Don't use long noun phrases. Often English speakers string together a series of nouns. "Damage recovery results," for instance, could be the results of damage recovery or the act of damaging those results. To avoid misinterpretation, rewrite the phrase for the non-native speaker: "results of the damage inspection."

Avoid using homophones. Homophones are two or more words that sound alike but have different meanings, and may have different spellings—like damage, which can be a noun or a verb. "Damage results" can mean "to damage the results" or "the results of the inspection of damage." To native speakers, the context often makes the meaning of these phrases clear, but non-natives often have trouble with the meaning.

Use clear modifier strings. Consider the phrase "black ergonomic keyboards and mouse pads." Does this mean that both the keyboards and mouse pads are black and ergonomic? Or just the keyboards? To help non-native speakers, you need to express the material in a more precise, though longer, form: "mousepads and black ergonomic keyboards" or "keyboards and mousepads that are black and ergonomic."

Write in clear subject–verb–object order. If speakers are not familiar with the rhythms of English language speech, they can become lost in the quickness and turns that sentences in English can take. Use the sentence order that it is likely non-native speakers learned in textbooks. Use "The director of the lab ordered new computers," rather than something like "Ordering lab computers was taken care of by the director."

If your text is to be translated, also be aware of these concerns: Leave space for expansion (Locke; Potsus). English phrases often expand in translation. Translated text can be as much as 30 percent longer in other languages. Even a simple Canadian highway sign illustrates this. The English text is *Chain-up area.* (13 spaces), and the French is *Attachez vos chaines ici.* (24 spaces). If you have pages designed so that text should fall at a certain spot, leave extra room in your English original so that after the translation and subsequent expansion, the text will still be relatively at the same spot.

Choose a simple font and avoid text effects (like boldface, italics, underlining) (Hoft; Locke). Many languages that use roman letters have diacritical marks that are not used in the United States (like Å or Ç). Custom fonts often do not include these letters, though "common" fonts, like Times New Roman, do. Many languages do use text effects, and these effects are simply not recorded in the translation, thus losing any emphasis they may have originally carried.

Web Copy

When it comes to writing for the Web, there are additional considerations for the technical communicator to address. Here is a helpful synopsis of many of the points made in this section as they pertain to website design (Gillette 17).

When designing a site for a professional, international audience, you must follow most of the standard international communication guidelines commonly used for printed documents, online help, and other forms of software design. In brief:

- Keep sentences short and to the point.
- Use simple subject–verb–object sentence structure.
- Avoid the use of embedded or dependent clauses.
- Use short paragraphs to allow for easier paragraph-by-paragraph interpretation.
- Avoid regional idioms or turns of phrase.

TIP ||||

Where Can You Find Assistance?

Many resources exist to help communicators in the international arena, including many sources easily accessible on the Web. Below are two sources that will be helpful in your intercultural work. Be careful, however. While various sites can give helpful tips, as a practitioner you need to carefully assess your own general and radical localizations in order to engage someone else's.

CIA World Factbook On-line. Web. 28 Feb. 2012 <https://www.cia.gov/library/publications/the-world-factbook/> "The World Factbook provides information on the history, people, government, economy, geography, communications, transportation, military, and transnational issues for 267 world entities."
Global Talk. STC International SIG News. Web. 28 Feb. 2012 <http://itcglobaltalk.org/>
Try Google www.google.com. A recent search of Google using the term *international communication* turned up 57, 500,000 sites.

- Avoid any visual, textual, or interactive metaphors based on a specific national or social context (e.g., mailboxes and envelopes vary from country to country, so a mailbox icon that indicates "send mail" in the United States may just look like a blue box to the international visitor).
- Define technical terms as directly as possible, avoiding elaborate metaphor.
- If you have any doubt about users' knowledge of a specific term, define it.
- Accompany all graphical buttons with a verb-based identifier (e.g., left-pointing arrow with "Go Back").

Exercises

▶ You Create

1. Make a list of several communities to which you belong (e.g., university students, this class, X corporation). Write a paragraph that explains how you used writing as a member of one of those communities to enable another member or members of the community to act. Specifically explain your word, format, and sequencing (which item you put first, which second, etc.) choices.

2. Write a paragraph that persuades a specific audience to act. Give two reasons to enroll in a certain class, to purchase a certain object, to use a certain method to solve a problem, or to accept your solution to a problem.

3. Write a paragraph that gives an audience information that they can use to act. For example, give them information on parking at your institution.

4. Create a visual aid to enable a reader to act. Choose one of these goals: show the location of an object in relation to other objects (machines in a lab; rooms in a building); show someone how to perform an act (how to print a document from a computer; how to hold a hammer, how to create a contacts list for e-mail or a social medium, how to sync contacts between mobile devices); show why one item is better than another (cost to purchase an object like an e-reader or a TV or class notebooks; features of two objects).

5. Interview a professional in your field of interest. Choose an instructor whom you know or a person who does not work on campus. Ask questions about the importance of writing to that person's job. Questions you might ask include

   - How often do you write each day or week?
   - How important is what you write to the successful performance of your job?
   - Is writing important to your promotion?
   - What would be a major fault in a piece of writing in your profession?
   - What are the features of writing (clarity, organization, spelling, etc.) that you look for in someone else's writing and strive for in your own?

Write a one-page report in which you present your findings. Your instructor may ask you to read your report to your classmates.

▶ You Analyze

6. Explain a situation in which you would write to a member of a community to enable him or her to act. Identify the community and detail the kind of writing you would do and what the reader would do as a result of your writing.

7. Bring to class a piece of writing that clearly assumes that you (or the reader) belong to a particular community (good sources include newspaper stories on social issues like taxes, editorials, letters that ask for contributions). Point out the words and design devices that support your analysis. Alternate assignment: For a piece of writing given to you by your instructor, determine the community to which the writer assumed the reader belongs.

8. Choose one of the models at the end of a chapter in this book or a sample of writing you find in your daily life. Write a paragraph that describes how you interact with that piece of writing to gather some meaning. Describe your expectations about the way this kind of writing should look or be organized; what features of the writing led you to the main point; and any reactions to the presentation language, visual aids, or context.

9. Research your library's online catalog. Write a paragraph that alerts your instructor to commands, screens, or rules that will give students trouble if they are not aware of them ("use AND to search for multiple terms"; "your user name must be typed in lowercase"; "the library closes at 9:00 p.m. on Fridays").

10. Research a database available through your library's online catalog. Tell students about at least two types of material in the database (abstracts of articles, U.S. demographic information) and explain how that material will help them.

11. Analyze the following paragraph to decide who the audience is and what their need is; then rewrite it for a different audience with a different need. For instance, you might recount it as a set of instructions or use it to tell a person what objects to buy for this step and why.

> The fixing solution removes any unwanted particles that may still be in the paper. This process is what clears the print and makes the image more "crisp." The photographer slips the print into the fixing solution, making sure it is entirely submerged. He or she will agitate the print occasionally while it is in the fixer. After two minutes, he or she may turn the room lights on and examine the print. The total fixing time should be no less than 2 minutes and no more than 30 minutes. After the fixing process is over, the print then needs to be washed.

▶ You Revise

12. Arrange the following block of information into meaningful chunks. Some chunks may contain only several sentences.

> You have expressed an interest in the process used to carve detailed feathers on realistic duck decoys. The tools used are the same ones needed to prepare the carving up to this point: flexible-shaft grinder, stone bits, soft rubber sanding disc, pencil, and knife. My intention is to explain the ease with which mastery of this process can be achieved. There are five steps involved: drawing, outlining, and concaving the feathers, stone carving the quill, and grinding the barbs. The key to drawing the feathers is research. Good-quality references have been used in getting the carving to this stage and they will prove invaluable here. When comfortable with the basic knowledge of placement and types of feathers, drawing can begin. As with the actual carving, drawing should be done in a systematic manner. All feathers should be drawn in from front to back and top to bottom. Drawing should be done lightly so that changes can be made if necessary. All of the other steps are determined by what is done here, so the carver must be satisfied before beginning. Outlining creates a lap effect

similar to fish scales or the shingles on a house. The carver uses the flexible-shaft grinder and a tapered aluminum-oxide bit to achieve the fish scale effect. The carver starts at the front and works toward the back using the lines that were drawn as guides. Each feather is tapered from a depth of about 1/32 inch at the lines up to the original height of the carving. This step also works toward a "shingle effect." Concaving by the artist is nothing more than using the soft rubber sanding disc and gently cupping each feather toward the quill (center) area. The carver only uses slight sweeping motions with the disc to achieve good results at this phase. Concaving starts at the back outside edge of each feather and proceeds toward the tips. In stone carving the quill, the carver uses the aluminum-oxide bit and flexible-shaft grinder to raise and outline the quill area. As with all of the other steps, none of the procedures should be exaggerated. The goal is to make everything as life-like as possible. The carver grinds the barbs to match as closely as possible the hair-like structures of the feather using the same stone and grinder that were used in carving the quill. Actual feathers are used to get the exact angles needed for realism. Gentle sweeping motions are used, starting at the quill and moving toward the outside edge. Using only the tip of the stone creates the desired effects. When finished, the carver uses a loose wire wheel to remove any unwanted hair-like matter on the surface of the carved areas.

▶ Group

13. In groups of three, ask each other if the writing you do as a student or as an employee enables other people to do something. As a group, create a paragraph in which you list the kinds of people and actions that your writing affects. Use the Magolan report (p. 37) as a guide.

14. Your instructor will assign groups of three or four to read any of the following documents that appear later in this book: Instructions (four examples, pp. 262–266); Introduction, Method, Results, Discussion (IMRD; Example 10.1. pp. 296–302); or Informal Recommendation (pp. 280–281, 303–304). After reading it, explain what made it easy or hard to grasp. Consider all the topics mentioned in this chapter. Compare notes with other people. If your instructor so requires, compose a report that explains your results.

15. In groups of three or four, analyze the following sample memo. Explain how the report enables a reader to act, demonstrates its purpose, and uses specific practices to help the reader grasp that purpose. If your instructor so requires, create a visual aid that would encourage a reader to agree with the recommendation (perhaps a table that reveals all the results at a glance) and/or create another report to show how the report is ethical. Your instructor will ask one or two groups to report to the class.

DATE: April 1, 2014
TO: Isaac Sparks
FROM: Keith Munson
SUBJECT: Recommendation on whether we should issue bicycles to the maintenance department

Introduction

As you requested, I have investigated the proposal about issuing bicycles to the maintenance department and have presented my recommendation in this memo. I consulted with a company that has already implemented this idea and with our maintenance department. The decision I made was based on five criteria:

- Would machine downtime be reduced?
- Is the initial cost under $5000?
- Will maintenance actually use them?
- Will maintaining them be a problem?
- Are the bicycles safe?

Recommendation

Through my investigation, I have found that the company could realize substantial savings by implementing the proposal and still sufficiently satisfy all the criteria. Therefore, I fully recommend it.

Would Machine Downtime Be Reduced? Yes. There would be less machine downtime if bicycles were used because maintenance could get to the machines faster and have an average of 2 hours more per day to work on them. This could save the company approximately $500 a week by reducing lost production time.

Is the Initial Cost Under $5000? Yes. The initial cost of approximately $1500 is well within our financial limitations.

Will Maintenance Actually Use Them? Yes. I consulted with the maintenance department and found that all would use the bicycles if it became company policy. The older men felt that biking, instead of walking, would result in their fatiguing more slowly.

Will Maintaining Them Be a Problem? No. The maintenance required is minimal, and parts are very cheap and easy to install.

Are the Bicycles Safe? Yes. OSHA has no problem with bicycles in the plant as long as each is equipped with a horn.

16. Perform this exercise individually or in a group, as your instructor requires. Assume that you work for a manufacturer of one of the following items: (a) electric motors, (b) industrial cranes, (c) processors, or (d) a product typical of the kind of organization that employs you now or that will when you graduate. Assume that you have discovered a flaw in the product. This flaw will eventually cause the product to malfunction, but probably not before the warranty period has expired. The malfunction is not life threatening. Write a report recommending a course of action.

17. In groups of three or four, react to the memos written for Exercise 16. Do not react to your own memo. If all individuals wrote reports, pick a report from someone not in your group. If groups wrote the reports, pick the report of another group. Prepare a report for one group (customers, salespeople, manufacturing division) affected by the recommendation. Explain to them any appropriate background and clarify how the recommendation will affect them. Your instructor will ask for oral reports of your actions.

18. You have just learned that the malfunction discussed in Exercises 16 and 17 is life threatening. Write new memos. Your instructor will ask for oral reports of your actions.

Web Exercise

Analyze a website to determine how it fills the characteristics of technical communication explained in this chapter. Use any site unless your instructor directs you to a certain type (e.g., major corporation, research and development site, professional society). Write a report or IMRD (see Chapter 10) in which you explain your findings to your classmates or coworkers.

Works Cited

Allen, Nancy J. "Community, Collaboration, and the Rhetorical Triangle." *Technical Communication Quarterly* 2.1 (1993): 63–74. Print.

Bacah, Walter. "Trends in Translation." *intercom* (May 2000): 22–23.

The Basics of Color Design. Cupertino, CA: Apple, 1992. Print.

Beamer, Linda. "Learning Intercultural Communication Competence." *Journal of Business Communication* 29.3 (1992): 285–303. Print.

———. "Teaching English Business Writing to Chinese-Speaking Business Students." *Bulletin of the Association for Business Communication* LVII.1 (March 1994): 12–18. Print.

Business Roundtable. "The Rationale for Ethical Corporate Behavior." *Business Ethics* 90/90. Ed. John E. Richardson. Guilford, CT: Dushkin, 1989. 204–207. Originally published in *Business and Society Review* 20 (1988): 33–36. Print.

Collins, Robert C. Letter to Art Muller, packaging concentration coordinator. The Dial Corp. Scottsdale, AZ. 8 March 1994.

Cunningham, Donald. Presentation. CCC Convention. Minneapolis. 20 March 1985. Print.

Daft, D. "Back to Classic Coke." *Financial Times.* (27 March 2000). Print.

deMooij, M. "Convergence and divergence in consumer behavior: Implications for global advertising." *International Journal of Advertising* 22 (2003):183–202. Print.

Dombrowski, Paul. *Ethics in Technical Communication.* Needham Heights, MA: Allyn & Bacon, 2000. Print.

———. "The Evolving Face of Ethics in Technical and Professional Communication: Challenger to Columbia." *IEEE Transactions on Professional Communication* 50.4 (2007): 306–319. Print.

Dragga, Sam. "A Question of Ethics: Lessons from Technical Communicators on the Job." *Technical Communication Quarterly* 6.2 (1997): 161–178. Print.

———. "'Is This Ethical?' A Survey of Opinion on Principles and Practices of Document Design." *Technical Communication* 43.3 (Third Quarter 1996): 255–265. Print.

Gatenby, Beverly, and Margaret C. McLaren. "A Comment and a Challenge." *Journal of Business Communication* 29.3 (1992): 305–307. Print.

Gillette, David. "Web Design for International Audiences." *intercom* (December 1999): 15–17. Print.

Green, Georgia M. "Linguistics and the Pragmatics of Language Use." *Poetics* 11 (1982): 45–76. Print.

Griffin, Jack. "When Do Rhetorical Choices Become Ethical Choices?" *Technical Communication and Ethics.* Ed. R. John Brockman and Fern Rook. Arlington, VA: Society for Technical Communication, 1989. 63–70. Originally published in *Proceedings of the 27th International Technical Communication Conference* (Washington, DC: STC, 1980). Print.

Hall, Dean G., and Bonnie A. Nelson. "Integrating Professional Ethics into the Technical Writing Course." *Journal of Technical Writing and Communication* 17.1 (1987): 45–61. Exercises 16–18 are based on material from this article.

Hall, E. T., and Mildred Reed Hall. *Understanding Cultural Differences.* Intercultural Press, Yarmouth, ME, 1990. Print.

Harcourt, Jules. "Developing Ethical Messages: A Unit of Instruction for the Basic Business Communication Course." *Bulletin of the Association for Business Communication* 53 (1990): 17–20. Print.

Hartley, Peter. "Writing for Industry: The Presentational Mode versus the Reflective Mode." *Technical Writing Teacher* 18.2 (1991): 162–169. Print.

Hofstede, Geert. *Culture's Consequences : Comparing Values, Behaviors, Institutions and Organizations Across Nations.* 2nd ed. Sage: Thousand Oaks, CA, 2001. Print.

Hoft, Nancy L. *International Technical Communication: How to Export Information About High Technology.* New York: Wiley, 1995. Print.

"How do they spend their time writing at work?" Purdue University. 2008. Web. 20 May 2012. <https://engineering.purdue.edu/MECOM/Assignments/2008.FALL/NCState.2008.time.at.work.html>.

Jeske, Holly. "Writing and Me." E-mail to Dan Riordan. 20 Aug. 2003. Print.

Killingsworth, M. Jimmie, and Michael Gilbertson. "How Can Text and Graphics Be Integrated Effectively?" *Solving Problems in Technical Writing.* Ed. Lynn Beene and Peter White. New York: Oxford, 1988. 130–149. Print.

———. *Signs, Genres, and Communities in Technical Communication.* Amityville, NY: Baywood, 1992. Print. The material on community and interaction is based on ideas developed in this book.

Levitt, T. "The Globalization of markets." *Global Marketers,* Harvard Business Review, 1983. pp. 99–102. Print.

Locke, Nancy A. "Graphic Design with the World in Mind." *intercom* (May 2003): 4–7. Print.

Martin, Jeanette S., and Lillian H. Chancey. "Determination of Content for a Collegiate Course in Intercultural Business Communication by Three Delphi Panels." *Journal of Business Communication* 29.3 (1992): 267–284. Print.

Mathes, J. C. "Assuming Responsibility: An Effective Objective in Teaching Technical Writing." *Technical Communication and Ethics.* Ed. R. John Brockman and Fern Rook. Arlington, VA: Society for Technical Communication, 1989. 89–90. Originally published in *Proceedings of the Technical Communication Sessions at the 32nd Annual Meeting of the Conference on College Composition and Communication* (Dallas, TX: NASA Publication 2203, 1981). Print.

Potsus, Whitney Beth. "Is Your Documentation Translation Ready?" *intercom* (May 2001): 12–17. Print.

Rasche, A., and Kell, G. *The United Nations Global Compact: Achievements, Trends and Challenges.* Cambridge University Press: Cambridge, 2010.

Redish, Janice. "What Is Information Design?" *Technical Communication* 47.2 (May 2000): 163–166. Print.

Rude, Carolyn D. "Format in Instruction Manuals: Applications of Existing Research." *Journal of Business and Technical Communication* 2 (1988): 63–77. Print.

Schriver, Karen. *Dynamics in Document Design: Creating Texts for Readers.* New York: Wiley, 1997. Print.

———. "Foreword." *Content and Complexity: Information Design in Technical Communication.* Ed. Michael J. Albers and Beth Mazur. Mahwah, NJ: Lawrence Erlbaum, 2003. ix–xii. Print.

Selzer, Jack. "Arranging Business Prose." *New Essays in Technical and Scientific Communication: Research, Theory, Practice.* Ed. Paul V. Anderson, R. John Brockman, and Carolyn Giller. Farmingdale, NY: Baywood, 1983. 37–54. Print.

Shimberg, Lee H. "Ethics and Rhetoric in Technical Writing." *Technical Communication and Ethics.* Ed. R. John Brockman and Fern Rook. Arlington, VA: Society for Technical Communication, 1989. 54–62. Originally published in *Technical Communication* 25.4 (Fourth Quarter 1988): 173–178. Print.

Sless, David. "Collaborative Processes and Politics in Complex Information Design." *Content and Complexity: Information Design in Technical Communication.*

Ed. Michael J. Albers and Beth Mazur. Mahwah, NJ: Lawrence Erlbaum, 2003. 59–80. Print.

Society for Technical Communication. "Certification." Web. 19 May 2012. <http://www.stc.org/education/certification/certification-main>.

Society for Technical Communication. "Defining Technical Communication." Web. 19 May 2012. <http://www.stc.org/about-stc/the-profession-all-about-technical-communication/defining-tc>.

Society for Technical Communication. "Ethical Principles." Web. 19 May 2012. <http://www.stc.org/about-stc/the-profession-all-about-technical-communication/ethical-principles>.

Society for Technical Communication. "General Information and Competition Rules." Web. 19 May 2012. <http://www.stc.org/images/stories/doc/2011_General_Information_and_Competition_Rules.doc>.

Southard, Sherry. "Practical Considerations in Formatting Manuals." *Technical Communication* 35.3 (Third Quarter 1988): 173–178. Print.

Swan, M. *Inside Meaning.* A Reprint of "Hamburgers and Hamlets" from The Observer. Cambridge University Press: Cambridge, 1992. Print.

"UN Global Compact." Web. 20 May 2012. <http://www.unglobalcompact.org/>.

"World Englishes," World Englishes—*Journal of English as an International and Intranational Language.* Wiley-Blackwell. Web. 20 May 2012. <http://onlinelibrary.wiley.com/journal/10.1111/%28ISSN%291467-971X>.

Yunker, John, E. "A Hands-On Guide to Asia and the Internet." *intercom* (May 2000): 14–19. Print.

Profiling Audiences

CHAPTER CONTENTS

Chapter 2 in a Nutshell

CHAPTER 2 **IN A NUTSHELL**

You write a different document based on how you define your audience. Because your understanding of your audience controls so many of your writing decisions, analyze the audience before you write. Create an audience profile by answering these questions:

▶ Who are they?

▶ How much do they know?

▶ What do they expect?

Find out who your audience is. Is it one person or a group or several groups? Are you writing a memo to a specific individual or instructions for "typical" workers?

Estimate how much they know. If they are advanced, they know what terms mean, and they understand the implications of sentences. If you are addressing beginners, you have to explain more.

Determine expectations. Expectations are the factors that affect the way in which the audience interprets your document. Will it conform to their sense of what this kind of document should look and sound like? Will it help them act in the situation? Does it reflect a sense of the history of the situation or the consequences of acting?

Chapter 1 showed how audience is a major concern in technical communication. Every piece of writing has an intended audience—the intended reader or readers of the document. Because your goal is to enable those readers to act, you must analyze the intended readers in order to discover the facts and characteristics that will enable you to make effective decisions as you write. The facts and characteristics that you discover will affect planning, organizing, and designing all aspects of the document, from word choice to overall strategy and structure.

This chapter explains the factors that writers investigate in order to analyze audiences. The chapter begins with an example of a technical memo, and then presents sections that help you answer these key questions:

- Who is the audience?
- What is the audience's task?
- What is the audience's knowledge level?
- What factors influence the situation?
- How do I create an audience profile?

An Example of Technical Writing

The following brief report (Figure 2.1) highlights the importance of audience in professional communication. Todd Magolan performs routine inspections of thick rubber cargo mats that fit into the beds of special hauling equipment in manufacturing facilities. He reports on their performance to Marjorie Sommers, his supervisor. Sommers uses the reports to determine whether or not her company has met the conditions of its contract and to decide whether or not to change manufacturing specifications. He sent the report as an e-mail attachment to Sommers with a "cover letter" e-mail. The body of the e-mail is very succinct: "Marjorie, attached is my report from my March 15, 2015, site visit to inspect mats at Oxbow Creek. If you have any concerns or questions, please contact me. Todd."

As you read the report, note the following points:

1. The writer states the purpose of the report for the audience (to deliver information on his impressions).

2. The writer uses unambiguous language to focus on the specific parts of the cargo mat (trim lines, holes, kick plate) and to point out specific problems ("the front edge of vinyl/maratex still needs to be evaluated with a base kick plate"). Note that the writer uses the word *good* to mean "implements the specification exactly," knowing that his audience understands that usage.

3. The information is designed, appearing in easy-to-scan chunks set off by heads. The writer sets up the document in the first paragraph by naming the three items—the 410, 430, and 480 mats—that he discusses in the body of the report. He repeats these keywords as section headings, presents information in a consistent pattern (vinyl/maratex, hole location, concerns) for each section, and numbers individual points within sections.

Clear title

Introduction "sets up" discussion

Purpose

Heading

Unemotional presentation

Review of Mats at Oxbow Creek Plant

After seeing the 410 and 430 Cargo mats, as well as the 480 Front mat, installed in the vehicles at Oxbow Creek, my impressions of each are as follows:

410 Cargo
1. The rear and side vinyl/maratex both fit. (The rear kick plate fits perfectly.)
2. All hole locations were good.
3. The front edge of vinyl/maratex was not evaluated because there was no base kick plate for the front.

Overall, I feel that the 410 Cargo mat fit was very good. However, the front edge of vinyl/maratex still needs to be evaluated with a base kick plate.

Words repeated as section heads

Chunking of information

430 Cargo
1. With our revised vinyl/maratex trim lines, I feel the mat fit is excellent. There was no pull out of the kick plate such as we noticed before.
2. All trim lines and holes are now good.

It is my feeling that our proposed design is much more functional than the original design and should be incorporated if feasible.

Words repeated as section heads

480 Cargo
1. All trim lines seemed good.
2. Hole locations were good. There was a little concern/suggestion that the rear group of holes (for the rear seat) be moved outward a few millimeters. The added lytherm seemed to bring them inward slightly.
3. The major concerns came in the B- and C-pillar areas (see attached sketch). There is much gapping between our mat and the molding. It is most evident in the C-pillar area. It is my opinion that our mat is correct in being molded to the sheet metal contour (in the C pillar) and that the pillars themselves are incorrect.

Chunking of information

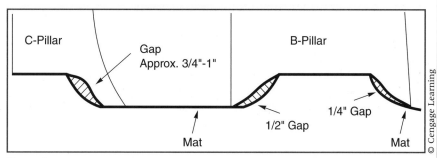

© Cengage Learning

Figure 1.
480 B-Pillar and C-Pillar Gap Problems

If you have any questions or would like to discuss these findings further, please let me know.

© Cengage Learning

Figure 2.1   Sample Report

4. The writer uses a visual aid—the drawing—to convey a problem discussed in the 480 section (Figure 1).

5. The writing is responsible. The writer tells the stakeholder (Sommers) all that she needs to know to be able to do her job. In addition, the report treats other stakeholders properly. Magolan's company, for instance, has informed individuals handling its affairs. Customers have received honest treatment of the problem, allowing them to interact with Magolan's company in an informed manner.

This brief report illustrates the skills and attitudes that technical writers employ. Although the report is a straightforward review of a site visit, it is nonetheless a well-crafted document that effectively conveys the information that writer and reader need to fulfill their roles in the organization.

Who Is the Audience?

The answer to this basic question dictates much of the rest of what you do in the document.

The audience is either someone you know (Sommers in the Magolan report) or a generalized group, such as "college freshmen" or "first-time cell phone users" (Coney and Steehouder). The audience could be a single person (your supervisor, a coworker), a small group (members of a committee), or a large group (the readers of a set of instructions). Sometimes the audience is multiple, that is, a primary audience who will act on the contents of your document and a secondary audience who read the document for information—to keep them in the loop—but who will not act on the information.

In order to communicate effectively with your audience, you have to engage them, that is, write in a way that makes it clear to them "that their knowledge and values are understood, respected, and not taken for granted" (Schriver 204). To create that engagement (see Schriver 152–163), you must answer these questions:

- What are the audience's demographic characteristics?
- What is the audience's role?
- How does the reader feel about the subject?
- How does the reader feel about the sender?
- What form does the reader expect?

What Are the Audience's Demographic Characteristics?

Demographic characteristics are basically objective items dealing with common ways of classifying people—age, ethnicity, and gender. The answers to these questions provide a base from which to act, but, in order to prevent stereotyping, have to be used with the answers to the other four questions listed earlier. For instance, in the Magolan example, the audience Marjorie Sommers is a 45-year-old white woman with a number of years' experience as a manager. However, the report is not written for a generic 45-year-old; it is written to Marjorie, an audience much more specific than a member of an age group. The report can be as individualized as it is because the writer also personally knows the audience and thus knows her attitudes and expectations.

Sometimes, however, if an audience is treated stereotypically, the message can fail. One researcher discovered that a brochure urging African-American teens to stay away from drugs failed because the writers used an image of a person who had an outdated hairstyle. The teens who read the brochure felt that the outmoded image indicated that the writer was someone who was out of touch; thus, they dismissed the brochure. In other words, the writers used demographic data but did not find answers to the other audience questions, and as a consequence failed to engage their audience (Schriver 171–189).

What Is the Audience's Role?

In any writing situation, your audience has a role. Like actors in a drama, audience members play a part, using the document as a "script." In the script, you "write a part" for department managers or tricycle assemblers or parents or students. For instance, the tricycle assembler could be a parent, on a Saturday morning, just home from the store, who needs to work quickly and efficiently to assemble the tricycle for an anxious three-year-old. The reader assumes that role as he or she reads. If the role in the document reflects the role the reader has in the real-life situation, the document will then engage the reader and thus make effective communication more likely. Readers' willingness to assume the role in the document often depends upon the way the role depicted in the writing is similar to the role they play in a real-life situation (Coney and Steehouder).

To create an effective role in the script, you have to understand the audience's role in real life. Because of their role, they have specific tasks, that is, specific responsibilities and actions. For instance, Marjorie Sommers is department manager. She evaluates data in relation to other aspects of the company. Her decisions have consequences because the corporation chooses to fund different actions based on her evaluations. She is professionally concerned that her clients be satisfied, that the company sell quality products. Personally, she is concerned that her work be judged as effective both by clients and by her supervisors. This matrix of characteristics defines her role. A writer must understand those characteristics in order to design an effective memo or any other document for her.

Role and task, in turn, are connected to need, those items necessary to fulfill the role. In order to fulfill her role in the organization, Marjorie Sommers must be assured that the mats serve the purpose for which they were sold. She must know of possible problems so that she can keep the customer happy

Personas Digging Deeper

Another method for understanding what motivates a particular type of audience is the creation of personas. Many writing professionals use this tool to understand who their customers are, and get in touch with their preferences, lifestyles, and beliefs (Duncan-Durst). A persona is "essentially a representative profile, which summarizes a key demographic target. It is usually accompanied by a photograph of a representative customer who is given a "real name" and assigned some basic demographic data (such as age, marital status, tangible occupation and income) as well as relevant information pertaining to personal behavior" (Duncan-Durst).

Creating a persona helps you understand whom you are talking to, so you know both how to talk to the audience and what to say. Using current slang to try and persuade an audience of senior citizens will likely not work. You need to think about what motivates and appeals to each type of audience. Then you need to tailor your messaging style to appeal to their beliefs, values, and lifestyles.

Although personas and profiles share a lot in common, personas tend to dig deeper. Some writers construct a "day in the life" story to imagine what their customers might do during the course of an ordinary day. As mentioned before, writers choose a stock photograph *of* their persona, so they can write *to* that person. Others write up imagined scenarios to determine what the persona would do in an imagined situation, like making a purchasing decision. Oftentimes, they base their personas on people they know.

Personas can be an effective tool when making any type of communication, whether print or Web. Having a visual representation of who your audience is can be extremely helpful. Keeping these ideas in mind when creating written messages will help you to make the best choices possible to ensure not only that readers understand your message, but also that the message appeals to them on a personal level.

with the product and with the company's service. She must be able to explain to her supervisor how her department is functioning. She has to decide whether she should talk to people in manufacturing about the fabrication of the part. In short, she needs the information to help her carry out her job responsibilities.

You can easily see the different effects of need by considering two audiences: operators of a machine and their department managers. Both groups need information but of different kinds. Operators need to know the sequence of steps that make the machine run: how to turn it on and off, how to set it to perform its intended actions, and how to troubleshoot if anything goes wrong. Managers need to know whether the machine would be a useful addition to the workstation and thus to purchase it. They need to know whether the machine's capabilities will benefit staff and budget. They need to know whether the machine has a variable output that can be changed to meet the changing flow of orders in the plant, whether the personnel on the floor can easily perform routine maintenance on the machine without outside help, and whether problems such as jamming can be easily corrected.

Because the audiences' needs differ, the documents directed at each are different. For the operator, the document would be a manual, with lots of numbered how-to-do-it steps, photos or drawings of important parts, and an index to help the operator find relevant information quickly. For the manager, the document would contain explanatory paragraphs rather than numbered how-to-do-it steps. Instead of photos, you might use a line graph that shows the effect of the variable rate of production or a table that illustrates budget, cost, or savings.

How Does the Reader Feel About the Subject?

The reader's feelings can be described as positively inclined, neutral, or negatively inclined toward the topic or the writer. Those feelings arise from many sources, which an author should analyze. For instance, any of these sources could affect the reader's attitudes: the topic (a pay cut, a meeting); the genre (a manual, a PowerPoint, a long e-mail, a Facebook message); the reader's opinion of the author (fair, clear, pushy, arrogant), the reader's level of knowledge (expert in this field, never heard of this issue, had an argument about this last week). If it is an e-mail from a committee chair announcing a meeting, the audience will probably feel neutral about the document and author. However, if the e-mail cancels a project led by the reader or announces a change in health insurance plans, the audience will likely feel negatively. If the e-mail announces that the advertising team won an award for its work, the audience will probably feel positively. Sometimes there is no way to predict the feelings. If the reader is having a "bad day," even a neutral message might be interpreted negatively.

If the audience is positively inclined, a kind of shared community can be set up rather easily. In such a situation, many of the small details won't make as much difference; the form that is chosen is not so important, and the document can be brief and informal. Words that have some emotional bias can be used without causing an adverse reaction. Marjorie Sommers is positively inclined toward the subject. Knowing about mats is part of her job; she is responsible for seeing that clients are satisfied with the product they purchased.

Much the same is true of an audience that is neutral. A writer who has to convey the minutes of a meeting that dealt with routine matters would assume that the audience will feel neutral about the topic and would just send an e-mail, if that is the usual practice. As long as the essential facts are present, the message will be communicated.

However, if the audience is negatively inclined, the writer cannot assume a shared community. The small details must be attended to carefully. Spelling, format, and word choice become even more important than usual because negatively inclined readers may seize upon anything that lets them vent their frustration or anger. Even such seemingly trivial documents as the announcement of a meeting can become a source of friction to an audience that is negatively inclined.

How Does the Reader Feel About the Sender?

A writer must establish a relationship with the reader, even if the reader has no previous relationship with the writer. Readers feel positively about a message if they feel that it is organized around their needs and if the writer "has taken the time to speak clearly, knowledgeably, and honestly to them" (Schriver 204). Documents that contain this sense of relationships tend to motivate the reader both to read and to act, a key requirement for effective communication (Carliner, 2000). To create this you-and-I-are-in-a-relationship sense, writers must create the belief that they are credible and authoritative and their documents must be "inviting" and "seductive."

Credibility means that you are a person who can be listened to. Credibility arises because of your role or your actions. If readers know that you are the quality control engineer, they will believe what you write on a quality issue. If readers know that you have followed a standard or at least a clear method of investigating a topic, they will believe you. The effect of demonstrating your credibility is that readers tend to think, "I will accept your message because I feel you are a credible person in the situation." Credibility grows out of competence and method.

Competence is control of appropriate elements. If you act like a competent person, you will be perceived as credible. The items discussed in this text will all improve your credibility—attention to formatting, to organization, to spelling and grammar, and to the audience's needs. Competence is also shown by tone. You will not seem credible if you sound casual when you should sound formal or facetious when you should sound serious.

Method includes the acts you have taken in the project. Simply put, audiences will view you as credible—and your message as believable—if they feel you have "acted correctly" in the situation. If the audience can be sure that you have worked through the project in the "right way," they will be much more likely to accept your requests or conclusions. If you have talked to the right people, followed the right procedures, applied the correct definitions, and read the right articles, the audience will be inclined to accept your results. For instance, if you tell the audience that package design A is unacceptable because it failed the Mullen burst test (an industry-wide standard method of applying pressure to a corrugated box until it splits), they will believe you, because you arrived at the conclusion the right way.

In all the informative memos and reports you write, you should try to explain your methodology to your reader. The IMRD report (Chapter 10) provides a specific section for methodology, and in other types of reports you should try to present the methodology somewhere, often in the introduction, as explained in this chapter. Sometimes one sentence is all you need: "To find these budget figures I interviewed our budget analyst." Sometimes you need to supply several sentences in a paragraph that you might call "background." However you handle it, be sure to include methodology both in your planning (what do I need to do to find this information?) and in your writing (be sure to add a reference or a lengthier section).

Authority means that you have the power to present messages that readers will take seriously (Lay). Basically, you have the right to speak because you have expertise, gained by either your role or your actions. Naturally, this authority is limited. Your report is authoritative enough to be the basis for company policy, even though you might not be the one who actually sets the policy.

Inviting documents cast the writer in a helpful role toward the reader. For instance, the writer could assume a role of guide who shows visitors the path through the forest of instructions in assembling a tricycle, a librarian who leads users through the information to find what they need. Often how inviting the document is depends on the wording. For instance, Coney and Steehouder suggest that on Web pages the phrase *e-mail us* is much more seductive than a link to an unnamed webmaster (332).

Seductive, in this usage, means "to attract our readers' attention and win their sympathy" (Horton 5). To create a seductive document is partly a matter of your attitude (and partly, as Horton explains, the way you design and present the information, concepts that will be dealt with in later chapters). Horton suggests that the key attitude is to present yourself as a person who will "guide and protect the reader" in order to stimulate him or her into action.

In our example, Sommers is positively inclined toward Magolan. Magolan knows that Sommers likes and trusts him because the two have worked together for a while. His past actions have generated a sense of authority. He is a person who has the right to speak because he has acted well in the past. Todd knows that Sommers is the supervisor and expects clear information without much comment.

Todd, however, creates a credible, inviting document. He establishes credibility in the first paragraph by explaining his methodology—he inspected the site. He clearly feels that he has the authority to make evaluative comments that suggest future actions (incorporate the proposed design, not the original one). His tone is informal, using "I" and "you"; notice, too, that he is comfortable enough to structure his communication as a short, no-nonsense list. In short, Todd presents himself as a person who will guide and protect the reader. He has created an easy-to-follow document that invites further action, should it be necessary. He is a person whom the reader can trust.

What Form Does the Reader Expect?

Many audiences expect certain types of messages to take certain forms. To be effective, you must provide the audience with a document in the form they expect. For instance, a manager who wants a brief note to keep for handy reference may be irritated if he gets a long, detailed business letter. An electronics expert who wants information on a certain circuit doesn't want a prose discussion because it is customary to convey that information through schematics and specifications. If an office manager has set up a form for reporting accidents, she expects reports in that form. If she gets exactly the form that she specified, her attitude may easily turn from neutral to positively inclined. If she gets a different form, her attitude may change from neutral to negatively inclined.

Marjorie Sommers expects an informal report that she can skim over easily, getting all the main points. She expects that this report, like all those she receives reporting on site visits, will have an e-mail "cover letter" and will be an attachment to that e-mail. It will have the main point first and heads to break up the text. Todd knows that the message must be brief (one to two pages), that its method of production must be a personal computer, and that it must appear as a typical Word document so that it can easily be printed on 8 1/2-by-11-inch paper if that becomes necessary.

What Is the Audience's Task?

What will the reader do after reading the document? Although fulfilling a need is why the audience is involved in the situation, the task is the action they must accomplish (Rockley). Tasks vary greatly, and can be nearly anything—to assemble a tricycle or a workstation, to say no to drugs, to agree to build a retail outlet at a site, to evaluate the sanitary conditions of a restaurant. The document must enable the reader to perform that task.

Marjorie Sommers's task is to act to protect the interests of the company. As a result, she will alert her superiors to the problem with the pillars. Because Magolan feels the problem is the customer's, Sommers will not ask manufacturing to change their process. But because she has been informed, she will have the facts she needs if she must act at a later date.

What Is the Audience's Knowledge Level?

Every audience has a knowledge level, the amount they know about the subject matter of the document. This level ranges from expert to layperson (or nonexpert). An expert audience understands the terminology, facts, concepts, and implications associated with the topic. A lay audience is intelligent but not well informed about the topic. Knowing how much the audience knows helps you choose which information to present and in what depth to explain it.

Adapting to Your Audience's Knowledge Level

You adapt to your audience's knowledge level by building on their schemata—that is, on concepts they have formed from prior experiences (Huckin). The basic principles are "add to what the audience knows," and "do not belabor what they already know." If the audience knows a term or concept (has a schema for it), simply present it. But if the audience does not know the term or concept (because they have no schema), you must help them grasp it and add it to their schemata.

Suppose for one section of a report you have to discuss a specific characteristic of a digitized sound. If the reader has a "digitized sound schema," you can use just the appropriate terminology to convey a world of meaning. But if the reader does not have this schema, you must find a way to help him or her develop it. The following two examples illustrate how writers react to knowledge level.

For a More Knowledgeable Audience

For a more knowledgeable audience, a writer may use this sentence:

That format allows only 8-bit sampling.

The knowledgeable reader knows the definitions of the terms *format* and *8-bit*. He or she also understands the implication of the wording, which is that the sound will not reproduce as accurately if it is sampled at 8 bits, but that the file will take up less disk space than, say, 16-bit sampling.

For a Less Knowledgeable Audience

A less knowledgeable audience, however, grasps neither the definitions nor the implications. To develop a schema for such readers, the writer must build on the familiar by explaining concepts, formatting the page to emphasize information, making comparisons to the familiar, and pointing out implications. You might convey the same information about sampling to a less knowledgeable audience in a manner as shown in Figure 2.2 (p. 46) (Stern and Littieri 146).

Finding Out What Your Audience Knows

Discovering what the audience knows is a key activity for any writer. It complements and is as important as discovering the audience's role. To estimate an audience's knowledge level, you can employ several strategies (Coney; Odell et al.; Selzer).

- Ask them before you write. If you personally know members of the audience, ask them in a phone call or brief conversation how much they know about the topic.
- Ask them after you write. Ask the audience to indicate on your draft where the concepts are unfamiliar or the presentation is unclear.
- Ask someone else. If you cannot ask the audience directly, ask someone who knows or has worked with the audience.

Explanation
of concept

In many ways, it helps to think of digitized sound as being analogous to digitized video.

Digitized sound is actually composed of a sequence of individual sound samples. The number of samples per second is called the *sample rate* and is very much like a video track's frame rate. The more sound samples per second, the higher the quality of the resulting sound. However, more sound samples also take up more space on disk and mean that more data need to be processed during every second of playback. (The amount of data that must be processed every second is called the *data rate*.)

Highlighted
text

Implication
of explanation

Analogy

Sound samples can be of different sizes. Just as you can reproduce a photograph more faithfully by storing it as a 24-bit (full-color) image than as an 8-bit image, 16-bit sound samples represent audio more accurately than 8-bit sound samples. We refer to the size of those samples as a sound's *sample size*. As with the sample rate, a larger sample size increases the accuracy of the sound at the expense of more storage space and a higher data rate.

Highlighted
text
Implications

© Cengage Learning

Figure 2.2 Writing for a Less Knowledgeable Audience

- Consider the audience's position. If you know the duties and responsibilities of the audience members, you can often estimate which concepts they will be familiar with.
- Consider prior contacts. If you have had dealings with the audience before, recall the extent of their knowledge about the topic.

What Factors Influence the Situation?

In addition to the personal factors mentioned earlier in relation to role, many business or bureaucratic situations have factors external to both the reader and the writer that will affect the design of a document. These factors can powerfully affect the way readers read. Common questions (based on Odell et al.) to answer are

- What consequences will occur from this idea?
- What is the history of this idea?
- How much power does the reader have?
- How formal is the situation?
- Is there more than one audience?

What Consequences Will Occur from This Idea?

Consequences are the effects of a person's actions on the organization. If the effect of your suggestion would be to violate an OSHA standard, your suggestion will be turned down. If the effect would be to make a profit, the idea probably will be accepted.

What Is the History of This Idea?

History is the situation prior to your writing. You need to show that you understand that situation; otherwise, you will be dismissed as someone who does not understand the implications of what you are saying. If your suggestion to change a procedure indicates that you do not know that a similar change failed several years ago, your suggestion probably will be rejected.

How Much Power Does the Reader Have?

Power is the supervisory relationship of the author and the reader. Supervisors have more power. Orders flow from supervisors to subordinates, and suggestions move in the reverse. The more powerful the reader, the less likely the document is to give orders and the more likely it is to make suggestions (Driskill; Fielden; Selzer).

How Formal Is the Situation?

Formality is the degree of impersonality in the document. In many situations, you are expected to act in an official capacity rather than as a personality. For an oral presentation to a board meeting concerning a multimillion-dollar planning decision, you would simply act as the person who knows about widgets. You would try to submerge personal idiosyncrasies, such as joking or sarcasm. Generally, the more formal the situation, the more impersonal the document.

Is There More Than One Audience?

Sometimes a document has more than one audience. In these situations, you must decide whether to write for the primary or secondary audience. The primary audience is the person actually addressed in the document. A secondary audience is someone other than the intended receiver who will also read the document. Often you must write with such a reader in mind. The secondary reader is often not immediately involved with the writer, so the document must be formal. The following two examples illustrate how a writer changes a document to accommodate primary and secondary audiences.

Suppose you have to write an e-mail to your supervisor requesting money to travel to a convention so that you can give a speech. This e-mail is just for your supervisor's reference; all he needs is a brief notice for his records. As an informal e-mail intended for a primary audience, it might read like Figure 2.3.

If this brief note is all your supervisor needs, neither a long, formal proposal with a title page and table of contents nor a formal business letter would be appropriate. The needs of the primary audience dictate the form and content of this e-mail.

Suppose, however, that your supervisor has to forward the e-mail to his manager for her approval. In that case, a brief, informal e-mail would be inappropriate. His manager might not understand the significance of the trip or

<table>
<tr><td>

Brief subject line

Informal use of name

No formatting of document

</td><td>

Upcoming Trip to SME

John

This is my formal request for $750 in travel money to give my speech about widgets to the annual Society of Manufacturing Engineers convention in San Antonio in May. Thanks for your help with this.

Fred

</td></tr>
</table>

© Cengage Learning

Figure 2.3 Informal E-mail for a One-Person Audience

<table>
<tr><td>

A more official format, including explanatory subject line:

Orients reader to background and makes request

Explains background of request

Adds detail that primary audience knows

</td><td>

Travel money for my speech to Society of Manufacturing Engineers convention

Hi John, As I mentioned to you in December, I will be the keynote speaker at the Annual Convention of the Society of Manufacturing Engineers in San Antonio. I would like to request $750 to defray part of my expenses for that trip.

  This group, the major manufacturing engineering society in the country, has agreed to print the speech in the conference proceedings so that our work in widget quality control will receive wide readership in M.E. circles. The society has agreed to pay $250 toward expenses, but the whole trip will cost about $1000.

  I will be gone four days, May 1–4; Warren Lang has agreed to cover my normal duties during that time. Work on the Acme Widget project is in such good shape that I can leave it for those few days. May I make an appointment to discuss this with you? Fred Paulson

</td></tr>
</table>

© Cengage Learning

Figure 2.4 More Formal E-mail Written for Multiple Audiences

might need to know that your work activities will be covered. In this new situation, your document might look like Figure 2.4.

  As you can see, this document differs considerably from the first e-mail. It treats the relationship and the request much more formally. It also explains the significance of the trip so that the manager, your secondary audience, will have all the information she needs to respond to the request.

Creating Audience Profiles

Before you begin to write, you need to use the concerns outlined in this chapter to create an audience profile, a description of the characteristics of your audience. The profile is an image of a person who lives in a situation. You use that description as the basis for decisions you make as you create your document (Lazzaro; Rockley).

In order to create a profile, you need to ask specific questions and use an information-gathering strategy.

Questions for an Audience Profile

To create the profile, ask the questions discussed in this chapter:

- ❯ Who is the audience?
  - ❯ What are their demographic characteristics?
  - ❯ What is the audience's role?
  - ❯ How does the reader feel about the subject?
  - ❯ How does the reader feel about the sender?
  - ❯ What form does the reader expect?
- ❯ What is the audience's task?
- ❯ What is the audience's knowledge level?
- ❯ What factors influence the situation?

Information-Gathering Strategies

To create these answers, use one of two methods—create a typical user or involve the actual audience at some point in your planning (Schriver 154–163).

Create a Typical User

Creating a typical user means to imagine an actual person about whom you answer all the profile questions. Suppose that you have to write a set of instructions for uploading a document to a Web server. You could create a typical user whom you follow in your mind as he or she enacts the instructions.

Here is such a creation: Marie Williamson, a sophomore and a Mac user (*demographics*) arrives in the lab (*situation*) with the assignment to upload several files to the Web using a PC (*task*). She is taking her first course that exclusively uses PCs and needs to upload the files to the Web so that she receives credit (*role*). She intensely dislikes PCs (*attitude toward the subject*) and has never really liked using manuals (*expectation about sender and form*). She has never done this before by herself (*knowledge level*). She is stressed because she has only 20 minutes in the lab to do this before she has to go to work, and she still has no babysitter lined up for her child (*other factors*) (see Garret 54–56).

To write the manual, you try to accommodate all the "realities" that you feel will affect "Marie's" or any user's ability to carry out the instructions. Your goal is to write an inviting, seductive manual, one that entices her to read and to act. If you can write a manual that would help "Marie," it is a good guess that it will help other people also. Using "Marie" you can make decisions that will help create a document that enables any reader to accomplish his or her task and enhance his or her relationships with teachers, coworkers, and children.

Involve the Actual Audience

You could interview actual members of the target audience. Instead of creating "Marie," you interview several people who have to upload files using a PC instead of a Mac. You ask them the profile questions and, in the best practice, later ask them to review the manual before the final draft is published. When you interview them, you ask the same profile questions as you would if you were creating a typical user, but of course you get their answers rather than your imagined ones. While this method is slower than creating an imaginary user, it is often more accurate in gauging real users' needs because it clarifies what the attitudes and experiences really are.

Exercises

▶ You Create

1. Write three different sentences in which you use a technical concept to explain three situations to a person who knows as much as you do. (Example: "You can't print that because the computer doesn't have the correct printer drivers installed." In this sentence, the writer assumes that the reader understands computers, printers, and their relationship to print drivers.)

2. Write a brief set of instructions (three to six steps) to teach an audience how to use a feature of a machine—for instance, the enlarge/reduce feature on a photocopier or changing the desktop wallpaper on a computer. Then exchange your instructions with a partner. After interviewing the partner to learn the procedure, rewrite the instructions as a paragraph that explains the value of the feature to a manager.

3. Write a brief memo in which you propose a change to a more powerful, positively inclined audience. Use an emotional topic, such as eliminating all reserved parking at your institution or putting a child care room in every building. Before you write, answer the questions on page 49. In a group of three or four, explain how your memo reflects the answers you gave to those questions.

▶ You Analyze

4. Analyze either the memo in Exercise 15, Chapter 1 (p. 29), or the report in Example 13.2 (pp. 399–401) in order to determine the audience's need and task. Be prepared to report your findings orally to the class.

5. Analyze this brief message to determine whether you should rewrite it. Assume it is written to a less powerful, negatively inclined audience

(Kostelnick). If you decide it needs revision, rewrite it. If your instructor so requires, discuss your revisions in groups of three.

> Thank you for your recent proposal on the 4-day, 10-hour-a-day week. We have rejected it because

> It is too short.
> It is too narrow-minded.
> It has too many errors.

▶ You Revise

6. Rewrite one of the three sentences in Exercise 1 into a larger paragraph that makes the same idea clear to someone who knows less than you do.

7. The Magolan report on page 37 is aimed at an audience of one. Rewrite it so that it includes a secondary audience who is interested in customer relations. If the subject matter of that memo is too unfamiliar to you, use a subject you know well (in-line fillers or a client's computer network).

▶ Group

8. In class, set up either of these two role-playing situations. In each, let one person be the manager, and let two others be employees in a department. In the first situation, the employees propose a change, and the manager is opposed to it. In the second, the employees propose a change, and the manager agrees but asks pointed questions because the vice president disagrees. In each case, plan how to approach the manager, and then role-play the situation. Suggestions for proposed changes include switching to a 4-day, 10-hour-a-day week; instituting recreational free time for employees; and having a random drawing to determine parking spaces instead of assigning the spaces closest to the building to executives. After you complete the role-playing, write a memo to the manager requesting the change. Take into consideration all you found out in the role-playing.

9. In groups of three or four, agree on a situation in which you will propose a radical change, perhaps that each class building at your college contain a child care room. Plan how to approach an audience that feels positive about this subject. For instance, which arguments, formats, and visual aids would be persuasive? Then write the memo as a group. For the next class, each person should bring a memo that requests the same change but addresses an audience that is neutral or negative (you choose) about this subject. Discuss the way you changed the original plan to accommodate the new audience. As a group, select the best memo and read it to the class.

Worksheet for Defining Your Audience

☐ **Who will read this document?**

- Name the primary reader or readers.
- Name any secondary readers.

☐ **Determine the audience's level of knowledge.**

- What terms do they know?
- What concepts do they know?
- To find your reader's knowledge level, you must (1) ask them directly, (2) ask someone else who is familiar with them, or (3) make an educated guess.
- Do they need background?

☐ **Determine the audience's role.**

- What will they do as a result of your document?
- Have you presented the document so they can take action easily?
- Why do they need your document? for reference? to take to someone else for approval? to make a decision?

☐ **Determine the audience's community attitudes.**

- What are the social factors in the situation?
- Is the audience negatively or positively inclined toward the message? toward you?
- What format, tone, and visuals will make them feel that you are focused on their needs?

☐ **What form does the audience expect?**

Writing Assignments

1. Interview one or two professionals in your field whose duties include writing. Find out what kinds of audiences they write for by asking a series of questions. Write a memo summarizing your findings. Your goal is to characterize the audiences for documents in your professional area. Here are some questions that you might find helpful:

   - What are two or three common types of documents (proposals, sets of instructions, informational memos, letters) that you write?
   - Do your audiences usually know a lot or a little about the topic of the document?

- Do you try to find out about your audience before you write or as you write?
- What questions do you ask about your audience before you write?
- Do you change your sentence construction, sentence length, or word choice to suit your audience? If so, how?
- Do you ever ask someone in your intended audience to read an early draft of a document?
- Does your awareness of audience power or inclination affect the way you write a document?
- Does your awareness of the history of a situation affect the way you write a document?
- Do you ever write about the same topic to different audiences? If yes, are the documents different?
- Do you ever write one document aimed at multiple audiences? If yes, how do you handle this problem?

2. Write two different paragraphs about a topic that you know thoroughly in your professional field. Write the first to a person with your level of knowledge. Write the second to a person who knows little about the topic. After you have completed these two paragraphs, make notes on the writing decisions you made to accommodate the knowledge level of each audience. Be prepared to discuss your notes with classmates on the day you hand in your paragraphs. Your topic may describe a concept, an evaluating method, a device, or a process. Here are some suggestions; if none applies to your field, choose your own topic or ask your instructor for suggestions.

Choosing a Web browser	The camera on a smart phone
Benefits of the most recent X-box	The food pyramid
A machining process used in wood	
Finding an address on the Internet	The relation of calories to grams
Speakers for an audio system	Registering for a class
Cranking amps of a battery	Antilock brakes
The capacity and speeds of external storage devices	Using a smart to receive and send e-mail (or to use a certain app, such as one that converts voice to text)
Creating a document using HTML code	Just-in-time manufacturing
	Securing an internship

3. Form groups of three or four. If possible, the people in each group should have the same major or professional interest. Decide on a short process (four to ten steps) that you want to describe to others. As a group, analyze the audience knowledge and then write the description.

Alternate: After writing the memo, plan and write the same description for an audience with different characteristics than your first audience.

Web Exercise

Analyze a website to determine how the authors of the site "envision" their readers. What have they assumed about the readers in terms of knowledge and needs, role, and community attitudes? Write an analytical report to your classmates or coworkers in which you explain strategies they should adopt as they develop their websites.

Works Cited

Carliner, Saul. (2000). "Physical, Cognitive, and Affective: A Three-Part Framework for Information Design." *Technical Communication* 47.4 (2000): 561–572. Print.

Coney, Mary. "Technical Communication Theory: An Overview." *Foundations for Teaching Technical Communication: Theory, Practice, and Program Design.* Ed. Katherine Staples and Cezar Ornatowski. Greenwich, CT: Ablex, 1997. 1–16. Print.

Coney, Mary, and Michael Steehouder. "Role Playing on the Web: Guidelines for Designing and Evaluating Personas Online." *Technical Communication* 47.3 (Third Quarter 2000): 327–340. Print.

Driskill, Linda. "Understanding the Writing Context in Organizations." *Writing in the Business Professions.* Ed. Myra Kogen. Urbana, IL: NCTE, 1989. 125–145. Print.

Duncan-Durst, Leigh. "The Power of Personas." *Marketing Profs.* 2006. Web. 28 Oct 2010. <http://www.mpdailyfix.com/the-power-of-personas>.

Fielden, John S. "What Do You Mean You Don't Like My Style?" *Harvard Business Review* 60 (1982): 128–138. Print.

Garret, Jesse James. *The Elements of User Experience: User-Centered Design for the Web.* Indianapolis, IN: American Institute of Graphic Arts, 2003. Print.

Horton, William. *Secrets of User-Seductive Documents.* Arlington, VA: Society for Technical Communication, 1997. Print.

Huckin, Thomas. "A Cognitive Approach to Readability." *New Essays in Technical and Scientific Communication: Research, Theory, and Practice.* Ed. Paul V. Anderson, R. John Brockman, and Carolyn Miller. Farmingdale, NY: Baywood, 1983. 90–108. Print.

Kostelnick, Charles. "The Rhetoric of Text Design in Professional Communication." *Technical Writing Teacher* 17.3 (1990): 189–203. Print.

Lay, Mary. *The Rhetoric of Midwifery: Gender, Knowledge, and Power.* New Brunswick, NJ: Rutgers University Press, 2000. Print.

Magolan, Todd. Personal memo, 15 March 1991. Used by permission.

Odell, Lee, Dixie Goswami, Ann Herrington, and Doris Quick. "Studying Writing in Non-Academic Settings." *New Essays in Technical and Scientific Communication: Research, Theory, and Practice.* Ed. Paul V. Anderson, R. John Brockman, and Carolyn Miller. Farmingdale, NY: Baywood, 1983. 17–40. Print.

Rockley, Ann. "Single Sourcing and Information Design." *Content and Complexity: Information Design in Technical Communication.* Ed. Michael J. Alebers and Beth Mazur. Mahwah, NJ: Lawrence Erlbaum, 2003. 307–335. Print.

Schriver, Karen. *Dynamics in Document Design: Creating Texts for Readers.* New York: Wiley, 1997. Print.

Selzer, Jack. "Composing Processes for Technical Discourse." *Technical Writing: Theory and Practice.* Ed. Bertie E. Fearing and W. Keats Sparrow. New York: MLA, 1989. 43–50. Print.

Stern, Judith, and Robert Littieri. *QuickTime and Moviemaker Pro for Windows and Macintosh.* New York: Peachpit, 1998. 146. Print.

The Technical Communication Process

CHAPTER CONTENTS

CHAPTER 3 **IN A NUTSHELL**

You have to plan, draft, revise, and finish your document, either by yourself or in a group.

Plan by establishing your relationship with your audience and your document. Situate yourself by determining your knowledge of the topic and your goal for the audience. Also determine whether legal, ethical, or global issues are involved. Create a clean document design, both for physical appearance and for content strategy. Develop a realistic production schedule. You want your audience to accept what you tell them. They have to accept you as credible—because they know "who you are" or because you have performed the "right action" to familiarize yourself with the topic.

Draft by carrying out your plan. Find your best production method. Some people write a draft quickly, focusing on "getting it out," whereas others write a draft slowly, focusing on producing one good sentence after another. Keep basic strategies—for instance, the top-down method of first announcing the topic and then filling in the details—in mind as you write. Almost no one gets the draft "right" on the first iteration. Think of it as your "alpha" version. Build in time to create a "beta" version. To do so, review your draft on different levels—look at whether the topics are in the most helpful order, consider whether examples or visual aids should be deleted or added, keep asking "Does this help my reader understand?" If the review causes you to see a new, better way to present the material, change.

Edit by making the document consistent. Look for surface problems, such as spelling, grammar, and punctuation. Make sure all the presentation elements—heads, captions, margins—are the same. Set and meet quality benchmarks.

Work in a group by expanding your methodology. For groups, add into your planning a method to handle group dynamics—set up a schedule, assign responsibilities, and, most important, select a method for resolving differences.

Like all processes, document production proceeds in stages. This chapter explains each of these stages and introduces you to writing as a member of a group, a practice common in industry and business.

An Overview of the Process

The goal of technical communication is to enable readers to act. To do so, you need to create a document that is helpful and appropriate, one in which the audience can find what they need, understand what they find, and use what they understand appropriately (Redish). In addition, you need to create a document that engages readers so that they focus their attention on the patterns in the document. In order to achieve these goals, you need to follow a process. This chapter explains that process, including special items you need to consider if you are composing a document as part of a group.

Many technical writers have discussed the process that they use to produce documents. These discussions have a remarkable consistency. Almost all writers feel that the process is both linear (following the sequence step by step) and recursive (returning to previous steps or skipping ahead as necessary).

Almost all use some version of this sequence:

▶ Plan by discovering and collecting all relevant information about the communication situation.
▶ Draft, test, and revise by selecting and arranging the elements in the document.
▶ Finish by editing into final form.

Figure 3.1, a flow chart of the process (adapted from Goswami et al. 38; Redish 164), indicates the steps and nature of the process. The blue arrows indicate the linear sequence: first plan, then write, and then edit. This path is the standard logical sequence that most people try to follow in any project. The black arrows, however, indicate the recursive nature of this process—you must be ready to return to a previous stage or temporarily advance to a subsequent stage to generate a clear document.

Planning Your Document

During the planning stage, you answer a set of questions concerning your audience, your message, your document's format, and the time available for the project. The answers to those questions will give you important information about your audience and a general idea of the document you will create.

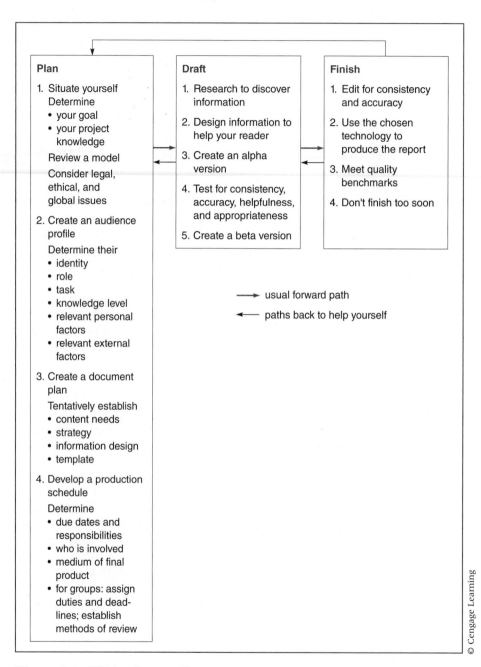

Plan

1. Situate yourself
   Determine
   • your goal
   • your project
     knowledge
   Review a model
   Consider legal,
   ethical, and
   global issues

2. Create an audience
   profile
   Determine their
   • identity
   • role
   • task
   • knowledge level
   • relevant personal
     factors
   • relevant external
     factors

3. Create a document
   plan
   Tentatively establish
   • content needs
   • strategy
   • information design
   • template

4. Develop a production
   schedule
   Determine
   • due dates and
     responsibilities
   • who is involved
   • medium of final
     product
   • for groups: assign
     duties and dead-
     lines; establish
     methods of review

Draft

1. Research to discover
   information

2. Design information to
   help your reader

3. Create an alpha
   version

4. Test for consistency,
   accuracy, helpfulness,
   and appropriateness

5. Create a beta version

Finish

1. Edit for consistency
   and accuracy

2. Use the chosen
   technology to
   produce the report

3. Meet quality
   benchmarks

4. Don't finish too soon

⟶ usual forward path

⟵ paths back to help yourself

© Cengage Learning

Figure 3.1   Writing Process Chart

Depending on the situation, planning can be either brief or lengthy. For a short report or e-mail, the planning step might be just making a few mental notes that guide you as you compose. For instance, to send out a routine meeting announcement via e-mail, you might only have to select the recipients from your address list; confirm the date, place, and agenda; and review previous routine announcements to get the correct page design. But for a lengthy proposal or manual, the planning sessions could yield a written document that specifies the audience and explains the way they will use the planned document, a detailed style sheet for format, and a realistic production schedule.

Regardless of the amount of time spent on this step, planning is a key activity in writing any document. Better writing results from better planning (Dorff and Duin; Flower).

To plan effectively, situate yourself, create an audience profile, create a document plan, design your information, design your template, and create a production schedule.

Situate Yourself

To situate yourself, you need to determine what you are trying to do. Take a moment (or several) to think through your answers to the questions discussed next.

What Is Your Goal?

Your goal is to direct the audience to a specific experience after reading your document. What should they know or be able to do after reading your document, and how should they be helped to that new condition (see Screven; Shedroff)? The answer depends on the situation, and often seems obvious, but specifically stating your goal will give you a mission statement that will guide your work. Examples include, "Through reading this document the audience will assemble the tricycle." "Through reading this document the audience will find compelling reasons to fund my research project." "Through reading this document the audience will understand the results and implications of this experiment" (see Horton, Chapter 1).

How Much Project Knowledge Do You Have?

You need to decide how much you know. The answer to this question depends on the relation of the project and the document. If the project is completed, you have the knowledge and need to perform little if any research. You know what happened in the experiment; you know what you found out when you investigated sites for a new retail store. On the other hand, in many situations you are handed an assignment that results in a document, but concerns a topic about which you know nothing. In this case, you need to plan how you will discover all the relevant information.

If you are assigned to create a set of instructions for software that you are unfamiliar with, then you will have to undergo some kind of training and perform research to find the information. If you have to write a proposal for funding a program, you will have to research all the information that the granting agency requires.

This information will eventually find its way into your production schedule, discussed later in this chapter.

Is There a Model to Help You Focus Your Thinking?

Most people find a document easier to create if they have some idea of what it ought to look like. As a college student, you no doubt have asked a teacher for specifications on a research paper's length and whether you need to include formal footnotes so you know how to shape the material. To find a model, consider any of these methods. Is there a look imposed by a genre? Many technical documents belong to a *genre,* a formalized way of handling recurring writing tasks. For instance, sets of instructions are a genre; most of them look roughly alike, with a title, introduction, and chronologically sequenced steps. Is there a company style sheet? Many companies have style guidelines to follow when creating a report—the title page, for instance, must contain specified information such as title, author, recipient, date, and project number; chapters must begin on a new page, using a font of a certain size. Look for an earlier document of the same kind to serve as a model. Many times, authors who have to write a report, say, a personnel evaluation or committee minutes, will find an example of previous work done acceptably, and use that approach.

All these are examples of the author finding out in the planning stage the "vessel" into which he or she will "pour" his or her ideas. Having this look in mind allows you to plan other elements effectively.

Are Global Issues Involved?

If the document will be translated for or used by people in other countries, there are global issues to be dealt with. In that case, you need to review the type of localization (as explained in Chapter 1, pp. 19–26) you will use—radical or general. You will then have to adjust the document's form and wording to include the audience who does not share your cultural background.

Are Legal or Ethical Issues Involved?

Ethical issues, as discussed in Chapter 1, deal with doing the right thing for the stakeholders involved. These issues can be very important, such as whether this project and the report that supports it actually break the law, or they can deal with issues of treating information clearly for all stakeholders. At times, issues may need to be reviewed by legal counsel, for instance, phrasing sentences so that you remain in compliance with terms of a contract.

Create an Audience Profile

To create an audience profile you need to answer the audience questions listed in Chapter 2. Rephrased slightly, those questions are

- Who is the audience?
- What is their role?
- What is their task?
- What is their knowledge level?
- What personal factors influence the situation (feelings about subject, sender, expected form of communication)?
- What external factors influence the situation?

As was explained in Chapter 2, planning for your audience is a key method for writing helpful and appropriate documents. Since these questions have all been dealt with there, they will not be discussed here. If the audience is located in another country, use ideas from Chapter 1 (see section "Technical Communication Is Global") in your planning.

Create a Document Plan

To create a document plan, establish your content needs, establish your strategy, decide whether to use a genre, decide whether to use an established pattern, and consider using a metaphor.

Establish Your Content Needs

To establish your content needs you must determine what you know about the topic and what the audience knows and needs to learn. If you know quite a bit, then you understand the content and have little need for more research. If you know little or nothing about the topic, then you have to set up a plan to discover all the information before you can decide what the audience needs.

To deal with the audience knowledge and needs, refer again to the schemata discussion (Chapter 2, pp. 42–45), where you will find the basic principle: Add to what the audience knows and don't belabor the obvious. If they know a lot, then you can assume that they know broad terminology and implications of actions (e.g., if you say, "Open a Word file," they will perform all the actions necessary to open a new file on the screen without further instruction). If they do not know a lot, then build on what they do know (if your audience is composed of computer novices, you would instruct them to turn on the computer, find the Word icon, double-click on it, and choose New from the File menu).

Establish Your Strategy

To establish your strategy, you determine how to carry out your goal. Strategy is your "creative concept," the way you present the material so that your reader can easily grasp and act upon it. What design will help the reader to do that? Essentially, you must create an experience for your reader, helping him or her to build a unique model of the information (Albers). In order to build the model, the reader must be able to see all the parts and then work to join them together into a final product. It is as if every writing situation were a box filled with parts that have to be assembled. As the writer, you have to lay out those parts and then give the reader what is needed to assemble them. For instance, you lay out the problem and then help the reader understand (and agree to) your solution. Or, you lay out the goal and then take the reader step by step through a process that leads to the goal. While establishing a strategy and building a model are often difficult to do, they are also a fun way to think about relating to your topic (Garret 40–59).

The idea of strategy is to help readers grasp the big picture so that they can interrelate all the details that you present to them. The organization of your document helps them find that big picture. Three common ways to organize are to follow a genre, to use established patterns, and to use a metaphor.

Decide Whether to Use a Genre

A genre is a standardized way to present information (Carliner 49). Various genres include documents such as sets of instructions, proposals, trip reports, or meeting agendas, documents so common that just by looking at them the reader knows what to expect. He or she understands what they are, what they are likely to include, and how they are likely to be structured. In other words, they know what experience they will have as they read the document. For instance, in instructions, lists of necessary materials appear in the introduction, and the steps appear in the body in sequential order. Readers know that they have to find the parts and then follow the steps to reach the goal.

Decide Whether to Use an Established Pattern

If no genre exists, you could try several other options. You could use a common rhetorical sequence: for instance, definition followed by example and analogy. You could establish an organizational progression, such as most vital to least important or top to bottom. To treat the components of a computer system from most to least important, you might start with the central processing unit and end with the electrical cord. To discuss it from top to bottom, you might start with the screen and end with the keyboard.

Design Your Information

In addition to choosing an overall organization, plan how you will present information within that organization. Several methods will help you. The key items are choosing an arrangement and using various cueing devices to help the reader pick out what is important and to "assemble" relationships. Since many of these devices become clearer to a writer as he or she actually writes, they are discussed at length in the sections on drafting, testing, and finishing your document.

Design Your Template

A *template* is a general guide for the look of your page. It is closely related to a style sheet, a list of specifications for the design of the page. The goal of a style sheet is to create a visual logic—a consistent way to visually identify parts of a paper. The parts of a style sheet are the width of margins, the appearance of several layers of heads, the treatment of visuals and lists, the position of page numbers, and the typeface. These items are described in detail in Chapter 6. Most sophisticated word processing programs (e.g., Microsoft Word) allow you to set these formats before you begin.

Figure 3.2 Shows a Sample Template.

Title—24 point bold Arial, set up and down (i.e., capitalize all important words)
Level 1 Heading—12 point Arial bold, set flush left, up and down
Level 2 Heading—12 point Arial bold, set flush left, up and down, followed by a period
Text font—12 point Arial
Spacing—single-space within paragraphs and double-space between
Visual aids—set flush left
Caption—10 point Arial italicized, set flush left, up and down; include the word Figure followed by a period, then the title and no period

Sample Report Template

Level 1 Heading

This is text font. It is Arial 12 point.

Level 2 Heading. This is more text font. Below is a visual aid with its caption.

Figure 1.
Caption

Figure 3.2 Sample Template

An Example of Developing a Style Sheet

Let's return again to the report (Figure 2.1, pp. 36–38) to see how the author developed a style sheet. He used headings to make the three sections of the report easy to locate, and he used numbering to set off each item. His lists use the "hanging indent" method, the second line starting under the first word of the first line. The visual aid is set at the left margin with the caption above it.

Create a Production Schedule

A production schedule is a chronological list of the activities required to generate the document and the time they will consume. Your goal is to create a realistic schedule, taking into consideration the time available and the complexity of the document. You need to answer these questions: How much time do I have? Who is involved in producing the document? What constraints affect production?

How Much Time Do I Have?

You have from the present to the deadline, be it one hour, two weeks, or four months. Determine the end point and then work backward, considering how long it will take to perform each activity. How long will it take to print the final document? How much time must you allot to the revision and review stage? How long will it take to draft the document? How much time do you need to discover the "gist" of your document?

A major problem with time management is "finishing too soon." Many people bring hidden time agendas to projects. They decide at the beginning that they have only so many hours or days to devote to a particular project. When that time is up, they must be finished. They do not want to hear any suggestions for change, even though these suggestions are often useful and, if acted upon, could produce a much better document. Another time management problem often occurs in research projects. A fascinated researcher continually insists on reading "just one more" article or book, consuming valuable time. When he or she begins to write the report, there is not enough time left to do the topic justice. The result is a bad report.

As your ability to generate good documents increases, you will get better at estimating the time it will take to finish a writing project. You will also develop a willingness to change the document as much as necessary to get it right. Developing these two skills is a sure sign that you are maturing as a writer. The worksheet on page 77 provides a useful editing checklist.

Also consider the time it will take to produce the final document. Who will type it? What kind of system will you use? Word processor? Professional typesetter and printer? E-mail? How well do you know the system? The less familiar you are with your production tools, the more time you are likely to spend.

For a short document, of a type you have created before, the answers are obvious. But for long, complex documents, these questions are critical. The creation of such a document is far easier if you answer these questions realistically and accurately in the planning stage.

Who Is Involved in Producing the Document?

The number of people involved in the procedure varies from one (you) to many especially if there is a review process. If it is only you who is involved, you need to consider only your own work habits as you schedule time to work through the document. If many people are involved—such as reviewers for technical accuracy, legality, and internal consistency—you will have to schedule deadlines for them to receive and return your document.

What Constraints Affect Production?

Constraints are factors that affect production of the document. They include time, length, budget, method of production, method of distribution, and place of use (Goswami et al.).

▶ *Time* is the number of hours or weeks until the date the document is due.

▶ *Length* is the number of pages in the final document.

▶ The *budget* specifies the amount of money available to produce the document. A negligible concern in most brief documents (those one to three pages in length), the budget can greatly influence a large, complex document. For instance, a plastic spiral binding could increase costs slightly but a glued, professional-type binding could be prohibitively expensive.

▶ *Method of production* is the type of equipment used to compose the document. If you plan to use two word processors, one in a public lab and one at home, both must run the same software so that you can use your disk in either system.

Worksheet for Planning—Short Version

☐ **What is the situation?**

  a. Who will do what? When? Where?
  b. Are there any special issues to be aware of? History? Personal ties? Physical limitations?
  c. Do you need to do any work to be comfortable with the topic?

☐ **Who is the audience?**

  a. How much do they know about the topic?
  b. What is their goal in reading your document?
  c. What do they need in order to accomplish their goal?

☐ **What will the final document look like?**

☐ **How will you present the information?**

☐ **How will you design your template?**

☐ **What is a realistic schedule for completing the document?**

Worksheet for Planning—Long Version

☐ **What is true of my audience?**

   a. Who will read this document?
   b. Why do they need the document?
   c. What will they do with it or because of it?
   d. How much do they know about the topic?
   e. What is their personal history with the topic?
   f. What expectations exist for this kind of document's appearance or structure?
   g. Will other people read this document? What do they need to know?
   h. What tone do they expect? Should I use personal pronouns and active voice?

☐ **What is true of me?**

   a. Why am I credible? Because of my role? Because of my actions? Do I need to tell the reader that role or those actions?
   b. What authority do I feel I have? Do I believe that I have the right to speak expertly and be taken seriously?
   c. What should I sound like? How formal or informal should I make myself appear?

☐ **What is my goal in this writing situation?**

   a. What basic message do I want to tell my audience?
   b. Why do I need to convey that message? To inform? To instruct? To persuade?

☐ **What constraints affect this situation?**

   a. How much time do I have?
   b. How long should the document be?
   c. Is budget an issue? If so, how much money is available? Can I produce the document for that amount?
   d. How will I produce the document? Hard copy? Web document? Do I really understand the potential problems with that process? (e.g., will the library printers be working that day?)
   e. Is distributing the document an issue? If so, do I understand the process?
   f. Where will the reader read the document? Does that affect the way I create the document?

☐ **What are the basic facts?**

   a. What do I already know?
   b. Could a visual aid clarify essential information?

c. Where can I read more?

d. With whom can I discuss the matter further?

e. Where can I observe actions that will reveal facts?

☐ **What is an effective strategy?**

a. Should I choose a standard genre (set of instructions, technical report)?

b. Should I follow one example throughout the document, or should I use many examples?

c. Should I develop a central metaphor?

d. Should I use definition followed by example and analogy?

e. Which organizational principle (e.g., top to bottom) is best?

☐ **What format should I use?**

a. What margins, heads, and fonts do I want?

b. What will my template for the document look like?

☐ **What is my production schedule?**

a. What is the due date of the document?

b. By what date can I complete each of the stages (research, alpha version, review, beta version—see p. 58)?

▶ *Method of distribution* is the manner in which the document is delivered to the reader. If it is to be mailed, for instance, it must fit into an envelope of a certain size.

▶ *Place of use* is the physical surroundings in which the audience reads the document. Many manuals are used in confined—even dirty—spaces that could require smaller paper sizes and bindings that can withstand heavy use.

Drafting and Revising Your Document

Drafting and *revising* are the actions you take as you create the document. This stage is concerned with "getting it down" and "making it easy for your reader to grasp." It is not the same as checking closely for spelling and other types of "surface" consistency. That stage is editing or finishing, discussed in a later section. As you draft you implement your plan, choosing words, paragraphs, examples, explanations that convey your topic to your reader. In this stage, you often have "discovery" moments in which you realize more about the topic and how to present it. You may suddenly think of new ideas or new ways to present your examples. Or you may discover an entirely new way to organize and approach the whole topic, and so you discard much of your tentative plan. This section explains strategies for drafting and revising.

Research to Discover Information

If you do not know the topic, obviously you must learn about it. To do so requires time spent researching. The basic methods used are keyword searching in a library or on the Web, interviewing users, and interviewing experts. These topics are covered in depth in Chapter 5.

Design Your Information to Help Your Reader

To design your information, remember that you are creating an experience for your readers. You are helping them build a model. An easy way to envision this concept is to think about how information is given along the interstate highway system. Large green signs, placed at crucial junctures, indicate such information as exit numbers and distances between points. You receive the information you need at the point where you have to act: Signage tells you where to exit, or enter, the system. As a result of the placement of the information, you perform the correct action. Your information design should accomplish the same thing for your readers because you engineer an experience for them (Shedroff).

The following sections contain many hints for engineering interactions and experiences. All of them are ways to design the kinds of "signs" needed to arrive at the "final destination." However, you must place them at the key points, depending on what you know about the audience and their goal in the situation.

General Principles

Researchers (Duin; Huckin; Slater; Spyridakis) have developed some specific guidelines to help you create the kind of interactive document your reader needs. For more detailed information on these topics, see Chapter 4.

1. For an audience with little prior knowledge about a topic, use the familiar to explain the unfamiliar. Provide examples, operational definitions, analogies, and illustrations. These devices invite your reader to become imaginatively involved with the topic and make it interesting.

2. For readers familiar with a topic, don't belabor the basics. Use accepted terminology.

3. For all readers, do the following:
   - State your purpose explicitly. Researchers have found that most readers want a broad, general statement that helps them comprehend the details.
   - Make the topic of each section and paragraph clear. Use heads. Put topic sentences at the beginning of paragraphs. This top-down method is very effective.
   - Use the same terminology throughout. Do not confuse the reader by changing names. If you call it *registration packet* the first time, don't switch to *sign-up brochure* later.

Hierarchy	List
Site A has two basic problems.	Site A has these problems.
Terrain	Water drains into the basement
Water in the basement	Hills block solar heating
Hills block solar heating	Roof leaks
Disrepair	Windows are broken
Roof leaks	Air conditioning system is
Windows are broken	broken
Air conditioning system is	
broken	

© Cengage Learning

Figure 3.3 Hierarchical and List Formats

- Choose a structuring method that achieves your goal. If you want your readers to remember main ideas, structure your document hierarchically; if you want them to remember details, use a list format. Figure 3.3 shows the same section of a report arranged both as a hierarchy and as a list. Writers typically combine both methods to structure an entire document or smaller units such as paragraphs (see Spyridakis). The list provides the details that fill out the hierarchy.
- Write clear sentences. You should try to write shorter sentences (under 25 words), rely on the active voice, employ parallelism, and use words the reader understands.
- Make your writing interesting (Duin; Slater). Use devices that help readers picture the topic. Include helpful comparisons, common examples, brief scenarios, and narratives. Include any graphics that might help, such as photographs, drawings, tables, or graphs.

Use Context-Setting Introductions

Your introduction should supply an overall framework so that the reader can grasp the details that later explain and develop it. You can use an introduction to orient readers in one of three ways: to define terms, to tell what caused you to write, and to explain the document's purpose.

Define Terms

You can include definitions of key terms and concepts, especially if you are describing a machine or a process.

> A closed-loop process is a system that uses feedback to control the movement of hydraulic actuators. The four stages of this process are position sensing, error detecting, controlling the flow rate, and moving the actuator.

Tell What Caused You to Write

Although you know why you are writing, the reader often does not. To orient the reader to your topic, mention the reason you are writing. This method works well in e-mails and business letters.

> In response to your request at the June 21 action group meeting, I have written a brief description of the closed-loop process. The process has four stages: position sensing, error detecting, controlling the flow rate, and moving the actuator.

State the Purpose of the Document

The purpose of the document refers to what the document will accomplish for the reader.

> This report defines the basic concepts related to the closed-loop system used in the tanks that we manufacture. Those terms are position sensing, error detecting, controlling the flow rate, and moving the actuator.

Place Important Material at the Top

Placing important material at the top—the beginning of a section or a paragraph—emphasizes its importance. This strategy gives readers the context so that they know what to look for as they read further. Put statements of significance, definitions, and key terms at the beginning.

The following two sentences, taken from the beginning of a paragraph, illustrate how a writer used a statement of significance followed by a list of key terms.

> A bill of materials (BOM) is an essential part of every MRP plan. For each product, the BOM lists each assembly, subassembly, nut, and bolt.

The next two sentences, also from the beginning of a paragraph, illustrate how a writer used a definition followed by a list of key terms.

> The assets of a business are the economic resources that it uses. These resources include cash, accounts receivable, equipment, buildings, land, supplies, and the merchandise held for sale.

Use Preview Lists

Preview lists contain the keywords to be used in the document. They also give a sense of the document's organization. You can use lists in any written communication. Lists vary in format. The basic list has three components: an introductory sentence that ends in a "control word," a colon, and a series of items. The *control word* (*parts* in the sample that follows) names the items in the list and is followed by a colon. The series of items is the list itself (italicized in this sample).

A test package includes three parts: *test plans, test specifications,* and *tests.*

A more informal variation of the basic list has no colon, and the control word is the subject of the sentence. The list itself still appears at the end of the sentence.

The three parts of a test are *test plans, test specifications,* and *tests.*

Lists can appear either horizontally or vertically. In a horizontal list, the items follow the introductory sentence as part of the text. In a vertical list, the items appear in a column, which gives them more emphasis.

A test package includes three parts:
- Test plans
- Test specifications
- Tests

Use Repetition and Sequencing

Repetition means restating key subject words or phrases from the preview list; *sequencing* means placing the keywords in the same order in the text as in the list. The author of the following paragraph first lists the three key terms: *test plans, test specifications,* and *tests.* She repeats them at the start of each sentence in the same sequence as in the list.

A test package includes three parts: test plans, test specifications, and tests. *Test plans* specify cases that technicians must test. *Test specifications* are the algorithmic description of the tests. The *tests* are programs that the technicians run.

Use Coordinate Structure

Coordinate structure means that each section of a document follows the same organizational pattern. Readers react positively once they realize the logic of the structure you are presenting. The following paragraphs have the same structure: first a definition, followed by details that explain the term.

CUTTING PHASE

The cutting phase is the process of cutting the aluminum stock to length. The aluminum stock comes to the cutoff saw in 10-foot lengths and the saw cuts off 6-inch lengths.

MILLING PHASE

The milling phase is the process of shaving off excess aluminum from the stock. The stock comes to the milling machine from the cutoff saw in exact lengths. The milling machine shaves the stock down to exact height and width specifications of 5 inches.

DRILLING/THREADING PHASE

The drilling/threading phase has two steps: drilling and threading. The drilling phase is the process of boring holes into the aluminum stock in specified positions. After the stock comes from milling, the drilling machine bores a hole in each of the four sides, creating tunnels. These tunnels serve as passageways for oil to flow through the valve.

The threading phase is the process of putting threads into the tunnels. After a tool change to enable threading, the drilling machine cuts threads into the tunnels. The threads allow valve inlets and outlets to be screwed into the finished stock.

Testing

Testing is asking other people to interact with your document in order to discover where it is effective and where it needs revision. The goal is to turn your alpha version into a finished, smooth beta version. There are two types of testing—formal and informal.

Formal Testing

Formal testing is conducted in a Usability Lab, which contains enough technology (e.g., one-way mirrors, video cameras, scan converters, banks of monitors) to allow an empirical rendering of what occurred as the reader interacted with the document (Barnum 14–15). This type of testing can occur only in a formal laboratory setting and is usually beyond the normal resources of a writer.

Informal Testing

Informal testing involves the writer asking members of the target audience, or sometimes people familiar with the target audience, to review the document (Barnum 14–15; Lazzaro; Schriver). Methods for performing informal tests vary, ranging from the author creating various questions for the tester, to the author simply getting from a tester feedback on what is effective. Two common methods of testing are soliciting opinions of quality (Schriver) and soliciting comments about the five dimensions of usability (the 5 e's) (Quesenbery).

The quality information that the writer hopes to find often falls into two broad areas: information on the quality of the written prose and on the quality of the page design and use of visual aids. Questions on quality can generate opinions on such items as weak sentence structure, poorly designed tables, and use of jargon. Questions on page design can generate comments on whether the design is consistent, whether the information is easy to scan, and whether text and visual aids act together to send meaning (Schriver 448–449).

The 5 e questions generate a slightly different kind of information. These questions ask, Is the document

effective? Does it help readers achieve their goals?

efficient? Can users complete their tasks (including understanding the concepts or information presented) with speed and accuracy?

engaging? Is the document pleasant to use?

easy to learn? How well does the document facilitate the readers' interaction with the process? (This question and the following are often used with sets of instructions.)

error tolerant? How well does the document help the reader avoid or recover from mistakes (Quesenbery 83–89)?

Worksheet for Drafting

☐ **Choose strategies that help your readers.**

a. For unspecialized readers, use comparison, example, and brief narrative to make the unfamiliar familiar.
b. For specialized readers, do not overexplain.
c. For all readers:
State your purpose.
Make topic sentences and headings clear.
Use consistent terminology.
Organize material hierarchically to emphasize main ideas.
Use a list format to emphasize details.
Write clear sentences by employing the active voice and parallelism.
Note: If these strategies inhibit your flow of ideas, ignore them in the first draft and use them later.

☐ **Be sure your introductory material accomplishes one of these three purposes:**

a. Defines terms.
b. Tells why you are writing.
c. States the purpose of the document.

☐ **Use preview lists as appropriate in the body.**

☐ **Repeat keywords from the preview list in the body.**

☐ **Use coordinate structure to develop the paragraphs in the body.**

☐ **Test your draft.**

a. Ask a member of the target audience or of people familiar with the target audience to read your draft.
b. Give them a list of specific items to check (from spelling to legal, contextual considerations)
c. Have them "talk through" the document with you. Record each time they have a question or comment. Revise those spots.

While these questions tend to refer to instructions, you should be able to adapt them to any document. If, for instance, the last two are not applicable to your document, delete them.

You can easily create a test by creating key questions, either of quality or related to the 5 e's, and asking one or several members of the target audience to read the document. After you receive the results, you need to revise your document. If you have time, you could ask for a second round of tests before you finalize the document.

Editing or Finishing

Editing (or *finishing*) means developing a consistent, accurate text. In this stage, you refine your document until everything is correct. You are looking for surface, consistency problems. You check spelling, punctuation, basic grammar, format of the page, and accuracy of facts. You make the text agree with various rules of presentation. When you edit, ask, Is this correct? Is this consistent? In general, you edit by constructing checklists.

Create a set of quality benchmarks. Widely used in industry, benchmarks are quality standards used to judge a product. In order to edit effectively, you must set similar benchmarks for your work. Typically, benchmarks are divided into categories with statements of quality levels. The following is a simple benchmark set for a Web document created early in a college technical writing course.

STYLE DESIGN

• No spelling or grammar errors.

INFORMATION DESIGN

• Title appears.
• Introduction appears. Introduction tells point of document and, if it is long, sections of the document.
• Body sections are structured similarly in type and depth of points.

PAGE DESIGN

• Fonts are standard roman, and large enough to be easily readable.
• Heads appear and indicate subject of their section.

VISUAL DESIGN

• Visuals appear to support a point in the text or provide a place for the text to begin.
• Visuals are effectively sized, captioned, and referenced.

NAVIGATION DESIGN

• Every link works.
• Link size and placement indicate the type of content (e.g., return to homepage or major section of document) that it connects to.
• Links provide helpful paths through the work.

Constructing checklists of typical problems is a helpful strategy. The key is to work on only one type of problem at a time. For example, first read for apostrophes, then for spelling errors, then for heading consistency, then for consistency in format, and so forth. Typical areas to review include paragraph indicators (indented? space above?); heads (every one of each level treated the same?); figure captions (all treated the same?); and punctuation (e.g., the handling of dashes).

Figures 3.4 and 3.5 demonstrate the types of decisions that you make when you edit. The goal is to correct errors in spelling, punctuation, grammar, and consistency of presentation. Figure 3.4 is the original; Figure 3.5 is edited.

Version 1

Unclear topic sentence	**TECHNICAL REPORTS**
Misused semicolon	The detailed technical report to upper management will be submitted at the end of the project. It must explain;
List elements are inconsistent	1. the purpose of the machine,
Vertical list misemphasizes content	2. its operation,
Sentence fragment	3. and the operation of its sub systems.
"An" indicates one example, but two appear	4. Assembly methods will also be presented.
Misused semicolon	It will also include all design calculations for loads, stresses, velocities, and accelerations. Justification for the choice of materials of subsystems. An example might be; the rationale for using plastic rather than steel and using a mechanical linkage as compared to a hydraulic circuit. The report also details the cost of material and parts.

© Cengage Learning

Figure 3.4   Technical Report: Original

Version 2

	TECHNICAL REPORTS
Clear topic sentence	The technical report to upper management, submitted at the end of the project, contains several sections. The report explains
List consistent	1. the machine's purpose and operation.
	2. the operation of its subsystems and methods to assemble it.
Two examples suggested	It also includes all design calculations for loads, stresses, velocities, and accelerations, as well as justification for the choice of materials in subsystems.
Complete sentence	Examples of this justification include the rationale for using plastic rather than steel and a comparison between a mechanical linkage and a hydraulic circuit. The report also details the cost of material and parts.

© Cengage Learning

Figure 3.5   Technical Report: Edited

TIP ||||

Editing with a Word Processor

As you work to achieve the consistency that is the goal of the finishing stage, you can use style aids, in particular, spell checker and grammar checker. A spell checker indicates any words in your document that are not in its dictionary. If you have made a typo, such as typing *wtih,* the checker highlights the word and allows you to retype it. Most spell checkers have an autocorrect feature. Once you engage it (check your program's instructions), it will automatically change every mistake, such as *teh* to *the.* However, these programs have problems. If your typo happens to be another word—such as *fist* for *first*—the program does not highlight it. Also, if you misuse a word—such as *to* instead of *too*—the program does not detect the error.

Grammar checkers detect such problems as subject–verb disagreement, fragments, and comma splices. Checkers also can detect features of your writing. For instance, the checker might highlight all the forms of *to be* in your paper, thus pointing out all the places where you may have used the passive voice. Checkers can also highlight words that could be interpreted as sexist or racist, that are overused, or that are easily confused. Thus the checker will highlight every *your* and *you're,* but you must decide whether you have used the correct form.

Follow these guidelines:

Use your spell checker.

Set your spell checker to "AutoCorrect."

Use your grammar checker for your key problems. (If you have trouble, for example, with fragments or passive voice, set the checker to find only those items.)

After using the spelling and grammar checkers, proofread to find any "OK, but wrong" words, for example, *their* used for *there* or *luck* used for *link.*

Producing the Document

Producing a document involves the physical completion—the typing or printing—of the final document. This stage takes energy and time. Failure to plan enough time for physical completion and its inevitable problems will certainly cause frustration. Many people have discovered the difficulties that can plague this stage when their hard drive crashes or their printer fails. Although physical completion is usually a minor factor in brief papers, in longer documents it often takes more time than the drafting stage.

Worksheet for **Editing**

☐ **Make a checklist of possible problems.**

a. Head format (does each level look the same?)
b. Typographical items (e.g., are all dashes formed the same way?)
c. Handling of lists
d. Handling of the beginning of paragraphs (e.g., are they all indented five spaces?)
e. References to figures
f. Spelling
g. Grammar
h. Consistent word use

Exercises

▶ You Create

1. Following the format of the on-site review (pp. 36–38), write a one-page plan for one of the following situations. Your instructor may ask you to form groups of three. In each situation, assume that you have completed the research; that is, the plan is for the document, not the entire project.

   a. Recommend a change in the flow of work in a situation. For example, change the steps in manufacturing a finished part or in moving the part from receiving to point of sale.
   b. Think of a situation at your workplace or at your college that you want changed. Write the plan for a document proposing that change.
   c. You have discovered a new opportunity for manufacturing or retail at your workplace. Plan a report to obtain permission to use this opportunity.
   d. You have been assigned to teach coworkers or fellow students how to create a simple Web page. Plan the set of instructions you will write. Your goal is to enable your audience to create a page with four or five paragraphs, heads, one hyperlink to another document, and one visual (optional).

2. Using the plan that you constructed in Exercise 1, draft the document.

3. Produce the final version of Exercise 1: the change in work flow, the situation change, the new opportunity, or the creation of a Web page. Then write a report in which you (1) describe the process you used, step by step, and (2) evaluate the strength of your process.

4. Create a brief top-down document. Choose a topic that you can easily break into parts (e.g., computer memory includes RAM and ROM). Use a preview list, coordinate structure, and repetition and sequencing. Use a highlighter or some other method to indicate the devices you used to achieve the top-down structure.

▶ You Revise

5. Revise the following paragraph. Use the strategies listed in the Worksheet for Drafting (p. 73 under "Choose strategies"). The new paragraph should contain sections on reasons for writing and on format. Revise individual sentences as well.

> Proposals are commonly used in the field of retail management. These recommendations are written in a standard format, which has a number of company parts that we often use. These are stating the problem, a section often used first in the proposal. Another one is providing a solution, of course, to the problem that you had. You write a proposal if you want to knock out a wall in your department. You also write a recommendation if you want to suggest a solution that a new department be added to your store. Another section of the proposal is the explanation of the end results. You could also write a proposal if in the situation you wanted to implement a new merchandise layout. A major topic involved with writing proposals is any major physical change throughout the store. A proposal can be structured so as to enable implementation of any major physical changes to the upper management level, who would be the audience for which the proposal is intended.

6. Revise this paragraph for effective structure.

Additional Writing

> Additional types of writing that may be encountered by a software developer include quality tips emails, trip reports, and development proposals. Other types of proposals are infrequently submitted since there are no moving parts or assembly lines in software development. In addition, since hardware is so expensive, changes in hardware are generally management directed. Consequently, proposals for a change in machinery are also rarely submitted. These types of suggestions and requests are generally incorporated into weekly status reports. Development proposals rise from recognizing the need for a software tool to do work more efficiently. These proposals are directed at management in areas that could utilize this tool. It describes the need for the tool and the advantages using it. The quality tips reports are usually for the department and are a summary of a quality roundtable. Trip reports contain an overview of the conference materials and the points that the attender found most interesting. It also makes the conference literature available to interested parties.

7. Edit (or finish) the following sections for consistency. Correct errors in spelling and grammar, as well as such details of presentation as indentation, capitalization, and treatment of heads.

For who will I write?

For quick correspondence, emails will be sent to entry level employees, co-workers, Immediate Supervisors and, occasionally, to chief executives. Proposals will be sent to potential clients in hopes of attracting their business. Response request forms are to be sent out following the Preposals in order to obtain a response regarding the proposal.

When I Will Write:

Emails will be written on a daily basis. Proposals and response request letters will be written on demand—once or twice a week.

Importance of Writing

Emails are crucial for effective inter-office communication. Proposals are the key for attrcating business to the organiztion—with successful proposals come revenue. The Response Request Forms are important in communication because they complete the process begun by the proposal. Because of all these forms are written, good writing is the core of good communion.

Importance to promotion—

Since writing is tangible, it is easily alleviated. Hence, the quality of writing is the primary source for promotional evaluation.

LONGER DOCUMENTS

I would be required to write a longer document if a proposal is excepted by a client. Once a proposal is excepted, I will be required to write longer documents—workbooks, reaserch reports, and needs analysis evaluations.

Workbooks which consist of guidelines for a workshop and appropriate information on the subject.

RESEARCH REPORTS—a closer and more detailed look at a particular topic or organization

Needs Analysis Elevations. These are documents we use in order to determine the needs of the organization and to assist in setting up a workshop

8. Analyze and revise the following paragraph. Create several paragraphs, and revise sentences to position the important words first.

The problem-solving process is started by the counselor and client working together to define the problem. The goal of defining the problem is getting the client to apply the problem in concrete and measurable terms. The client may have several problems, so the counselor should allow the client to speak on the problem he or she may want to talk about first. The client usually speaks about smaller problems first until the client feels comfortable with the counselor. Determining the desired goal of the client is the second step in the process. The main point of determining the desired goal of the client is making sure the client and counselor feel the patient's goal is realistic. A bad example of this would be "I don't want to be lonely anymore," but a measurable example that would be realistic is getting the client to spend two hours a day socializing with other people. A realistic goal is considered to be one in which the client can reach without overcoming huge obstacles. The next step is for the counselor to discover the client's present behavior. The goal of this step is for the counselor to rationalize where the client is in relation to the problem. The present behavior of the client is what they are currently doing in concrete terms. Following the previous example, the lonely client could state, "I spend about a half hour a week with other people outside of working." Working out systematic steps between the present behavior and the goal is the final step in the problem solving process. The object is for the counselor to break the process into small parts so the client can work toward their goal and to give the client concrete, real work. Making schedules is a common formality clients use to follow their goals. Counselors are to be supportive and give feedback when the client is not working toward the goal. Feedback would be pointing out to the client that he or she is procrastinating. The client is always open to redefine their goals and the counselor is to stay open and to renegotiate the process.

9. Exchange the document you created in Exercise 6 with a classmate, rewrite that document, and then compare results and report to the class.

▶ Group

10. In groups of three, critique the plans you wrote in Exercise 1. Your goal is to evaluate whether a writer could use the plan to create a document. Your instructor will ask for oral reports in which your group explains helpful and nonhelpful elements of plans.

11. From any document that explains technical information to lay readers, photocopy a page that illustrates several of the organization strategies explained on pages 67–72. In groups of three or four, read one another's photocopies to identify and discuss the strategies; list them in order of how frequently they occur. Give an oral report of your findings to the class or, as a group, prepare a one-page report of your findings for your instructor. In either case, base your report on the list you construct and use specific examples.

12. Your instructor will divide you into groups of three or four. Each member interviews three or four people on a relevant topic in your community. (Are you getting a good education at this college? Does the school district meet your children's needs?) Combine the results into a report to the relevant administrator or official, using one visual aid.

   First, as a group, create a report in which you answer the questions in the Worksheets for Drafting (p. 73) and for Editing (p. 77). Then give each member of your group and your instructor a copy, conduct the interviews, and create the final report. Your instructor will provide you with more details about constraints. After you hand in the report, be prepared to give an oral report describing the process you used and evaluating the strengths and weaknesses of the process.

13. In groups of three or four, develop the document you planned in Exercise 1 (p. 77). Plan the entire project, from collecting data and assessing audiences to producing a finished, formatted report. First, create a report in which you complete the Worksheet for Planning (pp. 65–67, choose either the long or short version) for Drafting (p. 73) and for Editing (p. 77). Give each member of your group and your instructor a copy. Then produce the report. Your instructor will provide you with more detailed comments about constraints.

Writing Assignments

1. Assume that you have discovered a problem with a machine, process, or form that your company uses. Describe the problem in a report to your supervisor. She will take your report to a committee that will review it and decide what action to take. Before you write the description, do the following:

   • Write your instructor a report that contains your plan, that is, the answers to all the planning questions about your document. Include a schedule for each stage of your process.
   • Hand in your plan with your description of the problem.

   Here are some suggestions for a problem item. Feel free to use others.

oscilloscope	deposit slip for a bank
oxyacetylene torch	application form for an internship
condenser/enlarger	black-and-white plotter
photoelectric sensing device	tool control software
speed drive system for a conveyor	bid form

layout of a facility (a computer lab, a small manufacturing facility)	compliance with an OSHA (Occupational Safety and Health Administration) or a DOT (United States Department of Transportation) regulation
instruction screen of a database	
plastic wrap to cover a product	
method of admitting patients	method for inspecting X (you pick the topic)
method of deciding when to order	
Web authoring program,	disposal of Styrofoam drinking cups
method for testing X (you pick the topic)	posting your resume on Monster.com or LinkedIn
tamper-evident cap to a pill bottle	

2. Interview three people who write as part of their academic or professional work to discover what writing process they use. They should be a student whose major is the same as yours, a faculty member in your major department, and a working professional in your field. Prepare questions about each phase of their writing process. Show them the model of the process (see Figure 3.1, p.58), and ask whether it reflects the actual process they undergo. Then prepare a one- to two-page report to your classmates summarizing the results of your interviews.

3. Your instructor will assign you to groups of three or four on the basis of your major or professional interest. As a group, perform Writing Assignment 2. Each person in the group should interview different people. Your goal is to produce a two-page report synthesizing the results of your interviews. Follow the steps presented in the "Focus on Groups" box of this chapter (pp. 85–87). In addition to the report, your teacher may ask you to hand in a diary of your group activities.

4. Write a one- to two-page paper in which you describe the process you use to write papers. Include a process chart. Then, as part of a group, produce a second paper in which your group describes either one optimal process or two competing models.

Web Exercise

Explain the strategy of a website that you investigate. Establish the ways in which the authors develop a starting point, a presentation "map," and an identity. Write an analytical report to your classmates or coworkers in which you explain strategies they should adopt as they develop their own websites.

Works Cited

Albers, Michael J. "Complex Problem Solving and Content Analysis." *Content and Complexity: Information Design in Technical Communication.* Ed. Michael J. Albers and Beth Mazur. Mahwah, NJ: Lawrence Erlbaum, 2003. 263–283. Print.

Barnum, Carol. *Usability Testing and Research.* New York: Longman, 2002. Print.

Burnett, Rebecca E. "Substantive Conflict in a Cooperative Context: A Way to Improve the Collaborative Planning of a Workplace Document." *Technical Communication* 38.4 (1991): 532–539. Print.

Carliner, Saul. "Physical, Cognitive, and Affective: A Three-Part Framework for Information Design." *Content and Complexity: Information Design in Technical Communication.* Ed. Michael J. Albers and Beth Mazur. Mahwah, NJ: Lawrence Erlbaum, 2003. 39–58. Print.

Debs, Mary Beth. "Recent Research on Collaborative Writing in Industry." *Technical Communication* 38.4 (1991): 476–484. Print.

Dorff, Diane Lee, and Ann Hill Duin. "Applying a Cognitive Model to Document Cycling." *Technical Writing Teacher* 16.3 (1989): 234–249. Print.

Duin, Ann Hill. "How People Read: Implications for Writers." *Technical Writing Teacher* 15.3 (1988): 185–193. Print.

Flower, Linda. "Rhetorical Problem Solving: Cognition and Professional Writing." *Writing in the Business Professions.* Ed. Myra Kogen. Urbana, IL: NCTE, 1989. 3–36. Print.

Garret, Jesse James. *The Elements of User Experience: User-Centered Design for the Web.* Indianapolis, IN: American Institute of Graphic Arts, 2003. Print.

Goswami, Dixie, Janice C. Redish, Daniel B. Felker, and Alan Siegel. *Writing in the Professions: A Course Guide and Instructional Materials for an Advanced Composition Course.* Washington, DC: American Institute for Research, 1981. Print.

Horton, William. *Secrets of User-Seductive Documents.* Arlington, VA: Society for Technical Communication, 1997. Print.

Huckin, Thomas. "A Cognitive Approach to Readability." *New Essays in Technical and Scientific Communication: Research, Theory, Practice.* Ed. Paul V. Anderson, R. John Brockman, and Carolyn R. Miller. Farmingdale, NY: Baywood, 1983. 90–108. Print.

Lazzaro, Heather. "How to Know Your Audience." *intercom* (November 2001): 18–20. Print.

McTeague, Michael. "How to Write Effective Reports and Proposals." *Training and Development Journal* (November 1988): 51–53. Print.

Quesenbery, Whitney. "The Five Dimensions of Usability." *Content and Complexity: Information Design in Technical Communication.* Ed. Michael J. Albers and Beth Mazur. Mahwah, NJ: Lawrence Erlbaum, 2003. 81–102. Print.

Redish, Janice. "What Is Information Design?" *Technical Communication* 47.2 (May 2000): 163–166. Print.

Schriver, Karen. *Dynamics in Document Design.* New York: Wiley, 1997. Print.

Screven, C. G. "Information Design in Informal Settings: Museums and Other Public Spaces." *Information Design.* Ed. Robert Jacobson. Cambridge, MA: MIT, 2000. 131–192. Print.

Shedroff, Nathan. "Information Interaction Design: A Unified Field Theory of Design Information Design." *Information Design.* Ed. Robert Jacobson. Cambridge, MA: MIT, 2000. 267–292. Print.

Slater, Wayne H. "Current Theory and Research on What Constitutes Readable Expository Text." *Technical Writing Teacher* 15.3 (1988): 195–206. Print.

Spyridakis, Jan H. "Guidelines for Authoring Comprehensible Web Pages and Evaluating Their Success." *Technical Communication* 47.3 (Third Quarter 2000): 359–382. Print.

Focus on
Groups

Once they are in business and industry, many college graduates discover that they must cowrite their documents. Often a committee or a project team of three or more people must produce a final report on their activities (Debs). Unless team members coordinate their activities, the give and take of a group project can cause hurt feelings and frustration, and result in an inferior report. The best way to generate an effective document is to follow a clear writing process. For each of the writing stages, not only must you perform the activities required to produce the document, you must also facilitate the group's activities.

Before you begin planning about the topic, you must:

Select a leader
Understand effective collaboration
Develop a method to resolve differences
Plan the group's activities
Choose a strategy for drafting and revising
Choose a strategy for editing
Choose a strategy for producing the document

Select a Leader

The leader is not necessarily the best writer or the person most informed about the topic. Probably the best leader is the best "people person," the one who can smooth over the inevitable personality clashes, or the best manager, the one who can best conceptualize the stages of the project. Good leadership is an important ingredient in a group's success (Debs).

Understand Effective Collaboration

Group members must understand how to collaborate effectively. Two effective methods are goal sharing and deferring consensus (Burnett).

Goal sharing means that individuals cooperate to achieve goals. As any individual strives to achieve a goal, he or she must simultaneously try to help someone else achieve a goal. Thus, if two members want to use different visual aids, the person whose visual is used should see that the other person's point is included in the document. This key idea eliminates the divisiveness that occurs when people see each issue competitively, so that someone wins and someone loses.

Deferring consensus means that members agree to consider alternatives and voice explicit disagreements. Groups should deliberately defer arriving at a consensus in order to explore issues. This process initially takes more time, but in the end it gives each member "ownership" of the project and document.

According to one expert, group success is greatly helped by the "ability to plan and negotiate through difficulties; failures may be caused by a group's inability to resolve conflict and to reach consensus" (Debs 481).

Develop a Method to Resolve Differences

Resolving differences is an inevitable part of group activity. Your group should develop a reasonable, clear method for doing so. Usually the group votes, reaches a consensus, or accepts expert opinion. Voting is fast but potentially divisive. People who lose votes often lose interest in the project. Reaching a consensus is slow but affirmative. If you can thrash through your differences without alienating one another, you will maintain interest and energy in the project. Accepting expert opinion is often, but not always, an easy way to resolve differences. If one member who has closely studied citation methods says that the group should use a certain format, that decision is easy to accept. Unfortunately, another group member may disagree. In that case, your group will need to use one of the other methods to establish harmony.

(Continued)

Plan the Group's Activities

Your group must also develop guidelines to manage activities and to clarify assignments and deadlines.

To *manage activities,* the group must make a work plan that clarifies each person's assignments and deadlines. Members should use a calendar to set the final due date and to discuss reasonable time frames for each stage in the process. The group should put everything in writing and should schedule regular meetings. At the meetings, members will make many decisions; for example, they will create a style sheet for head and citation format. Write up these decisions and distribute them to all members. Make—and insist on—progress reports. If group members evaluate one another, they should comment on the punctuality and helpfulness of these reports. Help one another with problems. Tell other group members how and when they can find you.

To *clarify assignments* and *deadlines,* answer the following questions:

What is the exact purpose of this document?
Must any sections be completed before others can be started?
What is each person's research and writing assignment?
What is the deadline for each section?
What is the style sheet for the document?

Each member should clearly understand the audience, the intended effect on the audience, any constraints, and the basic points that the document raises. This sense of overall purpose enables members to write individual sections without getting off on tangents (McTeague).

Worksheet for Group Planning

- *What considerations are necessary if this is a group planning situation?*
  a. Who will be the leader?
  b. How will we resolve differences?
  c. What are the deadlines and assignments?
  d. Who will keep track of decisions at meetings?
  e. Who will write progress reports to the supervisor?

Choose a Strategy for Drafting and Revising

To make the document read "in one voice," one person often writes it, especially if it is short. A problem with this method is that the writer gets his or her ego involved and may feel "used" or "put upon," especially if another member suggests major revisions. The group must decide which method to use, considering the strengths, weaknesses, and personalities of the group members.

Once drafting begins, groups should review each other's work. Generally, early reviews focus on large matters of content and organization. Reviewers should determine whether all the planned sections are present, whether each section contains enough detail, and whether the sections achieve their purpose for the readers. At this stage, the group needs to resolve the kinds of differences that occur if, for instance, one person reports that the company land filled 50,000 square yards of polystyrene and another says it was 5 tons. Later reviews typically focus on surface-level problems of spelling, grammar, and inconsistencies in using the style sheet.

Worksheet for Group Drafting and Revising

- *Ask and answer planning questions.*
  a. What is the purpose of this document?
  b. Who is the audience of this document?
  c. What is the sequence of sections?
  d. Must any sections be completed before others can be started?
  e. What is each person's writing assignment?
  f. Does everyone understand the style sheet?
  g. What is the deadline for each section?
- *Select a method of drafting.*
  a. Will everyone write a section?
  b. Will one person write the entire draft?

Choose a Strategy for Editing

Groups can edit in several ways. They can edit as a group, or they can designate an editor. If they edit as a group, they can pass the sections around for comment, or they can meet to discuss the sections. However, this method is cumbersome. Groups often "overdiscuss" smaller editorial points and lose sight of larger issues. If the group designates one editor, that person can usually produce a consistent document and should bring it back to the group for review. The basic questions that the group must decide about editing include:

Who will suggest changes in drafts? One person? An editor? The group?
Will members meet as a group to edit?
Who will decide whether to accept changes?

In this phase, the conflict-resolving mechanism is critical. Accepting suggested changes is difficult for some people, especially if they are insecure about their writing.

Choose a Strategy for Producing the Document

The group must designate one member to oversee the final draft. Someone must collect the drafts, oversee the inputting, and produce a final draft. In addition, if the document is a long, formal report, someone must write the introduction and attend to such matters as preparing the table of contents, the bibliography, and the visual aids. These tasks take time and require close attention to detail.

Questions for the group to consider at this stage include:

Who will write the introduction?
Who will put together the table of contents?
Who will edit all the citations and the bibliography?
Who will prepare the final version of the visual aids?

Who will oversee production of the final document?

The group writing process challenges your skills as a writer and as a team member. Good planning helps you produce a successful report and have a pleasant experience. As you work with the group, it is important to remember that people's feelings are easily hurt when their writing is criticized. As one student said, "Get some tact."

Worksheet for Group Editing

- *Select a method of editing.*
  a. Will the members meet as a group to edit?
  b. Who will determine that all sections are present?
  c. Who will determine that the content is complete and accurate?
  d. Who will check for conflicting details?
  e. Who will check for consistent use of the style sheet?
  f. Who will check spelling, grammar, punctuation, etc.?
- *Oversee the production process.*
  a. Who will write the introduction?
  b. Who will put together the table of contents?
  c. Who will edit the citations and bibliography?
  d. Who will prepare the final versions of the visual aids?
  e. Who will direct production of the final document?

Note: Each project generates its own particular editing checklist. Formulate yours on the basis of the actual needs of the document.

Technical Communication Style

CHAPTER CONTENTS

CHAPTER 4 **IN A NUTSHELL**

Develop a repertoire of strategies to use as you write and revise documents. Mastering just a few key strategies will make your ideas seem clearer to your audience and cause them to trust you.

The basic strategies are

▶ Write in the active voice.

▶ Use parallelism.

▶ Use *there are* sparingly.

▶ Write 12- to 25-word sentences.

▶ Change tone by changing word choice.

The first four principles are common writing advice. The last principle helps create an identity that is appropriate for the situation.

Managers, directors, and people who hire want employees who write clearly. The employers say things like they want "very concise writing which ensures that the idea being promoted is clear." They want proper spelling, proper grammar, and appropriate sentence structure. One manager said that understanding proper English is a "point of differentiation" in your favor. Appendix A (pp. 558–579) will help you with spelling, grammar, and especially issues with pronouns. This chapter will explain several strategies that will help your writing become clear, concise, and appropriately structured. Most of the sections deal with sentence strategies. The last section focuses on writing a clear paragraph.

Sentence Strategies

If you don't already know these strategies, how do you learn them? Read over your drafts, looking specifically for the areas where you should make changes in terms of the strategies. Be aware that it takes concentration to use these strategies. The various sections that follow show you what the "bad" constructions look like and provide several ways to fix them.

It helps, at the start of a discussion of style to recall two important concepts. (1) The normal word order in English sentences is subject–verb–object ("Jane threw the ball." "The Johnsons bought the house." "That machine pulverizes the grapes.") Using the subject–verb–object order usually makes the topic easier for the reader to grasp. (2) Readers tend to grasp information better if you arrange items in top-down order. Put the most important idea (the "top") first and the details (the "bottom") next. ("We rejected the proposal because it was too costly." "Her time in the dash was 10.5 seconds.") Both concepts underlie much of the following advice.

Probably the two most helpful sentence strategies are

- Write in the active voice
- Use parallelism

Of course there are a number of other helpful strategies. They include

- Use *there are* constructions carefully
- Avoid "nominalizations"
- Put the main idea first, if possible
- Try to keep sentence length between 12 and 25 words
- Use *you* correctly
- Don't use sexist constructions
- Eliminate common clarity errors

Write in the Active Voice

The active voice emphasizes the performer of the action rather than the receiver. The active voice helps readers grasp ideas easily because it adheres to the subject–verb–object pattern and puts the performer of the action first. When the subject acts, the verb is in the active voice ("I wrote the report"). When the subject is acted upon, the verb is in the passive voice ("The memo was written by me"). As you create your drafts watch for the passive voice; try to find an active way to phrase the idea. If you are unsure about identifying passive voice, your word processor will underline instances if you set the grammar checker to find them. But more than relying on the grammar checker, practice thinking in active voice.

Change Passive to Active

To change a verb from the passive to the active voice:

▶ Move the person acting out of a prepositional phrase.

Passive	The test was conducted *by the intern*.
Active	*The intern* conducted the test.

▶ Supply a subject (a person or an agent).

Passive	This method was ruled out.
Active	*The staff* ruled out this method.
Active	*I* ruled out this method.

▶ Substitute an active verb for a passive one.

Passive	The heated water is *sent* into the chamber.
Active	The heated water *flows* into the chamber.

Use the Passive If It Is Accurate

The passive voice isn't always wrong; sometimes it is more accurate; for instance, it is properly used to show that a situation is typical or usual or to avoid an accusation.

Typical situation needs no agent	Robots are used in repetitive activities.
Active verb requires an unnecessary agent (companies)	Companies use robots in repetitive activities.
Active accuses	You violated the ethics code by doing that.
Passive avoids accusing	The ethics code was violated by that act.

The passive voice can also be used to emphasize a certain word.

Use passive to emphasize *milk samples*	Milk samples are preserved by the additive.
Use active to emphasize *additive*	The additive preserves the milk samples.

Use Parallelism

If a sentence or a paragraph has several ideas, parallel structure indicates to the reader that the ideas are equal in value (*coordinate*). The following sentence contains three verbs. They are of equal value because they each tell you what Adam did at each step. They are parallel because each has the same form, past tense. "Adam *went* into the store, *bought* an ice cream cone, and *sat* on a bench eating it." Using parallelism means using similar structure for similar elements.

Another way to think about parallelism is that words and ideas in a series should have the same structure. Lists use parallelism because the items in the list make up a series. In the following sentence, the italicized words make up a series.

Technical writers create *memos, proposals,* and *manuals.*

If coordinate elements (items in a series) in a sentence are not treated in the same way, the sentence is awkward and confusing. In the following sentences the italicized words constitute a series.

Faulty	My duties included *coming* in early in the morning and doing preparation work, *to cook* on the front line, *trained* new employees, and *took* inventory.
Parallel	My duties included *coming* in early to do preparation, *cooking* on the front line, *training* new employees, and *taking* inventory.
Faulty	Typical writing situations include *proposals, the sending of* e-mail, and *how to update* the system.
Parallel	Typical writing situations include *editing* proposals, *sending e-mail,* and *updating* the system.

Use *There Are* Sparingly

Overuse of the indefinite phrase *there are* and its many related forms (*there is, there will be,* etc.) weakens meaning by "burying" the subject in the middle of the sentence. Most sentences are more effective if the subject is placed first. Watch for this construction in your writing. It typically appears regularly when people try to write about technical topics.

Ineffective	*There is* a change in efficiency policy that could increase our profits.
Effective	Our profits will increase if we change our efficiency policy.
Ineffective	*There are* three problems that this process has.
Effective	This process has three problems.
Ineffective	*There are* two reasons why we should talk about abandoning our current location.
Effective	We should abandon our current location for two reasons.

Avoid Nominalizations

Avoid using too many *nominalizations,* verbs turned into nouns by adding a suffix such as *-ion, -ity, -ment,* or *-ness.* Nominalizations weaken sentences by presenting the action as a static noun rather than as an active verb. These sentences often eliminate a sense of agent, thus making the idea harder for a reader to grasp. Express the true action in your sentences with strong verbs. Almost all computer style checkers flag nominalizations.

Static	The training policy for most personnel will have the *requirement* of the *completion* of an initial one-week seminar.
Active	The training policy will *require* most personnel *to complete* a one-week seminar.
Static	There will be costs for the *installation* of this machine in the vicinity of $10,000.
Active	We can *install* this machine for about $10,000. The machine will cost $10,000 *to install.*

Put the Main Idea First

To put the main idea first (at the "top") is a key principle for writing sentences that are easy to understand. Place the sentence's main idea—its subject—first. The subject makes the rest of the sentence accessible. Readers relate subjects to their own ideas and experiences and thus orient themselves. After readers know the topic, they are able to interact with the complexities you develop.

Note the difference between the following two sentences. In the first, the main idea, "two types of professional writing," comes near the end. The sentence is difficult to understand. In the second, the main idea is stated first, making the rest of the sentence easier to understand.

Main idea is last	The writing of manufacturing processes, which explain the sequence of a part's production, and design specifications, which detail the materials needed to produce an object, are two types of professional writing I will do.
Main idea is first	Two types of professional writing that I will do are writing manufacturing processes, which explain the sequence of a part's production, and design specifications, which detail the materials needed to produce an object.

Write Sentences of 12 to 25 Words

An easy-to-read sentence is 12 to 25 words long. Shorter and longer sentences are harder to read because they become too simple or too complicated. Often parallel structure helps sentences "flow" better. Combine short sentences into a longer one, connecting the parts with parallel structure. Longer sentences, especially those employing parallel construction, can be easy to grasp. The first of the following sentences is harder to understand, not just because it is long, but also because it ignores the strategy of putting the main idea first. The revision is easier to read because the sentences are shorter and the main idea is introduced immediately.

One sentence, 40 words long	The problem is the efficiency policy, which has measures that emphasize producing as many parts as possible, for instance, 450 per hour, compared to a predetermined standard, usually measured by the machine's capacity, say, 500, for a rating of 90%.
Two sentences, 20 and 21 words long	The problem is the efficiency policy, which calls for producing as many parts as possible compared to a predetermined standard. If a machine produces 450 per hour and if its capacity is 500 per hour, it has a rating of 90%.

Use *You* Correctly

Do not use *you* in formal reports (although writers often use *you* in informal reports). Use *you* to mean "the reader"; it should not mean "I," or a very informal substitute for "the" or "a" (e.g., "This is your basic hammer.").

Incorrect as "I"	I knew when I took the training course that *you* must experience the problems firsthand.
Correct	I knew when I took the training course that I needed to experience the problems firsthand.

Avoid Sexist Language

Language is considered sexist when the word choice suggests only one sex even though both are intended. Careful writers rewrite sentences to avoid usages that are insensitive and, in most cases, inaccurate. Several strategies can help you write smooth, nonsexist sentences. Avoid such clumsy phrases as *he/she* and *s/he*. Although an occasional *he or she* is acceptable, too many of them make a passage hard to read.

Sexist	The clerk must make sure that *he* punches in.
Use an infinitive	The clerk must make sure *to punch* in.
Use the plural	The clerks must make sure that *they* punch in.
Use the plural to refer to "plural sense" singulars	Everyone will bring their special dish to the company potluck.

In the last example, *their,* which is plural, refers to *everyone,* which is singular but has a plural sense.

Exercises

▶ Passive Sentences

1. Make the following passive sentences active.

   - When all work is completed, turn the blueprint machine off.
   - A link, such as products or specifications, was checked by me.
   - The OK button was clicked and the correct link was scrolled down to.
   - After these duplicate sites were eliminated there were still too many sites to be looked at.
   - Profits can be optimized by the manufacturer with the use of improved materials and the result can be better product value for the customer.
   - Revise the passive voice sentences in these paragraphs.

   Numerous problems with flexographic printing have to be considered. On very short runs of 1000 or less the cost can be excessive. In addition dot gain is another issue that can be a problem. Small print and reverse print can be other problems with flexography.

   The page is entitled features. All of the features of the Netmeeting software are listed down the center of this page. Under each feature a brief discussion of the feature is given so that the user is told what the feature is for. The uses of each feature in detail can be found by clicking on the icon.

▶ Parallel Structure

2. Revise the following sentences to make their coordinate elements parallel.

   - It serves the purpose of pulling the sheet off the coil and to straighten or guide the sheet through the rest of the machine.
   - The main uses of the data are to supply estimates of the workforce and estimating unemployment.
   - There were clear headings to each section, some links for navigating around the site, and it provided a glossary of terms.
   - During the internship I was responsible for merchandising, creating floor plans, and to train new employees.
   - Along with a case, the laptop package includes 18 feet of top-quality UGA cable, a surge protector with circuit checker, and there is also an extra battery.

3. Write a sentence in which you give three reasons why a particular Web search engine is your favorite. Alternate: Your instructor will provide topics other than Web search engines.

4. Create a paragraph in which your sentence from Exercise 5 is the topic sentence. Write each body sentence using active voice and parallelism.

▶ Use of *There Are*

5. Eliminate *there are,* or any related form, from the following sentences.

- There are several methods of research that I will use.
- There are slowness and inefficiency in the data processing and information retrieval.
- There is a need for some XML code to be learned by me.
- There is a reliance on business to provide product information on the Web.
- There is the necessity to modify packaging methods so that there can be fewer contaminants released into the environment.

▶ Nominalization

6. Correct the nominalizations in the following sentences.

- Insertion of the image into the document occurs when you click OK.
- Specification of the file as a Word document was necessary to make the conversion of the file from text to Word.
- The manipulation of the layout in order to cause the transformation of it into film is the process of color separation.
- The use of my two keywords resulted in the best results for me.
- The conversion of the screen captures to JPEGs was easy to perform.

7. Write a paragraph about a concept you know well. Use as many nominalizations, *there are*'s, and passive voice combinations as you can. Then rewrite it eliminating all those constructions.

▶ Use of *You*

8. Correct the use of *you* in the following sentences.

- To judge the site's accuracy I read the tutorial and evaluated the lessons. This is where you look for your clear technical explanations and your computer code examples.
- This evaluation is different than the change from your old remote control to the new laser ones.
- Two methods exist to enhance fiber performance. In the first you orient the fibers. The second is the alkali process.

▶ Sexist Language

9. Correct the sexist language in the following sentences.

- Each presenter must bring his own laptop.
- Every secretary will hand in her timecard on Friday.
- If he understands the process, the machinist can improve production.
- Details were very good up to a point, but then the author seemed to lose her focus.
- I am not sure who Dr. Jones is, but I am sure he is a good doctor.

Eliminate Common Clarity Errors

Three other common issues that affect clarity are choppy sentences, wordiness and redundancy, and strings of nouns. Learn to identify and change these issues in your drafts.

Avoid Strings of Choppy Sentences

A string of short sentences results in choppiness. Because each idea appears as an independent sentence, the effect of such a string is to deemphasize the more important ideas because they are all treated equally. To avoid this, combine and subordinate ideas so that only the important ones are expressed as main clauses.

Choppy	Both models offer safety belts. Both models have counterbalancing. Each one has a horn. Each one has lights. One offers wing-sided seats. These seats enhance safety.
Clear	Both models offer safety belts, counterbalancing, a horn, and lights. Only one offers wing-sided seats, which enhance safety.

Avoid Wordiness and Redundancy

Generally, ideas are most effective when they are expressed concisely. Try to prune excess wording by eliminating redundancy and all unnecessary intensifiers (such as *very*), repetition, subordinate clauses, and prepositional phrases. Although readers react positively to the repetition of keywords in topic positions, they often react negatively to needless repetition.

Unnecessary subordinate clause	I found the site *by the use of keywords that are* nanotechnology and innovation
Revised	I found the site using *the keywords* "nanotechnology" and "innovation."
Redundant intensifiers plus unnecessary subordinate clause	It is made of *very* thin glass *that is milky white in color.*
Revised	It is made of thin, milky white glass.
Redundant	The tuning handle is a metal protrusion that can be easily grasped *hold of by the hand* to turn the gears.
Revised	The tuning handle is a metal protrusion that can be easily grasped to turn the gears.
Unnecessary repetition plus over use of prepositions	*This search* was done by a *keyword search* of the *same words* using the *search function* of different *search engines.*
Revised	This investigation used the same keywords in different search engines.

Avoid Noun Clusters

Noun clusters are three or more nouns joined in a phrase. They crop up everywhere in technical writing and usually make reading difficult. Try to break them up.

Noun cluster	Allowing *individual input variance* of *data process entry* will result in *higher keyboarder morale.*
Revised	We will have higher morale if we allow the keyboarders to enter data at their own rate.

Exercises

▶ Choppiness

1. Eliminate the choppiness in the following sentences.

   - XYZ has introduced an LCD monitor. The monitor is 17 inches. The monitor has a Web camera. Web conferencing applications can use the Web camera.
   - Numerical control exists in two forms. CNC is one form. CNC is Computer Navigated Control. DNC is Distributed Numerical Control.
   - On-line registration is frustrating. It should make it faster to register. The words on the screen are not self-explanatory. Screen notices say things like "illegal command." Retracing a path to a screen is difficult.

▶ Wordiness and Redundancy

2. Revise the following sentences, removing unnecessary words.

   - Due to the fact that we have two computer platforms connected together, we must pay attention to basic fundamentals when we send a document such as an e-mail attachment.
   - This project will be presented in Web format with links to the resource sites as well as other links that are associated with the sites and links that are associated with the topic of researching a report.
   - In fact many sites are available on the Web where the viewer can have the actual experience of purchasing equipment, new and used, from the site.
   - In the printing business there are two main ways of printing. The first is by using offset and the second is by using flexography.

Write Clear Paragraphs for Your Reader

Put the Topic Sentence First

An easy way to achieve a clear paragraph is to use the top-down strategy. Put the topic sentence first and follow it by several sentences of explanation.

The *topic sentence* expresses the paragraph's central idea, and the remaining sentences develop, explain, and support the central idea. This top-down arrangement enables readers to grasp the ideas in paragraphs more quickly (Slater). In addition putting the topic idea in the first sentence makes it possible for readers to get the gist of your document by skimming over the first sentences.

Consider this example:

Topic sentence Supporting details	The second remarkable property of muskeg is that, like a great sponge, it absorbs and accumulates water. Water enters a muskeg forest through precipitation (rain and snow) and through the ground (rivers, streams, seeps). It leaves by evaporation (chiefly of vapor transpired by the plants) and by outflow through or over the ground. However, input and output are not in balance. Water accumulates and is held absorbed in the accumulating peat. One of the commonest plants of muskeg is sphagnum moss, otherwise known as peat moss; alive or dead, its water-holding capacity is renowned, and is what gives peat its great water-retaining power. (Pielou 97)

In a coherent paragraph each sentence amplifies the point of the topic sentence. You can indicate coherence by

arranging sentences by level
repeating terms in a new/old sequence,
placing key terms in the dominant position,
indicating class or membership,
using transitions (Mulcahy).

Arrange Sentences by Level

Here is another way to think about the top-down method—think of each sentence as having a "level." On the first level is the topic sentence. The second level consists of sentences that support or explain the topic sentence. The third level consists of sentences that develop one of the second-level ideas. Four sentences, then, could have several different relationships. For instance, the last three could all expand the idea in the first:

1 First level
 2 Second level
 3 Second level
 4 Second level

Or sentences 3 and 4 could expand on sentence 2, which in turn expands on sentence 1:

1 First level
　2 Second level
　　3 Third level
　　4 Third level

As you write, evaluate the level of each sentence. Decide whether the idea in the sentence is level 2, a subdivision of the topic, or level 3, which provides details about a subdivision. Consider this six-sentence example:

Hydraulic pumps are classified as either nonpositive or positive displacement units. Nonpositive displacement pumps produce a continuous flow. Because of this design, there is no positive internal seal against leakage, and their output varies as pressure varies. Positive displacement pumps produce a pulsating flow. Their design provides a positive internal seal against leakage. Their output is virtually unaffected as system pressure varies.

The sentences of this paragraph have the following structure:

(1) Hydraulic pumps are classified as either nonpositive or positive displacement units.
　(2) Nonpositive displacement pumps produce a continuous flow.
　　(3) Because of this design, there is no positive internal seal against leakage, as their output pressure varies.
　(2) Positive displacement pumps produce a pulsating flow.
　　(3) Their design provides a positive internal seal against leakage.
　　(3) Their output is virtually unaffected as the system pressure varies.

Globalization and Style

When writing for an audience whose first language is not English, you need to be as clear and concise as possible. You do not need to "dumb down" your style, but in order for your document to be effective, you will need to think about the way in which you write (Dehaas).

An important aspect of communicating with other cultures is understanding their way of "categorization": how they divide and perceive the world. Lakoff states that "Categorisation is not to be taken lightly; there is nothing more basic than categorization to our thought, perception, action, and speech." (Lakoff 5). In other words some cultures see it as a virtue to go "astray and revert" and pursue the central idea, whereas other cultures see it as virtue to follow a clear and narrow line of thought from beginning to end.

However, some guidelines for good writing can be given irrespective of the deeper levels of thought.

Put only one idea in each sentence.

Focus on making the subject and the action clear—you can accomplish this by keeping your sentences in natural order: subject first, verb in the middle,

(Continued)

and predicate last, as shown in the first sentence in the following example below. If you use a subordinate clause, put the clause where it makes the strongest focus and makes the best cohesive link, as in sentences 2 and 3 in the following example. Avoid too many subject holders and antecedents, which are pronouns; write the word for the real subject instead. In sentence 2, replace "they" with "the modules" and in sentence 3, replace "they" with "modules" or "power supply." By doing so, you avoid ambiguities in the sentences.

Our electronic circuit protection modules provide the preferred solid-state protection for secondary circuits of switched mode power supplies. Faster than the self-protection of the power supply, they are designed to accommodate inductive and capacitance loads, avoiding nuisance tripping. When they do trip, an indicator shows which of four circuits caused the fault, saving troubleshooting time. (Rockwell)

Stay away from examples that require readers to understand metaphors outside of their experience. For instance, the sentence "That approach hit a home run" depends for its meaning on the ability to translate "home run" into "very successful." Examples linking up with a certain culture often call for localized translations to other cultures, which take a lot of effort and time and thus add to the costs!

Consider your word choice. You can cause confusion without ever being aware of it. By using the word *outlet*, meaning a place into which an electric plug is inserted, readers from another culture might be unfamiliar with the American usage, asking "Is 'outlet' a door or window?" (The UK word is *socket*, for instance).

Use consistent vocabulary. Once you use a word for an object, continue to use the same word. If you use "screen," do not suddenly switch to a synonym, for example, to "page" or "Web page." English often has many words for the same object or phenomenon, and writers often like to vary their language, but variation in the specific vocabulary is confusing. Variation should be confined to common-core language and function words.

Use an adequate mix between the active voice and the passive voice. Use "The manager accepted the proposal" instead of "The proposal was agreed to by the manager." The choice of active and passive voice, however, depends on the genre. In some situations, for instance, a document that explains how a machine performs a certain action, it is sometimes difficult to avoid the passive form. Likewise, in legal writing, it is sometimes not desired to have an agent in the sentence.

Write words in full instead of using contractions, especially in negative statements. Say "Do not use" rather than "Don't use"; the use of the separate word *not* gives the statement more emphasis. However, consider revising the sentence to state the issue positively. For instance, avoid instructions like this: "Do not put away your old freezer without ruining the lock. Children might use it to play hide and seek and die from lack of air, because the lock can only be opened from the outside." A positive instruction would be phrased like: "Always ruin the lock of a freezer before putting it away to prevent accidents." Another example is: "Do not smoke cigars or a pipe near the lawn mower." The ironical question then is "Well, can you then smoke pot near a lawn mower"?

Use –ing and –ed forms carefully. These forms cause difficulty in understanding. Consider this sentence: *The car body is spray-painted by a robot passing through a room.* Notice that the subject of the sentence (*car body*) is not the subject of the –ing form "passing;" *robot* is. That slight change can make your sentences difficult to translate. The sentence can be revised so that the agent performs both acts: *Passing through a room, the car body is spray-painted by a robot.* Although the sentences contain exactly the same words, shifting of –ing form to the beginning from end solves the problem. The car body passes through, and it is spray painted (Mousten).

For further reference, check out these websites:

IBM "Guideline B: Writing for an International Audience" <http://www-01. ibm.com/software/globalization/guidelines/guideb.html> has articles on writing, communicating, and presenting for an international community. You'll find tips on writing style as well as articles on cultural sensitivity issues. A complete list of its guidelines for writing for global audiences can be found at "Guidelines Quick Reference" http://www-01.ibm.com/software/globalization/guidelines/outline.html>

Repeat Terms in a New/Old Sequence

Sentences in a paragraph can follow an alternating sequence of supplying new information, which in turn becomes old information as the next sentence add more new information. In the following example, notice that the *new* information, *collided* (sentence 1), becomes *old* information, *collision* (sentences 2 and 3), and that *mountain range* is *new* information in sentence 3 but *old* information in sentence 4.

> Subduction stopped when the continent *collided* with the island arc along its northern margin. This *collision* resulted in extensive deformation of the island arc as well as deformation of the sedimentary rocks on the continental margin described earlier. The *collision* produced a *mountain range* across northern Wisconsin. This ancient *mountain range* is called the Penokean Mountains. The eroded remnants of these *mountains* constitute much of the bedrock of Wisconsin, Minnesota, and Michigan. (LaBerge 111)

Use the Dominant Position

Placing terms in the dominant position means to repeat a key term as the subject, or main idea, of a sentence. In the earlier paragraph notice that in sentences 2 and 3 *collision* appears first in sentences 2 and 3. The repetition returns readers to the same topic where they find it developed in another way.

Maintain Class or Membership Relationships

To indicate class or membership relationships, use words that show that the subsequent sentences are subparts of the topic sentence. In the following sentences, *distributed media* and *online systems* are members of the class *paradigm*. *Notice that the two terms also appear in the dominant position and that, if you indicated their level, they would both begin level 2 ideas.*

> Interactive multimedia follows one of two paradigms. Distributed media, such as CDs, are self-contained and circulated to audiences in the same manner as books or audio recordings. Online systems, such as intranets and the Web, resemble broadcasting in that the content originates from one central location and the use accesses it from a distance. (Bonime and Pohlman 177)

Provide Transitions

Using transitions means connecting sentences by using words that signal a sequence or a pattern. The common transitions such as *first/second, not only/but also, however, therefore, and, and but* are well known. For easy reading stay consistent. If you say "first" follow later with "second." While the paragraph might not read quite as elegantly as one that used less obvious transitions, the clearly indicated sequence will keep your reader from getting lost. It is fine to begin a sentence with *but*. However, that position makes the word quite emphatic, calling attention to the oncoming sentence as very important.

Choose a Tone for the Reader

The strategies discussed thus far in this chapter make your documents easy to read. These strategies, however, assume that the reader and the writer are unemotional cogs in an information-dispensing system. It is as though reader and writer were the cut-and-paste commands in a word processing program. Create the idea on this page (the writer), cut and paste to a new page (the reader), and the new page has the information in exactly the same form as the old page.

Writing situations are not that predictable. The tone, or emotional attitude implied by the word choice, can communicate almost as much as the content of a

Focus on Ethical Style

Use Unambiguous Language

Suppose, for instance, that you are writing a manual for a machine that has a sharp, whirling part under a protective cover. This dangerous part could slice off a user's fingers. When you explain how to clean the part, you inform the reader of the danger in a manner that prompts him or her to act cautiously. It would be unethical to write, "A hazard exists if contact is made with this part while it is whirling." That sentence is not urgent or specific enough to help a user prevent injury. Instead write, "Warning! Turn off all power before you remove the cover. The blade underneath could slice off your fingers!"

However, the need for unambiguous language appears in other much less dramatic situations. Take, for instance, the phrase "When I click the 'Submit' button, it doesn't work." This phrasing is so imprecise that it does not allow another person to act in a helpful way. How can someone fix it if he or she does not know what is not working? But that phrasing also indicates a moral stance—"I am not responsible. It is your job. I will not take the time and effort to right this, whatever inconvenience it may cause you." This kind of ambiguous use of language certainly is not dangerous, the way the previous example was, but it is a refusal to take responsibility in the situation. As such, the language does not help other people achieve their goals. It is wrong, not just because it is imprecise, but because it does not help the stakeholders.

Use Direct, Simple Expression

Say what you mean in a way that your reader will easily understand. Suppose you had to tell an operator how to deal with a problem with the flow of toxic liquid in a manufacturing plant. A complex, indirect expression of a key instruction would look like this:

If there is a confirmation of the tank level rising, a determination of the source should be made.

A simple, direct expression of the same idea looks like this:

Determine if the tank level is rising. Visually check to see if liquid is coming out of the first-floor trench.

Clarity is the gold standard for all communication. Jargon, shop talk, or technobabble that marginalizes or excludes the reader or audience is not only confusing, it is unethical. It is both reasonable and desirable to create prose, be it technical or otherwise, that is written for its intended audience. Unfortunately, sometimes it is all too easy to slide into a vernacular that is common among those in-house. To use terms that are unique to a particular discursive community can create a boundary between the document and its intended audience. If you must use jargonistic terms, include a glossary, or define the term the first time you use it in the text. If your language can be misconstrued, it can cause problems. A good general rule is to guard against any use of terms that are common in-house when the audience for the document is "out of house." This is not only good practice; it is the ethical thing to do.

message (Fielden). To communicate effectively, you must learn to control tone. Let's consider four possible tones:

▶ Forceful
▶ Passive
▶ Personal
▶ Impersonal

The *forceful* tone implies that the writer is in control of the situation or that the situation is positive. It is appropriate when the writer addresses subordinates or when the writer's goal is to express confidence. To write forcefully,

▶ Use the active voice.
▶ Use the subject–verb–object structure.
▶ Do not use "weasel words" (*possibly, maybe, perhaps*).
▶ Use imperatives.
▶ Clearly indicate that you are the responsible agent.

I have decided to implement your suggestion that we supply all office workers with laptops and eliminate their towers. This suggestion is excellent. You have clearly made the case that this change will reduce eyestrain issues and will greatly enhance the flow of information in the department. Make an appointment with me so we can start to implement this fine idea.

The *passive* tone implies that the reader has more power than the writer or that the situation is negative. It is appropriate when the writer addresses a superior or when the writer's goal is to neutralize a potentially negative reaction. To make the tone passive,

▶ Avoid imperatives.
▶ Use the passive voice.
▶ Use "weasel words" (*very, several people, quite, fairly*).
▶ Use longer sentences.
▶ Do not explicitly take responsibility.

The proposal to implement laptops in our department has not been accepted because of a number of very difficult issues. To our surprise several people have indicated that the ergonomic benefits of the screens are not seen as not quite offsetting the potential disruption that will be caused by the migration of files to the new machines. The large footprint of the docking station has also been suggested as a possible problem for our employees due to their fairly restricted desk space. Because the need for action on computer replacement is necessary, a meeting will be scheduled next week to discuss this.

Compare this to a forceful presentation:

> The steering committee and I reject the laptop proposal. You have not included enough convincing data on morale or work flow, and you have not dealt with work flow disruption and the large size of the docking station. Make an appointment to see me if necessary.

The *personal* tone implies that reader and writer are equal. It is appropriate to use when you want to express respect for the reader. To make a style personal,

- ▶ Use the active voice.
- ▶ Use first names.
- ▶ Use personal pronouns.
- ▶ Use short sentences.
- ▶ Use contractions.
- ▶ Direct questions at the reader.

> Ted, thanks for that laptop suggestion. The steering committee loved it. Like you, we feel it will solve the eyestrain issue and will facilitate data flow. And we think it will also raise morale. I'd like you to begin work on this soon. Can you make an appointment to see me this week?

This tone is also appropriate for delivering a negative message when both parties are equal.

> Ted, thanks for the laptop suggestion, but we can't do it this cycle. The steering committee understands the ergonomic issue you raise, but they are very concerned about the disruption that migrating all those files will cause. In addition, they feel that we need to work out the entire issue of footprint—the model you suggested would cause a number of problems with current desk configurations. I know that this is a disappointment. Could we get together soon to discuss this?

The *impersonal* tone implies that the writer is not important or that the situation is neutral. Use this tone when you want to downplay personalities in the situation. To make the tone impersonal,

- ▶ Do not use names, especially first names.
- ▶ Do not use personal pronouns.
- ▶ Use the passive voice.
- ▶ Use longer sentences.

> A decision to provide each employee with a laptop has been made. Laptops will reduce the eye fatigue that some employees have experienced, and the laptops will increase data flow. Ted Baxter will chair the implementation committee. Donna Silver and Robert Sirabian will assist. The committee will hold its initial meeting on Monday, October 10, at 3:00 P.M. in Room 111.

Worksheet for Style

☐ **Find sentences that contain passive voice. Change passive to active.**

☐ **Look for sentences shorter than 12 or longer than 25 words. Either combine them or break them up.**

☐ **Check each sentence for coordinate elements. If they are not parallel, make them so.**

☐ **Read carefully for instances of the following potential problems:**

- Nominalizations
- Sexist language
- Too frequent use of *there are*

- Choppiness
- Incorrect use of *you*
- Wordiness

☐ *Sentences.* **Look for four types of phrasing. Change the phrasing as suggested here or as determined by the needs of your audience and the situation.**

- The word *this.* Usually you can eliminate it (and slightly change the sentence that is left), or else you should add a noun directly behind it. ("By increasing the revenue, this will cause more profit" becomes "Increasing the revenue will cause more profit.")
- The words *am, is, are, was, were, be,* and *been.* If these are followed by a past tense (*was written*) the sentence is passive. Try to change the verb to an active sense (*wrote*).
- Lists of things or series of activities. Put all such items, whether of nouns, adjectives, or verb forms, in the same grammatical form (*to purchase, to assemble,* and *to erect*—not *to purchase, assembling,* and *to erect*). This strategy will do more to clarify your writing than following any other style tip.
- The phrases *there are* and *there is.* You can almost always eliminate these phrases and a *that* which appears later in the sentence. ("There are four benefits that you will find" becomes "You will find four benefits.")

☐ **Use the top-down principle as your basic strategy.**

☐ **Make sure that each paragraph has a clear topic sentence.**

☐ **Check paragraph coherence by reviewing for**

a. Repetition of terms
b. Placement of key terms in a dominant position
c. Class or membership relationships
d. Transitions.

☐ **Evaluate the sentence levels of each paragraph. Revise sentences that do not clearly fit into a level.**

Exercises

❭ You Create

1. Write a report in which you reject an employee's solution to a problem. Give several reasons, including at least one key item that the employee overlooked.

❭ You Analyze

2 Analyze Example 4.1. Either in groups or individually, revise the document using the concepts outlined in this chapter. You may revise both wording and tone.

Example 4.1   Methods Statement

In this section of the report, I will discuss *www.flipdog.com*. I used the job search titles of Retail Analyst, Construction Project Manager, and Packaging Engineer. For each of these job search titles, I researched the entire United States, Minnesota, and Wisconsin.

When I first got to the website, I clicked on "Find Jobs." My first step was to then choose the area I wanted to search. I decided to start with the entire United States so I just left "Search: All of U.S." as the default search. With this option, there was a maximum of 264,941 jobs searched. I then had the choice to choose a category of search. You may search all job categories, for a category individually, or by "Ctrl+click" to search multiple categories.

I also had the choice of choosing the specific company of employment. Like the category search above, you have the choice to search all employers, multiple employers, or a single employer.

For this research project, I didn't use the job category search or the employer search. I skipped right to the keyword search and typed in "retail analyst."

I then clicked "Get Results." Within all of the United States, "retail analyst" received 179 results. After these results were listed, you are also given the option to search for specifics within results by entering keywords into this figure below. (For my research, I used total results only.)

You may also modify the dates. (For my research, I used all dates by default.)

I continued my research using the steps above. The tables below summarize my results. Each table demonstrates the area searched (all of the United States, Wisconsin, and Minnesota), as well as the three job titles **Retail Analyst, Construction Project Engineer, and Packaging Engineer.** *Note:* For results over 200, FlipDog reported as 200.

3. Review the following paper (Example 4.2) for tone. By indicating specific phrases and words, determine whether the author has adopted the correct tone. If your instructor requires it, rewrite the report with a different tone.

Example 4.2

HOW TO CHANGE THE BACKGROUND COLOR
IN PHOTOSHOP

INTRODUCTION

PhotoShop was always something that scared me a little when mentioned, but after exploring it for a couple of hours I got the hang of it. I am working with PhotoShop with other people in my class and they are doing examples of other projects to do in PhotoShop also. (I have links to their sites below.) My topic is how to change the background color and it was very simple to do.

METHOD

The first thing that I explored with ended up to be the right thing so I got a little lucky when doing this, but a little knowledge of Auto Cad helped too. I clicked on the magic wand, which looks just like it sounds, and then clicked onto the color in the background that I wanted to change. This produces a blinking outline around everything that touches that color in the picture.

I then knew that I needed to be able to select a new color so I clicked on "Window" and scrolled down to "Show Color." This brought up a new smaller window that had a palette full of colors. It asked me to select a color and I chose red. After clicking "Okay" absolutely nothing happened and this had me stumped for a while. Here is where my Auto Cad experience helped me out. I went to "Edit" and scrolled down to "Fill" which basically regenerates your picture the same as in Auto Cad. My background is now red.

RESULT

As you can see from the above photos [not shown here] I was able to change the background color to red. In the corners the gray background is still there because that is a slightly different shade and all I would have to do is go through the process again and click on the color and make sure I had the same shade of red. Another thing to notice is that when you look through the bottles near the top you can see a white background yet, which I could also click on and give a pinkish color.

DISCUSSION

This option in PhotoShop can allow you to do virtually whatever you can come up with in your head. I believe that I am going to take some photos of vacations where the sky is really dark and make it a little brighter day. (You won't remember the difference in 30 years anyway.) This feature could also help if you need to eliminate something in the background completely like an indecent sign or person. If you have any needs similar to this then this should be something that you look into.

4. Analyze Example 8.3 on pages 231–233 to determine the sentence strategies and tone used for the intended audience.

5. Compare strategies in Examples 4.1 (pp. 107–108) and 8.3 or 8.4 (pp. 231–233; 233–234). Write a brief report in which you give examples that illustrate how sentence tone creates a definition of an intended audience. Alternative: In groups of three or four, compare Examples 4.1 and 8.3. Rewrite one of them for a different audience. Read your new report to the class, who will identify the audience and strategies you used.

▶ You Revise

6. Rewrite two paragraphs of Examples 8.3 (pp. 231–233) or all of the text in Exercise 8 (Chapter 3, pp.79–80) in order to relate to a different audience. Keep the content the same, but change the sentence strategies and the tone.

7. Rewrite part or all of Example 8.3 (pp. 231–233) to apply to a different audience.

8. Revise the following paragraphs so that the sentences focus on the Simulation Modeling Engineer as the actor in the process of model building.

> The process of abstraction of the system into mathematical-logical relationships with the problem formulation is the Model Building Phase. The assumptions and decision variables are used to mathematically determine the system responses. Also, the desired performance measures and design alternatives are evaluated. The system being modeled is broken down into events. For each event, relevant activities are identified. The basic model of the system is now an abstraction of the "real" system.
>
> The Data Acquisition Phase involves the identification and collection of data. This phase is often the most time consuming and critical. If the data collected are not valid, the simulation will produce results that are not valid. The data to be acquired are determined by the decision variables and assumptions. All organizations collect large quantities of data for day-to-day management and for accounting purposes. To collect data for simulation, the SMEs need cooperation of management in order to gain knowledge and access to the information sources. When the existing data sources are inadequate, a special data-collection exercise is required or the data are estimated.

❯ Group

9. In groups of three or four, review the memos you wrote in Exercise 8 for appropriate (or inappropriate) tone. Select the most (or least) effective report and explain to the class why you chose it.

10. Break into groups of two to three. From each set select the sentence you like best. Explain what is wrong with the one you reject. Usually one sentence is an attempt to fix a style issue in the other. You can dislike both and present a different revision.

1. I felt that the text would be easier to read if the sections were combined.

2. I felt that there was needless repetition in what you wrote.

3. There are many errors and sloppy decisions in this.

4. This report contains many errors and is based on sloppy decisions.

5. I am finished with my work and I never received your contribution.

6. The work has been finished by the deadline we discussed. Because your section was not available at that time it has not been included.

7. To explain who will do it, you would tell who the focus group leader is and who the project manager.

8. To explain who will do it, you would describe the focus group leader and the project manager, the person who will arrange the ads and the assemble the group.

9. To explain who will do it, you would describe the focus group leader and the project manager, the person who will arrange calls to attend and choose the members of the group.

10. To review—the only way to fix splices and run-ons yourself is to know two things: what an independent clause is and what the appropriate way to join them is.

11. To review—you can only fix splices and run-ons if you understand independent clauses and the appropriate way to join them.

12. If group members evaluate one another, they should comment on the punctuality and helpfulness of these reports.

13. If group members evaluate one another, the punctuality and helpfulness of these reports should be commented on.

14. As I expected he is wrong because he has not investigated all the proper documents.

15. As I expected his data is inaccurate because he has not investigated all the proper documents.

16. Hi, I am trying to figure out how to use LinkedIn, hence this venue for my message.

17. Hi, I used this venue because I am trying to figure out how to use LinkedIn.

18. Can you pull all the info off the Qualtrics site? I think that we have as many as we will get.

19. Can you pull all the info off the Qualtrics site? I think that we have as many responses as we will get.

20. These problems are simply not easy to identify.

21. There is no easy way to identify these problems.

22. To put it another way between the capital letter of the first word and the period after the last word, there has to be a subject and predicate to make *it* a sentence.

23. To put it another way between the capital letter of the first word and the period after the last word, there has to be a subject and predicate to *make the group of words* a sentence.

24. The problems typically arise when you have several clauses in a sentence or when you shorten up the phrasing for emphasis.

25. The problems typically arise when the sentence has several clauses or when the phrasing is shortened up for emphasis.

26. That group of words are a fragment.

27. That group of words is a fragment.

28. There has to be some indicator, such as a semicolon, or else you have a "run-on."

29. You must supply some indicator, such as a semicolon, or else you have a "run-on."

30. The generic, or usual, method most presenters use is the text method because it makes creating a slide so easy.

31. Many presenters will continue to use the text method because it makes creating a slide so easy.

32. A series of slides could illustrate what happens at each step, for instance explaining the actions of each stage of moving a bottle through a bottle-filler machine.

33. A series of slides explaining the actions of each stage of moving a bottle through a bottle-filler machine could illustrate what happens at each step.

34. As you speak the words or phrases on screen will serve as an outline.

35. PowerPoint, for instance, has both an outline and a slide sorter function that allow you to see the entire sequence of your presentation.

36. PowerPoint, for instance, has two functions that allow you to see the entire sequence of your presentation: outline and slide sorter.

37. If you use of the software's templates, the logic is created for you.

38. Use one of the software's templates and it creates the logic for you.

39. Animation makes items move.

40. Animation is making items move.

41. For instance, a long italicized quote is difficult to read.

42. For instance, a long quote when italicized, is difficult to read.

11. Create a class collaboration site, such as a blog, a secret Facebook page, or wiki. Devote the site to Style. The goal is to have all members of the class contribute sentences that they rewrote to improve them. The examples should contain both the before and after versions. Individuals may post their comments concerning the effectiveness of the revision. Encourage classmates to use examples of the type found in this chapter and in Appendix A. Also encourage them to use examples of changing from inappropriate to appropriate wording (review the one-sentence example in the social appropriateness section of the "Technical Communication Is Appropriate," Chapter 1, p. 10). An example: *Original sentence:* "Your goal is to give the reader access to the screen so that they will feel comfortable using the site." *Changed to* "Your goal is to give the readers access to the screen so that they feel comfortable using the site." *Reason for change:* eliminated the faulty pronoun reference, also eliminated having to use "he or she" in the second sentence.

Writing Assignments

1. You are a respected expert in your field. Your friend, an editor of a popular (not scholarly) magazine, has asked you to write an article describing basic terms employed in a newly developing area in your field. Write a one- or two-page article using Example 8.3 (pp. 231–233) or Exercise 8 in Chapter 4 (p. 109) as a guide.

2. Write a learning report for the writing assignment you just completed. See Chapter 5, Assignment 8, page 137, for details of the assignment.

Web Exercise

Investigate any website to analyze the style and organizational devices used in the site. Use the principles discussed in this chapter: Are the sentences written differently from what they might be if they were in a print document? What devices are used to make organization obvious? Write an analytical report (see Chapter 10) to alert your classmates or coworkers to changes they should make as they develop their websites.

Works Cited

American Psyschological Association. *Publication Manual.* 6th ed. Washington, DC: APA, 2010. Print.

Bonime, Andrew, and Ken C. Pohlman. *Writing for New Media: The Essential Guide to Writing for Interactive Media, CD-ROMs, and the Web.* New York: Wiley, 1998. Print.

Dehaas, David. "Say What You Mean." *OHS Canada.* Web. 28 May 2012. <www.ohscanada.com/Training/saywhatyoumean.aspx>.

Fielden, John S. "What Do You Mean You Don't Like My Style?" *Harvard Business Review* 60 (1982): 128–138. Print.

LaBerge, Gene L. *Geology of the Lake Superior Region.* Tucson, AZ: Geoscience Press, 1994. Print.

Lakoff, G. *Women, Fire and Dangerous Things: What Categories Reveal about the Mind,* Chicago: University of Chicago Press, 1987. Print.

Maggio, R. *The Bias-Free Word Finder: A Dictionary of Nondiscriminatory Language.* Boston: Beacon, 1991. Print.

The Microsoft Manual of Style for Technical Publications. MicrosoftPress, 1995.

Mousten, B., Vandepitte, S. and Maylath, B. "Intercultural Collaboration in the Trans-Atlantic Project: Pedagogical Theories and Practices in Teaching Procedural Instructions across Cultural Contexts." Chapter 9. In Starke-Meyerring, D. and Wilson, M. (eds). 2008. Print.

Mulcahy, Patricia. "Writing Reader-Based Instructions: Strategies to Build Coherence." *Technical Writing Teacher* 15.3 (1988): 234–243. Print.

Pielou, E. C. *After the Ice Age: The Return of Life to Glaciated North America.* Chicago: University of Chicago Press, 1991. Print.

Rockwell Automation "Circuit Et Load Protection." 2012. Web. 20 June 2012. <http://ab.rockwellautomation.com/Circuit-and-Load-Protection>.

Slater, Wayne H. "Current Theory and Research on What Constitutes Readable Expository Text." *Technical Writing Teacher* 15.3 (1988): 195–206. Print.

Focus on
Bias in Language

Current theory has made clear just how much language and language labels affect our feelings. Biased language always turns into biased attitudes and actions that perpetuate demeaning attitudes and assumptions. It is not hard to write in an unbiased way if you apply a few basic rules. *The American Psychological Association (APA) publication manual* suggests that the most basic rule focuses on exclusion. A sentence that makes someone feel excluded from a group needs to be revised. It's rather like hearing yourself discussed while you are in the room. That feeling is often uncomfortable, and you should not write sentences that give that feeling to others. The APA manual (70–77; based heavily on Maggio) offers several guidelines.

Describe People at the Appropriate Level of Specificity

This guideline helps whenever you have to describe people. Technical writing has always encouraged precise description of technical objects. You should apply the same principle to the people who use and are affected by those objects. So, when referring to a group of humans of both sexes, say "men and women," not just "men."

Be Sensitive to Labels

Call people what they want to be called. However, be aware that these preferences change over time. In the 1960s, one segment of the American population preferred to be called "black"; in the 1990s, that preference changed to "African American."

Basically, do not write about people as if they were objects—"the complainers," "the strikers." Try instead to put the person first—"people who complain," "people who are striking." Because this can get cumbersome, you can begin by using a precise description and after that use a shortened form as long as

it is not offensive. The issue of what is offensive is a difficult question. How do you know that "elderly" is offensive but "older" is not? There is no easy answer. Ask members of that group. Listen to the words that national TV news applies to members of the group.

Acknowledge Participation

This guideline asks you to treat people as action initiators, not as the recipients of action. In particular, it suggests using the active voice to talk about people who are involved in large mass activities. So say, "The secretaries completed the survey," not "The secretaries were given the survey."

Avoid Ambiguity in Sex Identity or Sex Role

This guideline deals with the widespread use of masculine words, especially *he,* when referring to all people. This usage has been changing for some time but still causes much discussion and controversy. The basic rule is to be specific. If the referent of the word is male, use *he;* if female, use *she;* if generic, use *he or she* or, more informally, *they.*

Choose Correct Terms to Indicate Sexual Orientation

Currently the preferred terms are *lesbians* and *gay men, straight women* and *straight men.* The terms *homosexual* and *heterosexual* could refer to men or women or just men, so their use is not encouraged.

Use the Preferred Designations of Racial Groups

The preferred designations change and sometimes are not agreed upon even by members of the designated group. Be sensitive to the wishes of the group you are serving. At times, *Hispanic* is not a good choice because individuals might prefer *Latino* or *Latina,* *Chicano* or *Chicana,* or even a word related to a specific country, like *Mexican.* Similar issues arise when you discuss Americans of African, Asian, and Arabic heritage. If you don't know, ask.

Do Not Use Language That Equates a Person with His or Her Condition

"Disability" refers to an attribute of a person. Say, "person with diabetes" to focus in a neutral way on the attribute; do not say, "diabetic," which equates the person with the condition.

Choose Specific Age Designations

Use *boy* and *girl* for people up to 18; use *man* and *woman* for people over 18. Prefer *older* to *elderly.*

Researching

CHAPTER CONTENTS

CHAPTER 5 **IN A NUTSHELL**

When you conduct research, you are finding the relevant facts about the subject. Two strategies are *asking questions* and *using keywords*.

Ask questions. Start with predictable primary-level *questions:* How much does the research cost? What are its parts? What is the basic concept you need to know? The trick, however, is to ask secondary-level questions that help you establish relationships. Secondary questions include cause (Why does it do this? Why does it cost this much?) and comparison–contrast (How is this like that? Why did it act differently this time?).

Use keywords. Type in *keywords,* following search rules, to search all library databases and Web databases. The two basic skills are knowing how to use this database's "search rules" and knowing which *words* to use.

Spend time learning the database's "search rules." Typing in one word is easy, but how do you handle combinations—either phrases (French roast coffee beans) or strings (coffee, caffeinated, decaffeinated)? All search engines use logical connectors—*and, but, not, or*—in some fashion. "Italian" *and* "coffee" narrows the results to those that contain both terms; "Italian" *or* "coffee" broadens the results to all those that contain just one of the two terms.

Finding which *words* to use is a matter of educated guesses and observation. "Coffee" is too broad (i.e., it will give you too many choices—so many results that you cannot use them), so use "caffeinated," "decaffeinated," which will yield narrower results.

People research everything from how high above the floor to position a computer screen to how feasible it is to build a manufacturing plant. This chapter discusses the purpose of research, explores the essential activity of questioning, and suggests practical methods of finding information.

The Purpose of Research

The purpose of research is to find out about a particular subject that has significance for you. Your subject can be broad and general, such as roasting coffee, or narrow and specific, such as purchasing standing workstations for your office. The significance is the importance of the subject to you or your community. Will a particular method of roasting coffee make the flavor more robust? Will those new workstations make the office more productive?

Generally, the goal of research is to solve or eliminate a problem (Why are the current workstations not optimal?) or to answer a question (What differences are there between traditional and standing workstations?). You can use two strategies: talking to people and searching through information. To find out about standing workstations, for instance, you would talk to various users to discover features that they need, and you could read about other office setups to see what works best.

Questioning—The Basic Skill of Researching

Asking questions is fundamental to research. The answers are the facts you need. This section explains how to discover and formulate questions that will "open up" a topic, providing you with the essential information you and your readers need.

How to Discover Questions

To learn about any topic—such as the best standing workstations for your office—ask questions. Formulate questions that will help you investigate the situation effectively and that will provide a basis for a report. For instance, the question "In what ways does our staff use their workstation?" will not only produce important data but also be the basis for a section on "usage patterns" in a report. Several strategies for discovering helpful questions are to ask basic questions, ask questions about significance, consult the right sources, and interact flexibly.

Ask Basic Questions

Basic questions lead you to the essential information about your topic. They include

- What are the appropriate terms and their definitions?
- What materials are involved?
- What processes are involved?

Ask Questions About Significance

Questions about significance help you "get the big picture" and grasp the context of your topic. They include

- Who needs it and why?
- How is it related to other items?
- How is it related to current systems?
- What is its end goal?
- How do parts and processes contribute to the end goal?
- What controversies exist?
- What alternatives exist?
- What are the implications of those alternatives?
- What costs are involved?

Consult the Right Sources

The right sources are the people or the information that has the facts you need.

People who are involved in the situation can answer your basic questions and your questions about significance. They can give you basic facts and identify their needs. The basic facts about intranets can come from Web designers, computer programmers, or experienced users. Information about needs comes from people who use the system. They expect an intranet to provide certain resources, and they know the conditions that make using those resources effective.

Existing information also answers basic questions and questions of significance, often more thoroughly than people can. This includes everything from magazine articles to encyclopedias to blogs, online discussion groups, and social media posts. For the workstations, a health magazine could provide the health benefits of a standing workstation; bulletin boards could give you anecdotal evidence of increased productivity; social media posts could give you user testimony.

Interact Flexibly

To ask questions productively, be flexible. People have the information you need, and you must elicit as much of it as you can. Sometimes questions produce a useful answer, sometimes not. If you ask, "Which feature of your workstation is most important to you?" the respondent might say, "Adequate storage," which is a broad answer. To narrow the answer, try an "echo-technique" question, in which you repeat the key term of the answer: "Storage?" On the other hand, if you ask, "Is it important to have a printer at your workstation?" the respondent might say, "No, I don't print that frequently, but I do use dual monitors." That answer opens two lines of questioning for you. Why is a printer not important? Why is having dual monitors important?

You can also use questions to decide what material to read. If your question is "What are the benefits of using a standing workstation?" an article that would obviously interest you is "10 benefits of using a standing workstation." Carefully formulating your questions makes your reading more efficient. Read actively, searching for particular facts that answer your questions (Spivey and King).

While reading, take notes, constantly reviewing the answers and information you have obtained. Look for patterns in the material or gaps in your knowledge. If three articles present similar evaluations of standing workstations, you have a pattern on which to base a decision.

How to Formulate Questions

Basically, researchers ask two kinds of questions: closed and open (Stewart and Cash). You can use both types for interviewing and reading.

A closed question generates a specific, often restricted answer. Technically it allows only certain predetermined answers.

Closed question	How many hours per week do you spend at your workstation?

An open question allows a longer, more involved answer.

Open question	What aspects of your current workstation decrease focus and productivity?

In general, ask closed questions first to get basic, specific information. Then ask open questions to probe the subtleties of the topic.

Collecting Information from People

You collect information, or find answers to your questions, in a number of ways. You can interview, survey, observe, test, and read. This section explains the first four approaches. Collecting published information, especially in a library, is treated in a later section.

Interviewing

One effective way to acquire information is to conduct an information interview (Stewart and Cash). Your goal is to discover the appropriate facts from a person who knows them. To conduct an effective interview, you must prepare carefully, maintain a professional attitude, be willing to probe, and record answers.

Prepare Carefully

To prepare carefully, inform yourself beforehand about your topic. Read background material, and list several questions you think will produce helpful answers.

If you are going to ask about workstations, read about them before you interview anyone so that you will understand the significance of your answers. Listing specific questions will help you focus on the issue and discourage you and the respondent from digressing. To generate the list, brainstorm questions based on the basic and significance questions we have suggested. A specific issue to focus on could be workstation problems: What aspect of your current workstation is not ideal? Is it something that can be easily remedied? Is there an alternative solution?

Maintain a Professional Attitude

Schedule an appointment for the interview, explaining why you need to find out what the respondent knows. Make sure she or he knows that the answers you seek are important. Most people are happy to answer questions for people who treat their answers seriously.

Be Willing to Probe

Most people know more than they reveal in their initial answers, so you must be able to get at the material that's left unsaid. Four common probing strategies are as follows:

- Ask open-ended questions.
- Use the echo technique.
- Reformulate.
- Ask for a process description.

The basic probing strategy is to ask an *open-ended question* and then develop the answer through the echo technique or reformulation. The *echo technique* is repeating significant words. If an interviewee says, "I really lose focus in the afternoon," you respond with "Lose focus?" This technique almost always prompts a longer, more specific answer. *Reformulation* means repeating in your own words what the interviewee just said. The standard phrase is "I seem to hear you saying. . . ." If your reformulation is accurate, your interviewee will agree; if it is wrong, he or she will usually point out where. *Asking for a process description* produces many facts because people tend to organize details around narrative. As the interviewee describes, step by step, how he or she uses the workstation, you will find many points where you'll need to ask probing questions.

Record the Answers

Put the questions in a word processing document—like MS Word—leaving a space to record notes and answers. Ask people to repeat if you didn't get the whole answer written down. After a session, review your notes to clarify them so they will be meaningful later and to discover any unclear points about which you must ask more questions.

Surveying

To survey people is to ask them to supply answers to your questions. Distributing a survey allows you to receive answers from many people, more than you could possibly interview in the time you have allotted to the project. Surveys help you determine basic facts or conditions and assess the significance or importance of facts. They have three elements: a context-setting introduction, closed or open questions, and a form that enables you to tabulate all the answers easily.

A context-setting introduction explains (1) why you chose this person for your survey, (2) what your goal is in collecting this information, and (3) how you will use the information. The questions may be either closed or open. The answers to closed questions are easier to tabulate, but the answers to open questions can give you more insight. A good general rule is to avoid questions that require the respondent to research past records or to depend heavily on memory.

The form you use is the key to any survey. It must be well designed (Warwick and Lininger). Your goal is to design it so that it is both easy to read (so that people will be willing to respond) and easy to tabulate (so you can tally the answers quickly). For instance, if all the answers appear at the right margin of a page, you can easily transfer them to another page. Many surveys are Web-based, which makes for easy tabulation, collection, and analysis. Figure 5.1 shows a sample survey.

Observing and Testing

In both observing and testing, you are in effect carrying out a questioning strategy. You are interacting with the process yourself.

Observing

Observing is watching intently. You place yourself in the situation to observe and record your observations. When you observe to collect information, you do so with the same questions in mind as when you interview: What are the basic facts? What is their significance?

To discover more about problems with the workstations, you could simply watch people at their workstations. You would notice their posture, how well they maintain focus at different parts of the day, how they utilize the equipment, how often they stand up or move around, and so on. If you discover that after lunch productivity drops and people are slouched down in their chairs, the standing workstation might be a perfect solution. By observation—looking in a specific way for facts and their significance—you might find the data you need to solve the office problem.

Context-setting introduction	**SURVEY** Recently we have had increased interest in the benefits of standing workstations. In order to move forward with purchasing these stations we have decided to assess the current needs of our employees. To help us choose the most ideal workstations, please take a moment to fill in the attached survey. Please return it to Peter Arc, 150 M Nutrition Building, by Friday, January 30. Thanks.
Closed question	How many hours a week are you at your workstation? less than 20 ____ 20–30 ____ 30–40 ____ more than 40 ____
Closed question	Does your workstation require any of these features? Yes No Dual monitors ____ ____ Printer/scanner ____ ____ Docking station ____ ____
Closed question	Are the following aspects important to you? Yes No Ability to sit or stand ____ ____ Paper file storage area ____ ____ Built-in foot rest ____ ____ Desktop area for paperwork ____ ____
Open question	Please describe the issues with your current workstation. Use the back of this sheet if you need more space.
Open question	Indicate the two most important benefits motivating you to switch to a standing workstation: Increased productivity ____ Improved posture and circulation ____ Higher calorie burn ____ Increased focus ____
Open question	Please discuss any concerns you have about switching to a standing workstation. Use the back of this sheet if you need more space.

© Cengage Learning

Figure 5.1 Sample Research Survey

Testing

To test is to compare items in terms of some criterion or a set of criteria. Testing, which is at the heart of many scientific and technical disciplines, is much broader and more complex than this discussion about it. Nevertheless, simple testing is often a useful method of collecting information. Before you begin a test, you must decide what type of information you are seeking. In other words, what questions should the test answer?

In the case of deciding which standing workstation to buy, the questions should reflect the users' concerns. They become your criteria, the standards you will use to evaluate two workstations. Typical questions might be

- Which one accommodates all of the necessary equipment?
- Which one allows for sitting or standing?
- Which one contains adequate storage and workspace?

After determining suitable questions, you have people use both stations and then record their answers to your questions.

Collecting Published Information

This section discusses the basic techniques for gathering published information. As with all writing projects, you must plan carefully. You must develop a search strategy, search helpful sources, and record your findings.

Develop a Search Strategy

With its thousands of books and periodicals, the library can be an overwhelming place. The problem is to locate the relatively small number of sources that you actually need. To do so, develop a "search strategy" ("Tracking Information") by determining your audience, generating questions, predicting probable sources, and searching for "keys."

Determine Your Audience

As in any writing situation, determine your audience and their needs. Are you writing for specialists or nonspecialists? Do they already understand the concepts in the report? Will they use your report for reference or background information, or will they act on your findings? Experts expect to see information from standard sources. Thus mentioning articles from technical journals is more credible than citing material from the popular press. However, in some areas, particularly computing and also in subjects like photography, monthly magazines—primarily available online—are often the best sources of technical information. For computers, *Macworld* and *PC World* are excellent sources for technical decisions. Nonspecialists may not know standard sources, but they expect you to have consulted them.

Generate Questions

Generate questions about the topic and its subtopics. These questions fall into the same general categories as those for interviews: What are the basic facts? What is their significance? These questions include

- ▶ What is it made of?
- ▶ How is it made?
- ▶ Who uses it?
- ▶ Where is it used?
- ▶ What is its history?
- ▶ Do experts disagree about any of these questions?
- ▶ Who makes it?
- ▶ What are its effects?
- ▶ How is it regulated?

Such questions help you focus your research, enabling you to select source materials and to categorize information as you collect it.

Predict Probable Sources

All concepts have a growth pattern, from new and unusual to established and respected. Throughout the pattern they are discussed in predictable—but very different—types of sources. New and unusual information is available only from a few people, probably in the form of letters, conversations, e-mail, answers to listserv queries, and personal websites. More established information appears in conference proceedings and technical journal articles. Established information appears in textbooks, general periodicals, and newspapers ("Tracking Information").

If you understand this growth pattern, you can predict where to look. Two helpful ways to use the pattern are by age and by technical level. Use the following guidelines to help you find relevant material quickly.

Consider the Age of the Information.   If your topic demands information less than a year old, consult periodicals, government documents, annual reviews, and online databases. Write letters, call individuals, ask on a listserv, or search the Web (see section "Web Searching"). If your topic requires older, standard information, consult bibliographies, annual reviews, yearbooks, encyclopedias, almanacs, and textbooks.

Consider the Technical Level of the Information.   If you need information at a high technical level, use technical journals, interviews with professionals, and specialized encyclopedias or handbooks. On the job, also use technical reports from the company's technical information department. If you need general information, use popular magazines and newspapers. Books can provide both technical and general information.

Search for "Keys"

A helpful concept to guide your searching is the "key," an item that writers constantly repeat. Look for keywords and key documents.

Find Keywords.   Keywords are the specific words or phrases that all writers in a particular field use to discuss a topic. For instance, if you start to read about the Internet, you will quickly find the terms *navigation* and *hyperlink* in many sources. Watch for terms like these, and master their definitions. If you need more information on a term, look in specialized encyclopedias, the card catalog, periodical indexes, abstracts, and databases. Keywords can also lead you to other useful terms through cross-references and indexes.

Watch for Key Documents.   As you collect articles, review their bibliographies. Some works will be cited repeatedly. These documents— whether articles, books, or technical reports—are *key documents*. If you were searching for information about the Web, you would quickly discover that three or four books are the "bibles" of the Web. Obviously, you should find and review those books. Key documents contain discussions that experts agree are basic to understanding the topic. To research efficiently, read these documents as soon as you become aware of them.

Search Helpful Sources

To locate ideas and material, you can use Web searching, electronic catalogs, and electronic databases.

Web Searching

The Web contains stunning amounts of information—so much that, curiously, the problem is to cull out what is usable. The key is to know how to search effectively.

Search the Web as you would any other database. Choose a search engine (like Google or Yahoo!) whose purpose is to search a database for instances of the word you ask it to look for. Review the resulting list of sites, or hits, that contain the word you looked for. The list is the problem—you can generate a list that tells you there are a million sites. Which ones do you look at? Unfortunately, there is no standard answer. You will have to undergo a certain amount of trial and error to find which search engine and which method produce the best results for you.

You can help yourself by understanding how the search engines work, by developing strategies for finding sites, and by learning how to use keywords ("Searching"; Seiter).

How Search Engines Work.   For the most part, search engines function similarly. Using various algorithms, they crawl the Web looking for key terms in URLs, abstracts, titles, metadata, and the full text.

Most search engines have advanced search options—the terms vary—that tell you how to limit the number of hits you receive (see section "Using Keywords and Boolean Logic," below). However, search engines have grown to be much more powerful than they were 10 years ago, and can usually provide results for speech-based inquiries (e.g., "How do I make French press coffee?").

Strategies for Finding Sites.   The primary strategy for finding sites is to use a search engine. Search engines usually search just their database. Common search engines include Google, Yahoo!, Bing, and Ask.com. Each engine reports a list with a relevancy factor, the database's "guess" of which sites will give the best information. Google refers to this as "page ranking." Evaluate the effectiveness of this characteristic; it can be very helpful or misleading. Take the time to perform a comparative search—try the same keyword or phrase in three search engines and compare the results.

Using Keywords and Boolean Logic.   Most search engines allow you to join keywords with Boolean connectors—*and, or,* and *not.* The engine then reports results that conform to the restrictions that the connectors cause.

The basic guidelines are

▶ *Choose specific keywords.* Be willing to try synonyms. For example, if *coffee* gives an impossibly long list, try *Arabica* or *Robusta.*
▶ *Understand how to use Boolean terms.* Read the instructions in the custom search or help sections of the search engine.
  ▶ *And* asks the engine to report only sites that contain all the terms. Generally because *and* narrows a search, eliminating many sites, it is an excellent strategy.
  ▶ *Or* causes the engine to list any site that contains the term. Entering the words *French and wine* generates a list of just those sites that deal with French wine. *French or wine* generates a much larger list that includes everything that deals with either France or wine.
  ▶ *Not* excludes specific terms. Entering *French and wine not burgundy* generates a list that contains information about the other kinds of French wine.

Be willing to experiment with various strings of keywords.

Search the Electronic Catalog

The electronic catalog is your most efficient information-gathering aid. You can easily find information by subject, author, title, keyword, or call number. Many systems allow you to search periodical indexes. All of them allow you to print out an instant bibliography.

Because several major systems and many local variations exist, no textbook can give you all the information you need to search electronically. Take the time to learn how to use your local system. Each system differs, so take the time to learn it. Do several practice searches to find the capabilities of the system and the "paths," or sequences of commands, you must follow to produce useful bibliographies. This section focuses on a few items available in a typical catalog and offers information on using keywords, with electronic catalogs.

The Typical Catalog. The typical catalog presents you with screens for individual items, bibliographies of related items, and categories of searching. An individual item screen describes the book in detail: its authors; printing information; call number; and, most important, the subject keywords that can be used to locate it. If you use those words as keywords in the database, you will find even more books on this subject.

A bibliography screen appears in Figure 5.2. This screen lists the first 5 of 388 books contained in the library under the subject heading "sustainability." Note, however, that the books' topics are not all the same; the term *sustainability* is too broad to create a focused list. A researcher would have to use Boolean logic (sustainability and corporation) to narrow the 388 items down to a group on the same topic.

Obviously, these systems can generate a working bibliography on any subject almost instantly. You can search by subject, title, keyword, author, date, and call number. Use the method that agrees with what you know. For instance, if you know the author—perhaps he or she has written a

Figure 5.2 Subject Bibliography Screen

Source: University of Wisconsin–Stout.

key document, and you want to find what else he or she has written—use the author method. If you know little about the topic, start with a subject search.

Keywords.   By now you can see that the "trick" to using an electronic system is the effective use of *keywords,* any word for which the system will conduct a search. Figure 5.3 shows that, if you want to search for all the items that have *sustainability* in their title, enter your keyword ("sustainability") in the locator box and select "Title Words" in the pull-down menu. An effective initial strategy is to use the keyword category, which searches for the word anywhere on the individual item screen. If the word appears, for instance, in the item's title, subject heading, or abstract, the system includes the item in the bibliography.

Search Online Databases

Online databases are as efficient as the electronic catalog in generating resources for your topic. If you access the database correctly (by using the correct keyword), it produces a list of the relevant articles on a particular subject—essentially a customized bibliography.

Databases are particularly helpful for obtaining current information; sometimes entries are available within a day of when they appear in print, or even before. Most university and corporate libraries provide their patrons with

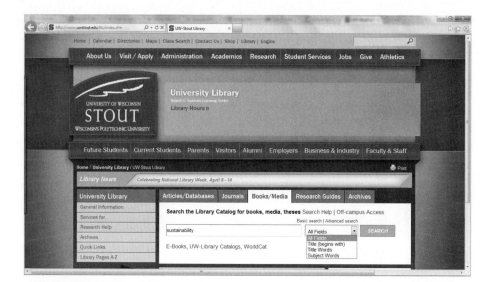

Figure 5.3   Search by Subject

Source: University of Wisconsin–Stout.

many databases for free. If your library offers databases such as EBSCOhost, use them.

To search a database effectively, you must choose your keywords carefully. Just as in the computerized catalog, if you pick a common term, like *technology* or *food* or *politics*, the database might tell you it has found 10,000,000 items. To narrow the choices, combine descriptors. For example, if you combine *food* with such descriptors as *organic* and *vegan*, the computer searches for titles that contain those three words and generates a much smaller list of perhaps 10 to 50 items. Figure 5.4 shows 50,265 results generated for the keyword *sustainability*. Obviously, this search would have to be narrowed (by adding more keywords in the available fields at the top). Figure 5.5 (p. 130) shows an item entry, giving all the relevant information. Notice especially the *Subject Terms* and the *Author-Supplied Keywords*. Type them into the locator to find articles close in topic to the one described. Also notice the abstract, which you can use to decide whether or not to read the full text of the article. (Most systems will let you retrieve full-text articles, an invaluable aid as you collect information for your report.)

Databases provide information on almost every topic. EBSCOhost offers many indexes, including Applied Science and Technology, ERIC, Hoover's Company Profiles, and Health Service Plus. Contact your library for a list of the services available to you.

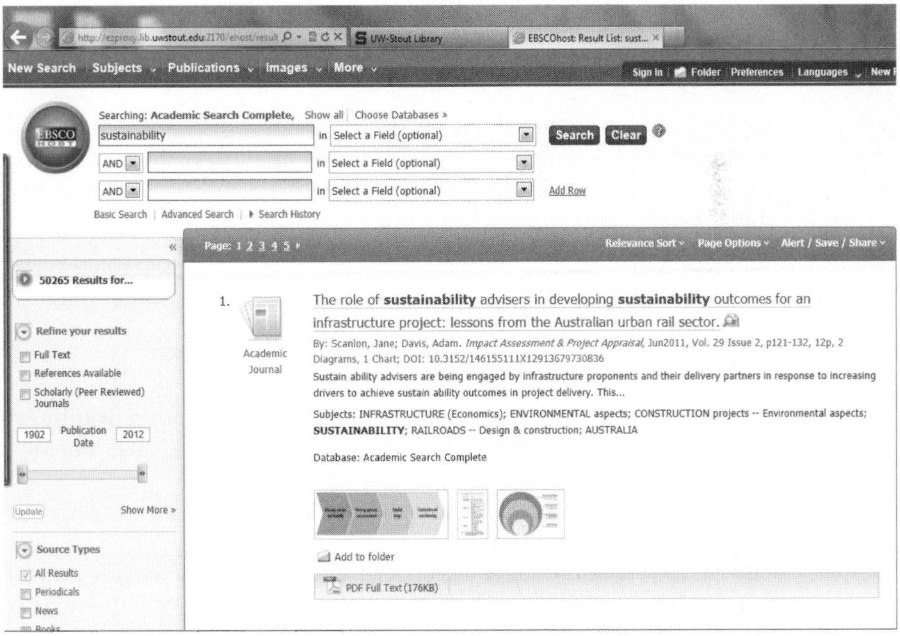

Figure 5.4 Generated List of Articles

Source: University of Wisconsin–Stout.

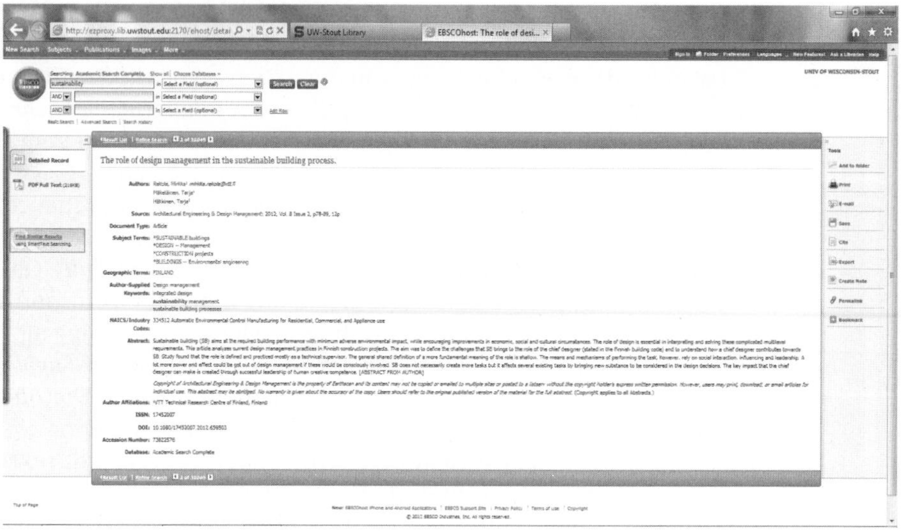

Figure 5.5   Online Bibliographic Entry with Abstract

Source: University of Wisconsin–Stout.

Focus on Ethical Citation

Credit Others

Suppose a new coworker has found a way to modify a procedure and save the company money. You are assigned to write the internal proposal that suggests the change. Your obligation is to present the facts so that your manager understands who conceived the idea—and who gets the credit. To do otherwise would be to deny your coworker proper credit for the idea.

Record Your Findings

As you proceed with your search strategy, record your findings. Construct a bibliography, take notes, consider using visual aids, and decide whether to quote or paraphrase important information.

Create a Working Bibliography

List potential sources of information in a Web-based document storage system ("the cloud"), such as Google Docs. Doing this allows you to access your sources from anywhere with an Internet connection. This document should contain the name of the author, the title of the article or book, and facts about the book or article's publication (Figure 5.6). Record this

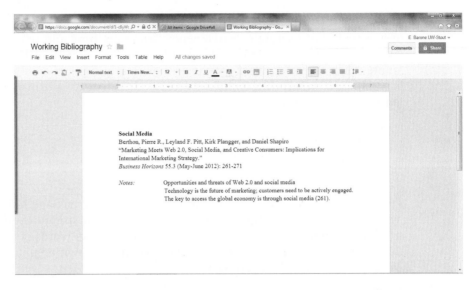

Figure 5.6 Working Bibliography with Notes

Source: Google.

information in the form that you will use in your bibliography. Also record the call number and any special information about the source (e.g., that you used EBSCOhost to find it). Such information will help you relocate the source later.

Take Notes

As you read, take notes and put these ideas in a *cloud* document as well (Figure 5.6). Create topic headings, and for each source include the name of the author and the page number from which you are recording information. Then write down the ideas you got from that source. This practice greatly simplifies arranging your notes when you finally organize the report.

Make Visual Aids

Visual aids always boost reader comprehension. You either find them in your research or create them yourself. If a key source has a visual aid that clarifies your topic, use it, citing it as explained in Appendix B (pp. 580–600). As you read, however, be creative and construct your own visual aids. Use flow charts to show processes; tables, charts, or graphs to give numerical data; and diagrams to explain workstations—whatever will help you (and ultimately your readers) grasp the topic.

Quoting and Paraphrasing

It is essential in writing research reports to know how and when to quote and paraphrase. *Quoting* is using another writer's words verbatim. Use a quote when the exact words of the author clearly support an assertion you have made or when they contain a precise statement of information needed for your report. Copy the exact wording of

▶ Definitions
▶ Comments about significance
▶ Important statistics

Paraphrasing means conveying the meaning of the passage in your own words. Learning to paraphrase is tricky. You cannot simply change a few words and then claim that your passage is not the exact words of the author. To paraphrase, you must express the message in your own original language. Write paraphrases when you want to

▶ Outline processes
▶ Give illustrative examples
▶ Explain causes, effects, or significance

The rest of this section explains some basic rules for quoting and paraphrasing. Complete rules for documenting sources appear in Appendix B.

Consider this excerpt from J. C. R. Licklider's 1968 essay on the benefits of the Internet, written after one of the first "technical meeting[s] held through a computer" (276).

When people do their informational work "at the console" and "through the network," telecommunication will be as natural an extension of individual work as face-to-face communication is now. The impact of that fact, and of the marked facilitation of the communication process will be very great—both on the individual and on society.

First, life will be happier for the online individual because the people with whom one interacts most strongly will be selected more by commonality of interests and goals than by accidents of proximity. Second, communication will be more effective and productive, and therefore more enjoyable. Third, much communication and interactions will be with programs and programmed models, which will be (a) highly responsive, (b) supplementary to one's own capabilities, rather than competitive, and (c) capable of representing progressively more complex ideas without necessarily displaying all the levels of their structure at the same time—and which will therefore be both challenging and rewarding. And fourth, there will be plenty of opportunity for everyone (who can afford a console) to find his calling, for the whole world of information, with all its field and disciplines, will be open to him. . . .

For the society, the impact will be good or bad, depending mainly on the question: Will "to be on line" be a privilege or a right? If only a favored segment of the population gets a chance to enjoy the advantage of "intelligence amplification," the network may exaggerate the discontinuity in the spectrum of intellectual opportunity (Licklider 277).

To quote, place quotation marks before and after the exact words of the author. You generally precede the quotation with a brief introductory phrase.

According to Licklider, "If only a favored segment of the population gets a chance to enjoy the advantage of 'intelligence amplification,' the network may exaggerate the discontinuity in the spectrum of intellectual opportunity" (277).

Note that the quotes around *intelligence amplification* are single, not double, because they occur within a quotation.

If you want to delete part of a quotation from the middle of a sentence, use ellipsis dots (. . .).

Licklider notes that "The impact of that fact . . . will be very great—both on the individual and on society" (277).

If you want to insert your own words into a quotation, use brackets.

Licklider points out that "For the society [certainly the global society], the impact will be good or bad, depending mainly on the question: Will 'to be on line' be a privilege or a right?" (277).

To paraphrase, rewrite the passage using your own words. Be sure to indicate in your text the source of your idea: the author and the page number on which the idea is found in the original.

After one of the first technical meetings held via a computer, Licklider theorizes that the new possibilities of interaction will have enormous benefits to individuals, who will no longer be limited by physical proximity. The benefits include the possibility to interact with people of like minds from many different locations, to develop complex ideas in greater depth, and to explore one's interests more fully.

While the benefits are clear, the potential information and communication explosion also throws the issue of parity into the spotlight. Those who don't have access to a computer will not be able to participate fully in this intellectual challenge. Current inequalities will be made even more pronounced (277).

Remember when you quote or paraphrase that you have ethical obligations both to the original author and to the report reader.

1. When in doubt about whether an idea is yours or an author's, give credit to the author.

2. Do not quote or paraphrase in a way that misrepresents the original author's meaning.

3. Avoid stringing one quote after another, which makes the passage hard to read.

Worksheet for Research Planning

☐ **Name the basic problem that you perceive or a question that you want answered.**

☐ **Determine your audience. Why are they interested in this topic?**

☐ **List three questions about the topic that you feel must be answered.**

☐ **List three or four search words that describe your topic.**

☐ **Determine how you find information about this topic. Do you need to read? Interview? Survey? Search library databases? Search the Web? Perform some combination of these?**

☐ **List the steps you will follow to find the information. Include a time line on which you estimate how many hours or days you need for each step.**

☐ **Name people to interview or survey, outline a test, or list potential sources (technical or nontechnical) of information.**

☐ **Create a form on which to record the information you discover. If you use interviews or surveys, create this form carefully so that you can later collate answers easily.**

Exercises

❱ You Create

1. Develop a research plan and implement it.

   a. Create a list of five to eight questions about a topic you want to research. For each question, indicate the kind of resource you need (book, recent article, website) and a probable search source (library, electronic catalog, Google, Bing). Explain your list to a group of two or three. Ask for their evaluation, changing your plan as they suggest.

   b. Based on the list you created in Exercise 1a, create a list of five to ten keywords. In groups of two or three, evaluate the words. Try to delete half of them and replace them with better ones.

   c. Select a database (e.g., EBSCOhost) or a Web search engine (e.g., Google).

   d. Go on to Writing Assignment 3.

2. Conduct a subject search of your library's catalog. Start with a general term (*sustainability*) and then, using the system's capabilities, limit the search in at least three different ways (e.g., sustainability not corporate, sustainability

and urban development, sustainability or conservation). Print out the bibliography from each search. Write a description of the process you used to derive the bibliographies and evaluate the effectiveness of your methods.

Alternate: Write a report to your classmates on at least two tips that will make their use of the catalog easier.

3. Select a topic of interest. Generate a list of three to five questions about the topic. Read a relevant article in one standard reference source. Based on the article, answer at least one of your questions and pose at least two more questions about the topic.

▶ You Analyze

4. Write a report in which you analyze and evaluate an index or abstracting service. Use a service from your field of interest, or ask your instructor to assign one. Explain which periodicals and subjects the service lists. Discuss whether it is easy to use. For instance, does it have a cross-referencing system? Can a reader find keywords easily? Explain at what level of knowledge the abstracts are aimed: beginner? expert? The audience for your report is other class members.

5. Write a report in which you analyze and evaluate a reference book or website in your field of interest. Explain its arrangement, sections, and intended audience. Is it aimed at a lay or a technical audience? Is it introductory or advanced? Can you use it easily? Your audience is other class members.

▶ Group

6. Form into groups of three. One person is the interviewer, one the interviewee, and one the recorder. Your goal is to evaluate an interview. The interviewer asks open and closed questions to discover basic facts about a technological topic that the interviewee knows well. The recorder keeps track of the types of questions, the answers, and the effectiveness of each question in generating a useful answer. Present an oral report that explains and evaluates your process. Did open questions work better? Did the echo technique work?

7. In groups of three, review material on the Web related to "crowd sourcing" or using Twitter as a method of researching a topic. Select a topic you wish to research. It could be current events, developments in your major, or hobbies. Each group member creates a Twitter account, and then all of you use the same hash tag (#) (for more details on hash tags see "Microblogs" in Chapter 11, p. 345). After a reasonable amount of time (a week) collect your results. Report your findings to the class, either orally or in writing, as your instructor requires. Report both the information you received on your topic and the effectiveness you find in this research method.

Writing Assignments

1. Write a research report on the value of a Career Networking site such as (at this writing) LinkedIn. If possible, interview placement officers or human resources people in order to determine its value. If interviews are not possible, conduct a Web search. Your goal is that your readers will feel comfortable either using or not such a site.

2. Write a short research report explaining a recent innovation in your area of interest. Your goal is to recommend whether your company should become committed to this innovation. Consult at least six recent sources. Use quotations, paraphrases, and one of the citation formats explained in Appendix B (pp. 580–600). Organize your material into sections that give the reader a good sense of the dimensions of the topic. The kinds of information you might present include

   - Problems in the development of the innovation and potential solutions
   - Issues debated in the topic area
   - Effects of the innovation on your field or on the industry in general
   - Methods of implementing the innovation

   Your instructor might require that you form groups to research and write this report. If so, he or she will give you a more detailed schedule, but you must formulate questions, research sources of information, and write the report. Use the guidelines for group work outlined in Chapter 3.

3. You (or your group, if your instructor so designates) are assigned to purchase a Customer Relationship Management (CRM) program for your company. Research three actual programs and recommend one. Investigate your situation carefully. Talk to users, discover the capabilities, and address the pros and cons of each system. Investigate cost and site licenses. Read several reviews.

4. Following one of the documentation formats, write a brief research paper in which you complete the process you began in Exercise 1.

   a. In your chosen database or Web search engine, use your search words and combinations of them to generate a bibliography.

   b. Read two to five articles that will answer one of your questions.

   Alternate: Write a brief report in which you explain the questions you asked, the method you used, and the results you achieved.

5. Divide into groups of three or four. Construct a three- or four-item questionnaire to give to your classmates. Write an introduction, use open and closed questions, and tabulate the answers. At a later class period, give an oral report on the results. Use easy topics, such as demographic inquiries (size of each class member's native city, year in college, length of employment) or inquiries into their knowledge of some common area in a field chosen by the group (e.g., using search engines on the Web).

6. Interview four people in a workplace to determine their attitude toward a technology (smart phones, tablets, wireless printing, social media). Present to an administrator a report recommending a course of action based on the responses. One likely topic is the need for training.

7. Write the report your instructor assigns.

   a. Describe your actions, the number of items and type of information, the value of the entries.

   b. Answer the question in several paragraphs. See Appendix B for listing sources.

   c. Describe your database or search engine. Explain why you selected it, whether it was easy to use, and whether it was helpful.

   d. Describe your article. Summarize it and explain how it relates to your topic.

   e. List two questions that you can research further as a result of reading your article.

8. After you have completed your writing assignment, write a learning report, a report to your instructor. Explain, using details from your work, what new things you have learned or old things confirmed. Use some or all of this list of topics: writing to accommodate an audience, presenting your identity, selecting a strategy, organizing, formatting, creating and using visual aids, using an appropriate style, developing a sense of what is "good enough" for any of the previous topics. In addition, explain why you are proud of your recent work and tell what aspect of writing you want to work on for the next assignment.

Web Exercise

Decide on a topic relevant to your career area. Using the Web, find three full-text professional articles that previously appeared in print and three documents that have appeared only on a website. Usually, access through a major university library will achieve the first goal; access to a corporate site will usually achieve the second goal.

Do either of the following, whichever your instructor designates:

a. In an analytical report (see Chapter 10) compare the credibility and the usability of the information in the two types of sources.

b. Write a research paper in which you develop a thesis you have generated as a result of reading the material you collected.

Works Cited

"Advanced Scholar Search Tips." *Google Scholar.* Google, 2011. Web. 16 Apr. 2012.

Licklider, J. C. R. "The Computer as a Communication Device." *Science and Technology* (April 1968). Rpt. as "The Internet Primeval." *Visions of Technology.* Ed. Richard Rhodes. New York: Simon & Schuster, 1999. 274–282. Print.

Seiter, Charles. "Better, Faster Web Searching." *Macworld* (December 1996): 159–162. Print.

Spivey, Nancy Nelson, and James R. King. "Readers as Writers Composing from Sources." *Reading Research Quarterly* (Winter 1989): 7–26. Print.

Stewart, Charles J., and William B. Cash, Jr. *Interviewing Principles and Practices.* 8th ed. Boston: McGraw-Hill, 1997. Print.

"Support for Libraries." *Google Scholar.* Google, 2011. Web. 16 April 2012.

"Tracking Information." *INSR* 33. Menomonie: University of Wisconsin–Stout Library Learning Center, January 2000. Print.

Warwick, Donald P., and Charles A. Lininger. *The Sample Survey: Theory and Practice.* New York: McGraw-Hill, 1975. Print.

Focus on
Google Scholar

Google Scholar is a search engine developed specifically for scholarly literature, providing sources from numerous disciplines—including

- Articles
- Theses
- Books
- Abstracts
- Court opinions

These sources come from various academic publishers, professional societies, online repositories, universities, and other websites. Figure 5.7 shows the results of a query on *social media*.

Google Scholar ranks the documents using full-text analysis, the publisher, the relevance (based on the date of publication), and the frequency with which the document is cited within other literature. Through careful searching you can find documents quite relevant to your research. One way to fine-tune your search results is to use the "Advanced Scholar Search" feature within Google Scholar. Through the addition of various "Operators," it is possible to pare down your search significantly. Some of these "Operators" include

- Author Search
- Publication Restrict
- Date Restrict
- Legal Opinions and Journals
- Jurisdiction Restrict

Figure 5.8 shows the "Advanced Scholar Search Page" where a search on *social media* has been refined by restricting the date of publication to 2011 through 2012 only.

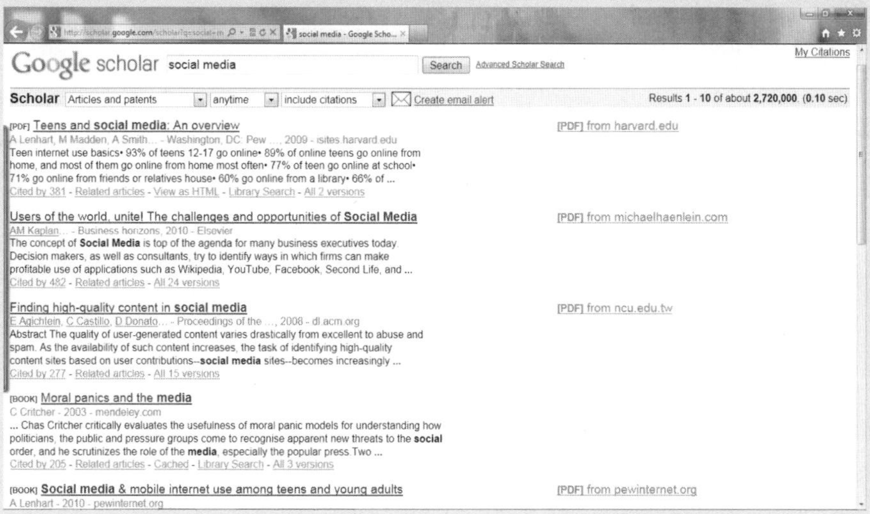

Figure 5.7   Google Scholar Search

Source: Google.

(Continued)

Figure 5.8   Advanced Google Scholar Search

Source: Google.

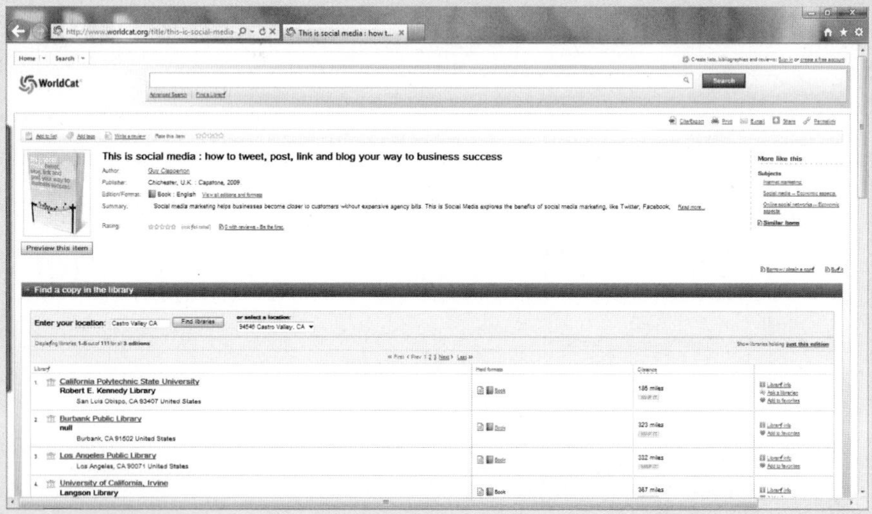

Figure 5.9   Google Scholar Library Search

Source: Google.

In order to provide full access to the sources listed, Google Scholar has linked up with libraries through a feature called "Support for Libraries." This works in two ways:

1.On-campus students, at universities participating in the "Library Links Program," are able to see additional links within Google Scholar, giving them access to their campus's library resources. This feature allows the students to easily access full-text articles from search queries on Google Scholar.

2. General users have the ability to access local library resources while searching Google Scholar. Libraries that are linked with OCLCs open World Cat system work with Google Scholar to provide users with local libraries housing the texts that they are seeking ("Support"). Users need to only click on the "Library Search" link on the query results page to access this. Figure 5.9 shows the "Library Search" page.

Designing Pages

CHAPTER CONTENTS

CHAPTER 6 **IN A NUTSHELL**

Design is the integration of words and visuals and all of the elements on the page or screen in ways that help readers achieve their goals for using the document. The key idea is to establish a *visual logic*—the same kind of information always looks the same way and appears in the same place (e.g., page numbers are italicized in the upper-right corner).

Visual logic establishes your credibility, because you demonstrate that you know enough about the topic and about communicating to be consistent. Visual logic helps your audience to see the "big picture" of your topic, and as a result they grasp your point more quickly. Both visual and textual features establish visual logic. Two key visual features are *heads* and *chunks*.

Heads tell the content of the next section. Heads should inform and attract attention—use a phrase or ask a question; avoid cryptic, one-word heads.

Heads have levels—one or two are most common. The levels should look different and make their contents helpful for readers.

Chunks are any pieces of text surrounded by white space. Typically, readers find a topic presented in several smaller chunks easier to grasp than one longer chunk.

A key textual feature is *highlighting*—changing the look of the text to draw attention, for instance, by using boldface or italics. In addition, *standardization* and *consistency* are effective ways to orchestrate textual design. Standardization means that each feature, such as boldface, has a purpose. For instance, in instructions for using software, boldfaced words could indicate which menu to access. Consistency means that all items with a similar purpose have a similar design; for instance, all level-one heads in the document have the same look (e.g., Arial 12 point boldfaced).

A *style sheet* and *template* are effective methods to plan design. A style sheet lists the specifications of the design (e.g., "All level-one heads appear in all caps, Arial 12 point, flush left"). A template is a representative page that indicates the correct look of each item of design.

Technical communicators design their document pages to produce what Paul Tyson, a designer, describes as a document "from which readers can quickly get accurate information" (27). Although *design* is a word with many meanings, in this book *design* means the integration of words and visuals in ways that help readers achieve their goals for using the document (Schriver). This definition implies two major concerns. First, design is about the look of the page—its margins, the placement of the visual aids, the size of the type. The pattern of these items is called a *template*. Second, design is about helping readers relate to the content. This aspect of design is called *visual logic.*

The second concern is actually more important than the first. Karen Schriver, a document design expert, indicates the relationship of the two concerns by saying that the look or design must reveal the structure of the document, and in order to achieve that goal communicators must orchestrate the look to achieve a visual logic, which makes structural relationships clear. Tyson explains the relationship this way: "In a well-designed document, the writing and formatting styles expose the logical structure of the content so readers can quickly get what they want from the document" (27).

This chapter will familiarize you with both goals. You will learn how to reveal a document's logical structure by using both text and visual features. This chapter covers using visual features to reveal content, using text features to convey meaning, and developing a style sheet and template.

Using Visual Features to Reveal Content

The visual features that reveal content are white space and chunks; bullets; head systems; and headers, pagination, and rules.

White Space and Chunks

The key visual feature of a document is its white space. *White space* is any place where there is no text or visual aid. White space creates *chunks*—blocks of text—and chunks reveal logical structure to readers, thus helping them grasp the meaning. The rule for creating chunks is very simple: Use white space to make individual units of meaning stand out. You can apply this rule on many levels. The contrast in the two examples in Figures 6.1 and 6.2 gives you the basic idea.

Figure 6.1 shows an email message produced as one chunk, which makes it seem there is only one message. The number of points and the content of the message are not at all clear.

Now, Figure 6.2 shows the same report with the content units turned into chunks. Notice that you can see that the content really has three parts—two

Hi John, The reports that were presented at the meeting won't be as effective on the company website as the people in the meeting suggested. The tables are too complicated and the actual explanations are unclear and not positioned near enough to refer back and forth easily. These reports will not make it much easier for our intended audience to use our data. This group has members who don't really belong, and a few who do belong are missing. I would like to be able to remove two of the marketing people and add the director of library services. Will you call me with your suggestions?

© Cengage Learning

Figure 6.1   One-Chunk Report

Hi John,

The reports that were presented at the meeting won't be as effective on the company website as the people in the meeting suggested. The tables are too complicated and the actual explanations are unclear and not positioned near enough to refer back and forth easily. These reports will not make it much easier for our intended audience to use our data.

This group has members who don't really belong, and a few who do belong are missing. I would like to be able to remove two of the marketing people and add the director of library services.

Will you call me with your suggestions?

© Cengage Learning

Figure 6.2   Three-Chunk Report

issues and a request for advice. In addition because the chunks divide the large chunk into units that look the same, you can tell that the three points have equal importance.

However, chunks can also show hierarchy; that is, they can indicate which material is subordinate to other material. Let's take the report and chunk it so that the design clarifies the main idea and the support idea (Figure 6.3). Notice that indenting the two reasons makes them appear subordinate to the main objection. As a result, the design shows the reader the logical structure of the chunk.

Report Bullets

Another visual feature that facilitates conveying meaning is introductory symbols—either *numbers* or *bullets,* dots placed in front of the first word in a

The reports that were presented at the meeting can't go on the website as easily as the people in the meeting suggested.

The tables are too complicated.
The actual explanations are unclear and not positioned near enough to refer back and forth easily.

These reports will not make it much easier for our intended audience to use our data.

© Cengage Learning

Figure 6.3   Hierarchy in Chunks Report

The reports that were presented at the meeting can't go on the website as easily as the people in the meeting suggested.

- The tables are too complicated.
- The actual explanations are unclear and not positioned near enough to refer back and forth easily.

These reports will not make it much easier for our intended audience to use our data.

© Cengage Learning

Figure 6.4   Bulleted List

unit. Figure 6.4 shows the paragraph with bullets added; the resulting list is called a *bulleted list*. Notice that the bullets emphasize the list items, causing the reader to focus on them. For even more emphasis, the author could have used numbers instead, emphasizing that there are two reasons.

Also notice that the second reason has two lines; the second line starts under the first letter of the first line, not under the bullet. This strategy of indenting is called a *hanging indent* and is commonly used in lists to make the parts stand out.

Head Systems

A *head* is a word or phrase that indicates the contents of the section that follows. A *head system* is a pattern of heads (called *levels*) to indicate both the content and the relationship (hierarchy) of the sections in the document. With chunks, heads are a key way to help readers find information and also to see the relationship of the parts of the information (Figure 6.5).

All Caps (each letter is capitalized) Is Superior to "Up and Down" Style (capitals and small letters mixed)

GOOD NEWS FOR WIDGETS
Production Doubles
Sales Increase

Big Is Superior to Little

Good News for Widgets
Production Doubles
Sales Increase

Dark Is Superior to Light

Good News for Widgets
Production Doubles
Sales Increase

Far Left Is Superior to Indented

Good News for Widgets
Production Doubles
Sales Increase

© Cengage Learning

Figure 6.5 Ways to Indicate Hierarchy

Figure 6.6 (p. 147) illustrates this idea. Note that each head summarizes the contents of the section below it. Also note that there are two levels. Level 1 condenses the overall topic of the section into a few words, and the two level 2 heads show that the topic has two subdivisions. But in particular notice the design of the two levels. Level 2 is indented, and the clear content of the phrases reveals the logical structure of the document—a claim (Good News) and the reasons for the claim (Production Doubles; Sales Increase).

Head systems vary. The goal of each variation is to indicate the hierarchy of the contents of the document—the main sections and the subsections. Head systems are subject to certain norms.

Head systems also have two basic styles: open and closed. An open system uses only the position and size of the heads to indicate hierarchy. Figure 6.7 (p. 148) illustrates an open system. A closed system uses a number

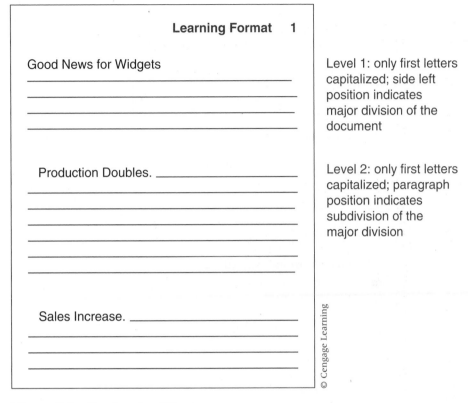

Learning Format   1

Good News for Widgets
_____

Level 1: only first letters capitalized; side left position indicates major division of the document

Production Doubles. _____

Level 2: only first letters capitalized; paragraph position indicates subdivision of the major division

Sales Increase. _____

© Cengage Learning

Figure 6.6 Two Levels of Heads

arrangement to indicate hierarchy. Level 1 is preceded by 1, a subsection is 1.1, and a sub-subsection 1.1.1. Figure 6.8 (p. 148) shows a closed system.

Headers or Footers, Pagination, and Rules

Three other features of visual layout are headers or footers, pagination, and rules. *Headers* or *footers* appear in the upper or lower margin of a page. They usually name the section of the document for the reader. *Page numbers* usually appear at the top right or top left (depending on whether the page is a right-hand or left-hand page), or bottom center of the page. Usually both headers and footers and page numbers are presented in a different type size or font from that of the body text. *Rules,* or lines on the page, act like heads—they divide text into identifiable sections and they can indicate hierarchy. A thinner rule is subordinate to a thicker rule. Figure 6.9 (p. 149) shows headers and footers and page numbers.

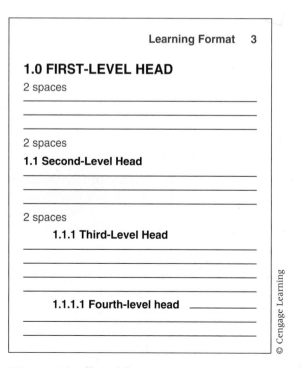

Figure 6.7   Open System

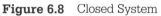

Figure 6.8   Closed System

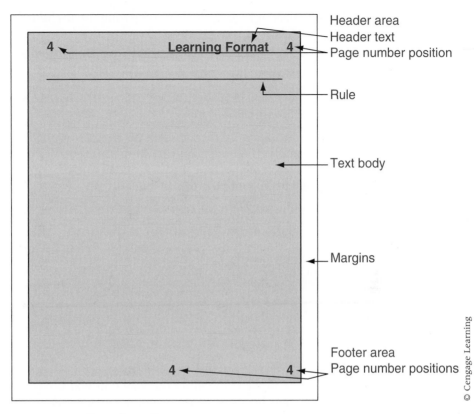

Header area
Header text
Page number position

Rule

Text body

Margins

Footer area
Page number positions

© Cengage Learning

Figure 6.9 Basic Page Parts

Using Text Features to Convey Meaning

Text features that are used to convey meaning are highlighters, font, font size, leading, columns and line length, and justification. You can use text features to emphasize words or groups of words and to give the text a certain personality.

Highlighters

Highlighters focus the reader's attention on an idea by making a word or phrase stand out from other words.

Types of Highlighters

Common highlighters are

> **Boldface**
> *Italics*
> ALL CAPS
> Vertical lists
> Quotation marks

You can see the effect of highlighters by comparing the use of boldface in the following two sentences:

> Your phone comes from the factory set to "700 msec." The talk indicator must be off before programming.

> Your phone comes from the factory set to "700 msec." **The TALK indicator must be off before programming.**

In the first example, the two sentences look the same. Nothing is emphasized. In the second example, the important condition stands out, and the keyword (*talk*) stands out even more because it appears in all caps.

Here is a second example of the use of boldface:

> Another must is **Brazos Bend State Park,** one of the best places in Texas to photograph alligators in their natural habitat (Miller).

In this example, the boldface focuses the reader on the important name in the sentence.

Here is another example, from the instructions for a scanner:

> Click **Scan.**

The boldfaced word is the one found on the screen.

Use Highlighters to Help Your Readers

The key to effective usage is to define the way you will use the highlighter. Give it a function. For instance, in the scanner instructions, the highlighting is used to signal that the word in the text is the one to look for on the screen. As soon as you use formatting to indicate a special use or meaning once, you

Focus on Ethical Design

Design Honestly. Suppose that in a progress report you must discuss whether your department has met its production goal. The page-formatting techniques you use could either aid or hinder the reader's perception of the truth. For instance, you might use a boldfaced head to call attention to the department's success:

> **Widget Line Exceeds Goals.** Once again this month, our widget line has exceeded production goals, this time by 18%.

Conversely, to downplay poor performance, you might use a more subdued format, one without boldface and a head with a vague phrase:

> Final Comments. Great strides have been made in resolving previous difficulties in meeting monthly production goals. This month's achievement is nearly equal to expectations.

If reader misunderstanding could have significant consequences, however, your use of "Final Comments" is actually a refusal to take responsibility for telling the stakeholder what he or she needs.

set up a convention that readers will look for: You have defined a style guide rule for your document. Once you establish a convention, maintain it. In the scanner manual, every time the writer uses boldface, readers know that they will find that word on their computer screen.

Other Ways to Use Highlighters

Use italics to emphasize a word that you will define:

> Each element on your form will have a *name* and *value* associated with it. The name identifies the data that is being sent (Castro 178).

Use quotation marks to introduce a word used ironically or to indicate a special usage:

> That was a "normal" sale in their opinion.
> The "dense page" issue affects all designers.

Use all caps as a variant of boldface, usually for short phrases or sentences. All caps has the written effect of orally shouting.

> Your phone comes from the factory set to "700 msec." THE TALK INDICATOR MUST BE OFF BEFORE PROGRAMMING.

Use vertical lists to emphasize the individual items in the list and to create the expectation in the reader that these are important terms that will be used later in the discussion. Notice in the following list that commas are not used after the items. The indentation heightens the sense that these words are different from the words in a usual sentence.

> Highlighters include boldface, italics, quotation marks, and all caps.

> Highlighters include
> • Boldface
> • Italics
> • Quotation marks
> • All caps

Font, Font Size, Leading, Columns and Line Length, and Justification

Use text features such as fonts, font size, leading, columns and line length, and justification to affect the reader's ability to relate to the text. Features can seem appropriate or inappropriate, helpful or not helpful.

Font

Font, or typeface, is the style of type. Fonts have personality—some seem frivolous, some interesting, some serious, some workaday.

Consider this sentence in four different typefaces. Alleycat and Sand seem frivolous; Shelley and Alien Ghost are illegible.

The TALK indicator must be off before programming.
The TALK indicator must be off before programming.

The TALK indicator must be off before programming.
THE TALK INDICATOR MUST BE OFF BEFORE PROGRAMMING.

Typefaces that routinely appear in reports and letters are Times, Helvetica, and Palatino, all of which appear average or usual, the normal way to deliver information.

This is Times
This is Helvetica
This is Palatino

Fonts belong to one of two major groups: serif and sans serif. The letters in serif faces have extenders at the ends of their straight lines. Sans serif faces do not. Serif faces give a classical, more formal impression, whereas sans serif faces appear more modern and informal. There is some evidence that serif faces are easier to read. However, some tests have indicated the readers prefer serif for longer, continuous text (like this chapter), and prefer sans serif for shorter, more telegraphic text—manuals, for instance. In addition, sans serif fonts are preferable for use in online material (Schriver 298, 508).

Many designers suggest that you use the same font for heads and text. Some designers suggest that you use a sans serif font for heads and titles (display text) and a serif font for body text.

Font Size

Font size is the height of the letters. Size is measured in points: 1 point equals $\frac{1}{72}$ inch. Common text sizes are 9, 10, and 12 points. Common heading sizes are 14, 18, and 24 points. Most magazines use 10-point type, but most reports use 12 point.

Font size affects characters in a line; the larger the point size, the fewer characters in a line.

18-point type allows this many characters in this line.

9-point type allows many more characters in a line of the same length, causing a different sense of width (Felker).

As Figure 6.10 shows, type size affects the appearance, length, and readability of your document.

Leading

Related to size is *leading,* or the amount of space between lines. Leading is also measured in points and is always greater than the font size. Word processing

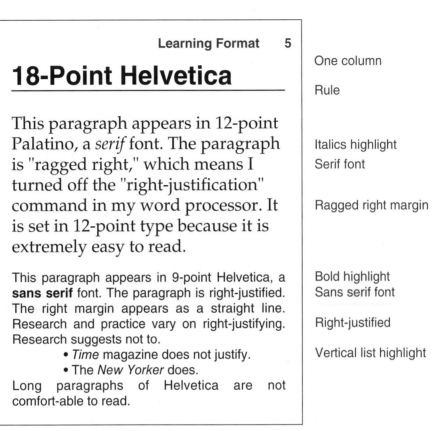

Figure 6.10   Text Features

programs select leading automatically so it is not usually a concern. However, too much or too little leading can cause text to look odd. Notice the effects of leading on the same sentence:

▶ 12-point text with 12-point leading:

Technical communication is "writing that aims to get work done, to change people by changing the way they do things."

▶ 12-point text with 18-point leading:

Technical communication is "writing that aims to get work

done, to change people by changing the way they do things."

Columns

Columns are vertical lines of type; a normal typed page is just one wide column. Many word processing programs allow multiple columns (12 or more); in practice, however, reports seldom require more than two columns. In general, use a single column for reports. To achieve a contemporary design, consider using a 2- or 2-½-inch-wide left margin. In other cases, two columns are especially useful for reports and manuals if you plan to include several graphics. For various column widths, see Figures 6.17 and 6.18, pages 161–163.

Column width affects line length, the number of characters that will fit into one line of type. Line length affects readability (Felker; Schriver). If the lines are too long, readers must concentrate hard as their eyes travel across the page and then painstakingly locate the next correct line back at the left margin. If the lines are too short, readers become aware of shifting back and forth more frequently than normal. Short lines also cause too much hyphenation.

Typographers use three rules of thumb to choose a line length and a type size:

Use one and a half alphabets (39 characters) or 8 words of average length per line.
Use 60 to 70 characters per line (common in books).
Use 10 words of average length (about 50 characters).

Unfortunately, no rule exists for all situations. You must experiment with each situation. In general, increase readability by adding more leading to lines that contain more characters (White).

Justification

Justification (see Figure 6.10) means aligning the first or last letters of the lines of a column. Documents are almost always *left-justified;* that is, the first letter of each line starts at the left margin. *Right-justified* means that all the letters that end lines are aligned at the right margin. Research shows that ragged-right text reads more easily than right-justified text (Felker).

Combining Features to Orchestrate the Text for Readers

Given all the possibilities for combining features in order to help readers quickly get accurate information, what are some guidelines to help with that task? The goal of design is twofold: to help readers easily find the information and to reveal the logical structure of the document. This section will give you several guidelines to help you with your orchestration (Schriver).

Analyze: Identify the Rhetorical Clusters in Your Document

Rhetorical clusters are visual and verbal elements that help the reader interpret the content in a certain way. Every document has many rhetorical clusters: titles, heads, visuals, captions, paragraphs, warnings, numbers, and types of links (in online reports). You must be aware of all of these items and treat them appropriately.

Standardize: Give Each Text or Visual Feature a Purpose

In the phone example (p. 150), a boldfaced word in the text indicated a word that appeared on the computer screen. Thus, boldfaced words are a rhetorical cluster. They tell the reader how to interpret a word that is treated differently, from the main text. Readers will quickly interpret your purpose and cluster design, counting on it to help them with the contents of the document.

Be Consistent: Treat All Like Items Consistently Throughout the Document

In effect, you repeat the design of any item and that repetition sets up the expectation of readers. Once the expectation is set up, readers look for the same item to cue them to interpret the content. They know that all-cap heads indicate the start of a new section, that boldface indicates a special item, that indented lists are important, and that 12-point links go to major sections of a document and 9-point links lead to other sites.

Be Neat: Align Items

Aligning basically means to create a system of margins and start similar features at the same margin. Figure 6.11 shows items haphazardly related and aligned. Notice that the left edges of all the visuals align with one another, as do the left

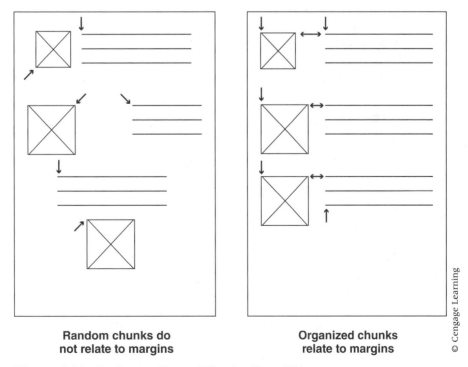

**Random chunks do
not relate to margins**　　　　**Organized chunks
relate to margins**

© Cengage Learning

Figure 6.11　Ineffective Versus Effective Use of Edges

edges of all the text chunks. In addition, the top edge of each visual is aligned with the top edge of each text chunk. Alignment creates meaningful units.

Learn: Use the Design Tips of Experts

Designers have researched many features to determine what is most effective:

1. Use top-to-bottom orientation to gain emphasis. Typically, readers rank the item at the top as the most important. Put your most important material near the top of the page (Sevilla).

2. Use brightness to gain emphasis. Readers' eyes will travel to the brightest object on the page. If you want to draw their attention, make that item brighter than the others (Sevilla).

3. Use larger-to-smaller orientation (Figure 6.12). Readers react to size by looking at larger items first (Sadowski). Put important material (e.g., main heads) in larger type. *Note:* Boldfacing causes a similar effect. Boldfaced 12-point type seems larger than normal (roman) type.

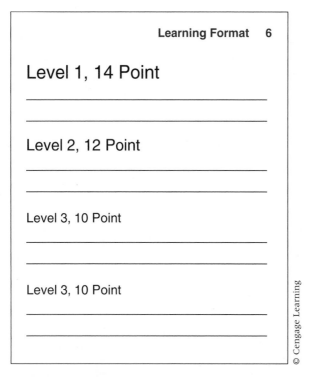

Figure 6.12 Larger-to-Smaller Orientation

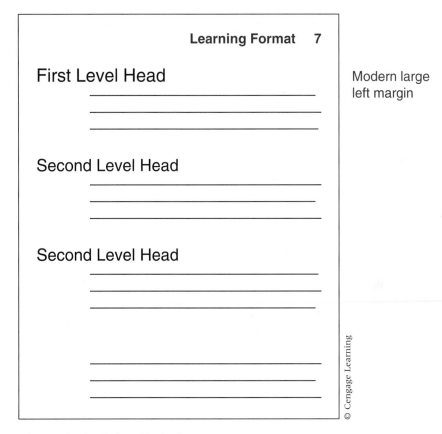

Figure 6.13 Left-to-Right Orientation

4. Use left-to-right orientation to lead your readers through the text (Rubens). Place larger heads or key visuals to the left and text to the right in order to draw readers into your message. See Figure 6.13, where the large left margin and heads perform this function.

5. Place visuals so that they move readers' attention from left to right (Rubens; *Xerox*). In a two-column format (Figure 6.14, p.158), you can place the visuals to the left and the text to the right or vice versa, depending on which you want to emphasize. Whatever you do, always anchor visuals by having one edge relate to a text margin (Sadowski).

6. In a multiple-page document, "hang" items from the top margin. In other words, keep a consistent distance from the top margin to the top of the first element on the page (whether head, text, or visual) (Cook and Kellogg). See Figure 6.15 (p. 158).

7. Learn to use color effectively. Guidelines for the effective use of color can be found in "Focus on Color," pages 171–178.

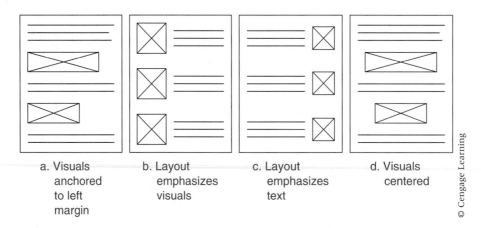

a. Visuals anchored to left margin

b. Layout emphasizes visuals

c. Layout emphasizes text

d. Visuals centered

© Cengage Learning

Figure 6.14 Placement of Visuals

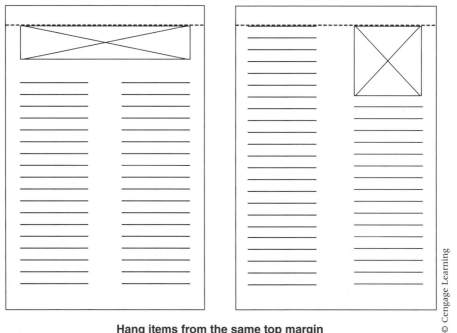

Hang items from the same top margin

© Cengage Learning

Figure 6.15 Items Hanging from Top Margin

Developing a Style Sheet and Template

In order to remain consistent, especially if you are working as part of a group, develop a *style sheet,* a list of specifications for each element in your document. You develop this list as part of your planning process. For brief documents you may not need to write it out, but you do need to think it through. Longer documents or group projects require a written or electronic style sheet. For instance, for a two-page memo, the style sheet would be quite short:

- Margins: 1-inch margin on all four sides
- Line treatment: no right justification
- Spacing within text: single-space within paragraphs, double-space between paragraphs
- Heads: heads flush left and boldfaced, triple-space above heads, and double-space below
- Footers: page numbers at bottom center

For a more complicated document, you need to make a much more detailed style sheet. In addition to margins, justification, and paragraph spacing, you need to include specifications for

- A multilevel system of heads
- Page numbers
- Rules for page top and bottom
- Rules to offset visuals
- Captions for visuals
- Headers and footers—for instance, whether the chapter title is placed in the top (header) or bottom (footer) margins
- Lists

Figure 6.16 (p. 160) shows a common way to handle style sheets. Instead of writing out the rules in a list, you make a *template* that both explains and illustrates the rules. (For more on planning style sheets, see the section "Format the Pages," in Chapter 15, pp. 453–455.)

The electronic style sheet is a particularly useful development. Many word processing and desktop publishing programs allow you to define specifications for each style element, such as captions and levels of heads. Suppose you want all level 1 heads to be Helvetica, 18-point, bold, flush left, and you want all figure captions to be Palatino, 8-point, italic. The style feature allows you to enter these commands into the electronic style sheet for the document. You can then direct the program to apply the style to any set of words.

Usually you can also make global changes with an electronic style sheet. If you decide to change all level 1 heads to Palatino, 16-point, bold, flush left, you need only make the change in the style sheet, and the program will change all the instances in the document.

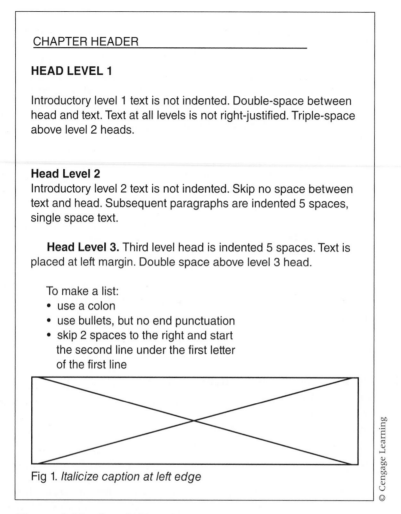

CHAPTER HEADER

HEAD LEVEL 1

Introductory level 1 text is not indented. Double-space between head and text. Text at all levels is not right-justified. Triple-space above level 2 heads.

Head Level 2
Introductory level 2 text is not indented. Skip no space between text and head. Subsequent paragraphs are indented 5 spaces, single space text.

　Head Level 3. Third level head is indented 5 spaces. Text is placed at left margin. Double space above level 3 head.

　To make a list:
- use a colon
- use bullets, but no end punctuation
- skip 2 spaces to the right and start the second line under the first letter of the first line

Fig 1. *Italicize caption at left edge*

© Cengage Learning

Figure 6.16　Sample Template

Worksheet for a Style Sheet

☐ **Select margins.**

☐ **Decide how many levels of heads you will need.**

☐ **Select a style for each level.**

☐ **Select a location and format for your page numbers.**

☐ **Determine the number of columns and the amount of space between them.**

☐ **Choose a font size and leading for the text.**

☐ **Place appropriate information in the header or footer area.**

☐ **Establish a method for handling vertical lists.**

Determine how far you will indent the first line. Use a bullet, number, letter, or some other character at the beginning of each item. Determine how many spaces will follow the initial character. Determine where the second and subsequent lines will start.

☐ **Choose a method for distinguishing visuals from the text.**

- Will you enclose them in a box or use a rule above and below?
- Where will you place visuals within the text?
- How will you present captions?

Figures 6.17 and 6.18 present the same report section in two different formats, each the result of a different style sheet.

DISCUSSION

High-Protein Diets

Introduction   The goal of this search was to determine if the Internet was a valuable source of information regarding high-protein diets. To define my information as usable, it must meet three criteria.

The three criteria are as follows: The information must be no older than 1997, the sites must be found to be credible sites, and it must take no longer than 10 minutes to find information pertinent to the topic on each site.

Findings   I used the Dogpile search engine to find my sites on high-protein diets. The keywords I used were *protein, high protein,* and *fad diets.* These keywords led me to the sites listed in Table 1.

As seen in Table 1, most of the sites fit the criteria. Using Dogpile to search for nutrition information yielded mixed results of commercial and professional sites. The Internet provided a vast amount of information regarding high-protein diets.

In my results I determined that the information from the website cyberdiet.com was credible even though it was a commercial site. The Tufts University Nutrition Navigator, a well-known, credible website that evaluates nutrition websites, recommended Cyberdiet and gave it a score of 24 out of 25.

Conclusion   The credibility and recency of the information did not all meet the criteria. Because of this, I conclude that the Internet

(Continued)

does have valuable information regarding high-protein diets, but that the Web user must use caution and be critical in determining the validity of each website.

TABLE 1
Standards of High-Protein Diet Search

	Less Than 10 Min.?	Recency (>1997)	Credibility
Cyberdietcenter.com "High-Protein, Low-Carbohydrate Diet"	Yes	1999 Yes	Yes
Heartinfo.org "The Reincarnation of the High-Protein Diet"	Yes	1997 Yes	Yes—Professional
more.com "Information on High-Protein Diets"	Yes	———	No—Commercial
Prevention.com "A Day in the Zone"	Yes	1995 No	No—Consumer
Eatright.org (ADA) "In the News: High-Protein/ Low-Carbohydrate Diets"	Yes	1998 Yes	Yes—Professional

Figure 6.17 Two-Column Design

DISCUSSION

High-Protein Diets

Introduction.   The goal of this search was to determine if the Internet was a valuable source of information regarding high-protein diets. To define my information as usable, it must meet three criteria.

The three criteria are as follows: the information must be no older than 1997, the sites must be found to be credible sites, and it must take no longer than 10 minutes to find information pertinent to the topic on each site.

Findings.   I used the Dogpile search engine to find my sites on high-protein diets. The keywords I used were *protein, high protein,* and *fad diets.* These keywords led me to the sites listed in Table 1.

As seen in Table 1, most of the sites fit the criteria. Using Dogpile to search for nutrition information yielded mixed results of commercial and professional sites. The Internet provided a vast amount of information regarding high-protein diets.

In my results I determined that the information from the website cyberdiet.com was credible even though it was a commercial site. The Tufts University Nutrition Navigator, a well-known, credible website that evaluates nutrition websites, recommended Cyberdiet and gave it a score of 24 out of 25.

Conclusion.   The credibility and recency of the information did not all meet the criteria. Because of this, I conclude that the Internet does have valuable information regarding high-protein diets, but that the Web user must use caution and be critical in determining the validity of each website.

TABLE 1
Standards of High-Protein Diet Search

	Less Than 10 Min.?	**Recency (>1997)**	**Credibility**
Cyberdietcenter.com "High-Protein, Low-Carbohydrate Diet"	Yes	1999 Yes	Yes
Heartinfo.org "The Reincarnation of the High-Protein Diet"	Yes	1997 Yes	Yes—Professional
more.com "Information on High-Protein Diets"	Yes	——	No—Commercial
Prevention.com "A Day in the Zone"	Yes	1995 No	No—Consumer
Eatright.org (ADA) "In the News: High-Protein/ Low-Carbohydrate Diets"	Yes	1998 Yes	Yes—Professional

Figure 6.18   One-Column Design

Exercises

▶ You Create

1.   For a nonexpert audience, write a three- to five-paragraph description of a machine or process you know well. Your goal is to give the audience a general familiarity with the topic. Create two versions. In version 1, use only chunked text. In version 2, use at least two levels of heads, a bulleted list, and a visual aid. Alternate: Using the same instructions, create a description for an expert audience.

2.   Write a paragraph that briefly describes a room in which you work. Create and hand in at least two versions with different designs (more if your

instructor requires). Alter the size of the type, the font, and the treatment of the right margin. For instance, produce one in a 12-point sans serif font, right-justified format, and another in a 10-point serif font, ragged-right format. Label each version clearly with a head. Write a brief report explaining which one you like the best.

3. Create a style sheet and template for either Figure 6.17 or 6.18.

4. Create a layout for the following instructions and visual aids. Correct any typographical or formatting errors. Develop a style sheet to submit to your instructor before you redo the text.

SOLDERING PROCESS

Introduction

The following information will show you how to make a correct solder joint between two wires. A solder joint is needed to electrically and physically connect two wires. A correct solder joint is one that connects the wire internally as well as externally (Figure 1).

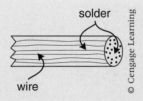

Figure 1.
Correct Joint

An incorrect solder joint does not maximize the electrical connection and also will probably break if handled. Due to the wide varieties of wire and soldering irons, this demonstration will be done using a 35-watt soldering iron and 16-gauge insulated wire. Generally, the thicker the wire, the more energy it takes to heat the wire. This means that any wire larger than 16 gauge would require a larger-wattage soldering iron.

Process

Step 1—Preheat Iron
The soldering iron must be plugged in and allowed to heat for at least 5 minutes. This will assure that the iron has reached its heating potential. This means that the iron will properly heat the iron and solder.

Step 2—Prepare Wire
Prepare both wires by cutting the ends to be soldered. Be sure to leave no frayed wire or insulation hanging.

Step 3—Strip Wire
Using a wire stripper, strip off approximately ½" of insulation from both wires (Figure 2). Visually check to be sure there is no hanging or frayed wire. If there is trim them off.

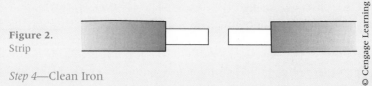

Figure 2.
Strip

Step 4—Clean Iron
Using a damp cloth, wipe off the end of the soldering iron. Be sure the tip is clear of any residue. This will assure that the solder will be free of impurities.

Step 5—Tin Iron
Using Rosin Core solder, apply a small amount of solder to the iron tip (Figure 3). Rosin Core solder is a special solder that has cleaning solvents internally in the solder. This makes soldering possible without the need of using other solvents such as Flux. Tinning the iron tip will allow maximum heat transfer between the iron and the wire.

Figure 3.
Tinning

Step 6—Twist Wires
Setting wires end-to-end, wrap them around each other. Be sure they are twisted together tight to be sure that they are making a good physical connection. The solder will join the wires only where they are touching (Figure 4).

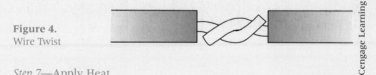

Figure 4.
Wire Twist

Step 7—Apply Heat
Touch the soldering iron tip firmly to the middle of the joined wire. Set the tip at a 45-degree angle to the wire (Figure 5). Heat focuses to a point so the most energy is at the tip of the iron.

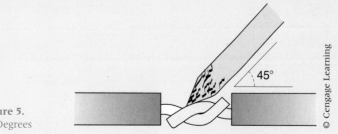

Figure 5.
45-Degrees

Step 8—Apply Solder

Apply solder where tip meets the wire (Figure 6). Observe solder to see if it is flowing into the wire. If the solder is globbing to the tip, either the tip is dirty and must be cleaned or the iron is not being firmly pressed to the wire. Watch insulation to be sure it does not melt. Apply solder until the wire appears to be full. Do not apply too much solder for it will begin to glob on the surface.

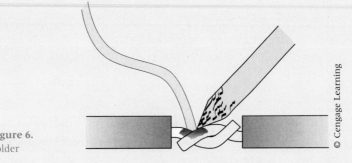

Figure 6.
Solder

© Cengage Learning

Step 9—Let Cool

Allow the soldered joint to cool slowly. Rapid cooling will cause the solder to become brittle.

Step 10—Inspect Solder Joint

Visually inspect the solder joint. Check to see if the solder has flowed into the wire. Make sure that it has not globbed around one point in the joint. Also check to be sure that the solder has made a solid physical connection (Figure 7). If the solder joint fails either of these qualifications, you must cut the wire and start over.

Good   Bad

Figure 7.
Inspect

© Cengage Learning

Step 11—Tape Joint

Using electrical tape, wrap joint to be sure that no metal is exposed (Figure 8).

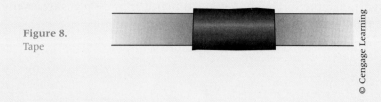

Figure 8.
Tape

© Cengage Learning

5. Use the following information to create a poster on your desktop system or word processor.

 a. The campus "golden oldies" club will hold a workshop for interested potential members on Wednesday, November 4, from 6:30 to 8:30 p.m. at the main stage in the University Student Center. No prior experience is necessary.

 b. The local Digital Photo Club will hold a benefit auction for the area United Way Food Pantry on Sunday, May 8, at 1:00 p.m. at the local fairgrounds. Admission is $10.00. The event will consist of photos for sale and workshops on buying digital cameras and creating digital photos.

6. Create a poster for an event that interests you.

▶ You Revise

7. Use principles of design and visual logic to revise the following paragraph. Also eliminate unnecessary information.

> **Capabilities of System.**   The new system will need to provide every capability that the current system does. I spoke with Dr. Franklin Pierce about this new system. Dr. Pierce has a vast amount of experience with Pascal (the old system is written in it) and in Ada (the new system is to be written in it). He also developed many software systems, including NASA's weather tracking system. After looking at the code for the current system, Dr. Pierce assured me that every feature in the old system can be mapped to a feature written in Ada. He also said that using Ada will allow us *multiple versions* of the program by using a capability of Ada to determine what type of computer is used for the menu. Furthermore, he said that using Ada will allow us to *improve the performance* of some of the capabilities, such as allowing the clock to continually be updated instead of stopping while another function is being performed and then being updated after that function has been completed. Thus, the first criterion could be met.

8. Rearrange the following layout so that it is more pleasing. Write a report that explains why you made your changes.

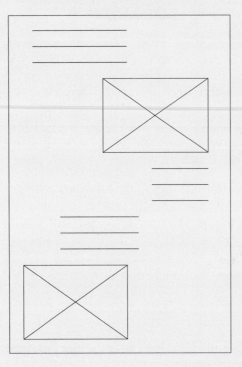

▶ Group

9. In groups of three or four, analyze Figures 6.17 and 6.18, pages 161–163. Decide which one you prefer, and explain your decision in a group report to the class.

10. In groups of two, create a design for the following text, which recommends that a student center purchase a particular sound board. Your instructor will require each group to use one of the various design strategies discussed in this chapter. Be prepared to present an oral report explaining your format decisions.

> This is a recommendation on whether to purchase a Goober sound board or a Deco sound board for the Technical Services Crew at the Student Center. Over the past few years, we have rented a sound board when bands or other large speaking groups come to the Student Center. The rental costs run from $200 to $500, and now that money has been allocated to buy a sound board, we should seriously consider purchasing one. The use we will get out of the board will make it pay for itself in two or three years. This report will detail my recommendation as to which board to purchase.
>
> I recommend we purchase the Goober XS-2000 24-channel sound board. Greg Newman, the Tech Crew Chief, and I have compared the Goober XS-2000 to the Deco TXS-260 24-channel board in audio magazines and by talking to people who have used the Goober board and the Deco board. We found the Goober to

be an overall better board. We have also experienced using both boards on many occasions and have liked the Goober board better. We compared the boards in terms of these criteria: features, reliability/experience, and cost.

The first criterion is the features. This is the most important factor in deciding which sound board to purchase. We are looking for a sound board that provides 24 channels in and 8 lines out. The lines out are used for sound effects and equalization. The more "outs" you have, the better the sound will be. This is what the Goober XS-2000 has: 24 channels in and 8 lines out. The Deco TXS-260 has 24 channels in and only 6 lines out. We find it necessary to have 8 lines out because we want to offer bands the best-quality sound possible.

Reliability is another important factor. Goober and Deco are both reputable companies that make good sound boards, but when it comes down to operating at maximum efficiency, we feel the Goober board can offer us better reliability. I had the opportunity to speak with Will Hodges of Southern Thunder Sound, a sound rental company from which we have rented sound boards. He said that most bands that come in to rent his equipment rent the Goober boards because they are easier to work with and don't break down as often as the Deco boards. Mr. Hodges's opinion lends further credibility to my recommendation that we purchase the Goober board.

Greg and I have had experience working with both boards. Once when we rented a Deco board, a loud buzzing sound began to be emitted from the speakers halfway through the concert. We did everything we could, but the buzzing continued throughout the concert. On the occasions when we have worked with a Goober board, it has operated without any problems.

The cost is the last criterion. We have allocated an amount of $3500 to purchase a sound board. The Deco TXS-260 is $2800 with a two-year limited warranty. The Goober XS-2000 is $3600 with a five-year limited warranty. We have dealt with this Goober dealer before, and she said she would throw in a 100-foot 24-channel snake (a $700 value) free and would knock $300 off the total price—a $4300 value for $3300. A $1000 savings. Although the Deco board is less expensive, we should spend the extra $500 for a better board with a longer warranty, greater reliability, and a free 100-foot snake.

Writing Assignments

1. In one page of text (just paragraphs), describe an opportunity for your firm, and ask for permission to explore it. Your instructor will help you select a topic. Your audience is a committee that has resources to take advantage of the opportunity. Use a visual aid if possible. Then use the design guidelines in this chapter to develop the same text into an appropriate document for the committee. In groups of three or four, review all the documents to see where the format best conveys the message. Report to the class your conclusion about which was best and why it works.

2. Write a learning report for the assignment you just completed. See Chapter 5, Writing Assignment 8, page 137, for details of the assignment.

Web Exercises

1. Analyze two Web pages in order to explain to a beginner audience how to lay one out—where to place titles, how large to make them, and so on. To give your advice a range of examples, use a homepage/index and an information link by following a particular path through several links, for example, About Us/Our Products/Cameras/FX20MicroZoom/Technical Spec.

2. Analyze the color scheme in a website. Use a major corporation like AT&T or Sun Micro. Write a report to your classmates that explains how the site uses color to make its message clear.

Works Cited

The Basics of Color Design. Cupertino, CA: Apple, 1992, Print.

Castro, Elizabeth. *HTML4 for the World Wide Web.* Berkeley, CA: Peachpit, 1998, Print.

"Color Wheel Pro—See Color Theory in Action." 10 April, 2004 <www.color-wheel-pro.com/color-meaning.html>.

Cook, Marshall, and Blake R. Kellogg. *The Brochure: How to Write and Design It.* Madison, WI: privately printed, 1980, Print.

Felker, Daniel B., Frances Pickering, Veda R. Charrow, V. Melissa Holland, and Janice C. Redish. *Guidelines for Document Designers.* Washington, DC: American Institutes for Research, 1981, Print.

Horton, William. "Overcoming Chromophobia: A Guide to the Confident and Appropriate Use of Color." *IEEE Transactions of Professional Communication* 34.3 (1991): 160–171, Print.

Jones, Scott L. "A Guide to Using Color Effectively in Business Communication." *Business Communication Quarterly* 60.2 (1997): 76–88, Print.

Keyes, Elizabeth. "Typography, Color, and Information Structure." *Technical Communication* 40.4 (1993): 638–654, Print.

Mazur, Beth. "Coming to Grips with WWW Color." *intercom* 44.2 (1997): 4–6, Print.

Miller, Brian. "The Uniquely Southern Landscape." *Outdoor Photographer* (June 2003): 66, Print.

Rubens, Phillip M. "A Reader's View of Text and Graphics: Implications for Transactional Text." *Journal of Technical Writing and Communication* 16 (1986): 73–86, Print.

Sadowski, Mary A. "Elements of Composition." *Technical Communication* 34 (1987): 29–30, Print.

Schriver, Karen A. *Dynamics in Document Design: Creating Texts for Readers.* New York: Wiley, 1997, Print.

Sevilla, Christine. "Page Design: Directing the Reader's Eye." *intercom* (June 2002): 6–9.

Tyson, Paul. "Designing Documents." *intercom* (December 2002): 27–29, Print.

White, Jan. *Color for the Electronic Age.* New York: Watson-Guptill, 1990, Print.

White, Jan V. *Graphic Design for the Electronic Age.* New York: Watson-Guptill, 1988, Print.

Xerox Publishing Standards: A Manual of Style and Design. New York: Xerox Press, 1988, Print.

Focus on
Color

Color invites us in; it entices. Cheaper technology that allows individuals to add color to documents has opened a new world for communicators. Now anyone can produce a multiple-color document. Working effectively with color means knowing

- How color relationships cause effects
- How color can be used in documents

Effects Produced by Color Relationships

When colors are placed next to each other, their relationships cause many effects. The concepts of the color wheel and value illustrate how color affects visibility. Various hues cause emotional and associational reactions.

The Color Wheel and Visibility The key concept is the color wheel, which provides a way to see how colors relate to one another (*Basics*). Figure 1 shows the six basic rainbow colors, from which all colors can be made.

The colors that occur across from one another are called *complementary* (e.g., red and green), those that touch are called *adjacent*

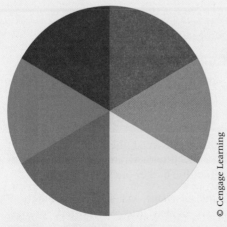

Figure 1. *The Basic Color Wheel*

© Cengage Learning

(e.g., red and orange or violet), and those that are two apart are called *contrasting* (e.g., red and yellow, red and blue, orange and green, orange and violet). The relationships are shown in Figures 2 through 4. Each of the relationships affects visibility.

Maximum Visibility Figure 2 shows that sets of complementary colors cause high contrast or maximum visibility. This relationship strongly calls attention to itself. Notice, however, that the colors tend to "dance." Most people find them harsh and can view them for only a short time.

Minimum Visibility Figure 3 shows that sets of adjacent colors cause low contrast or minimal visibility. These relationships tend to have a relaxing, pleasing effect. The colors, however, tend to blend in with each other, making it difficult to distinguish the object from the background.

Pleasing Visibility Figure 4 shows that sets of contrasting colors cause medium contrast or pleasing visibility. These relationships have strong contrast, but they do not dance. Most people find them bold and vivid.

Value Affects Visibility of Individual Hues
In addition to having relationships with other colors, any color has relationships with itself. So a basic color, like blue, is called a *hue*.

However, you can mix white or black with the hue and so create a value—a *tint* (a hue and white) or a *shade* (a hue and black). Figure 5 shows blue as a hue in the center, but as a tint to the left and a shade to the right. The more white you add, the lighter the tint; the more black, the darker the shade.

Reduce Harsh Contrast by Changing Value Understanding value allows you to affect the impact of color relationships. Although complementary colors of equal value are quite jarring, as shown in Figure 2, notice how the jarring decreases when the background

(Continued)

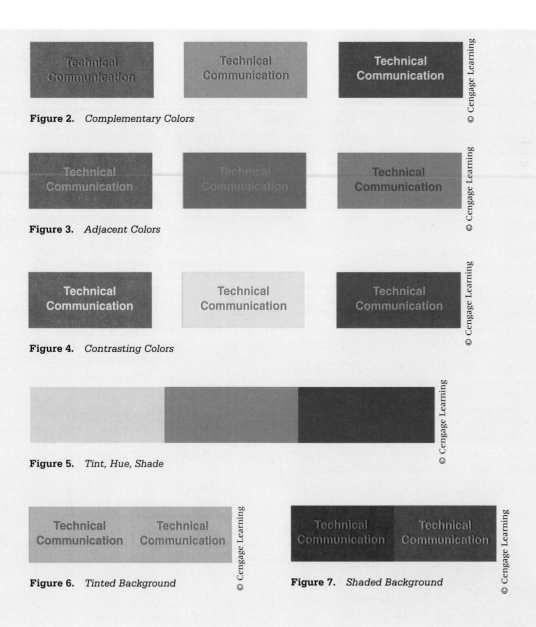

Figure 2.   *Complementary Colors*

Figure 3.   *Adjacent Colors*

Figure 4.   *Contrasting Colors*

Figure 5.   *Tint, Hue, Shade*

Figure 6.   *Tinted Background*

Figure 7.   *Shaded Background*

color changes value either as a tint (Figure 6) or a shade (Figure 7).

Increase Visibility by Using Innate Value   Colors have innate value. In other words, the basic hue of some colors is perceived as brighter than the basic hue of others. The brightest basic hue is yellow, and the darkest is violet (see Figure 8, p. 173). The brighter hues are more difficult to read (because they do not contrast as well with white, the color of most pages) than the darker ones.

Colors and Emotions   As shown in Figure 9 colors are divided into warm (red, orange, yellow) and cool (green, blue, violet). Warm colors appear to most people to be soft; cozy; linked to passions, celebrations, and

© Cengage Learning

Figure 8.   *Innate Values*

excitement. Cool colors appear to most people as harder; icier; linked to rational, serious, reliable decorum.

Colors and Associations   Colors have traditional associations. In the United States, red, for instance, implies stop, hot, desire, passion, energy. Blue implies peace, heavenliness, space, cold, calm. Green implies jealousy, nature, safety. Yellow implies joy, happiness, intellect (Color Wheel Pro). Using colors appropriately in context makes your work more effective. Consider:

> That pipe is hot and That pipe is hot.
>
> Try to remain calm and Try to remain calm.

Colors have different associations in different cultures. See "Globalization and Visual Aids" in Chapter 7.

How Color Can Be Used in Documents

Color has four important functions in documents (Horton; Jones; Keyes; Mazur; White). Use color to

- Make text stand out
- Target information
- Indicate organization
- Indicate the point in a visual aid

Use Color to Make Text Stand Out   As Figures 2 through 7 illustrate, one common use of color is to make text clearly visible, or legible. Legibility is a matter of contrast. If the text contrasts with its background, it will be legible. Two principles are helpful.

- The best contrast is black type on a white background.
- Colored type appears to recede from the reader. Higher values (yellow) appear farther away and less legible than lower values (blue).

Color alone will not cause individual words to stand out in black text. To make

Figure 9a.   *Warm Colors*

Figure 9b.   *Cool Colors*

(Continued)

Figure 10.   *Minimal Emphasis via Size*

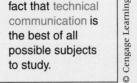

Figure 11.   *Minimal Emphasis via Color*

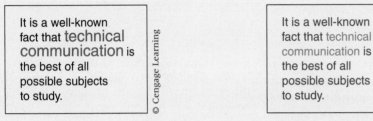

Figure 12.   *Emphasis via Size and Color*

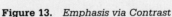

Figure 13.   *Emphasis via Contrast*

colored text stand out from black text, you must also change its size. Figures 10 and 11 show that size and color cause minimal emphasis in text, but Figure 12 shows that colored words made larger do show more emphasis. Figure 13 shows that words will stand out if the color contrasts with the color of the surrounding text.

Use Color to Target Information   Color focuses attention. It does this so strongly that color creates "information targets." In other words, people see color before they see anything else. As a consequence, you should follow these guidelines:

- Use color to draw attention to "independent focus" text—types of text that readers must focus on independently of other types.
- Make each type of text look different from the other types. Use different value (tint or shade, not hue) and different areas or shape to cause the difference.

Common examples of independent focus texts are

- Warnings
- Hints
- Cross-references
- Material the reader should type
- Sidebars

Figure 14 (p. 175) shows a page that has each of these items. Notice that the difference is achieved by differences in value, shape, and location.

Use Color to Indicate Organization   Color creates a visual logic. Readers quickly realize that color indicates a function. How strongly a color indicates that function is increased when that color is combined with shape and area.

Getting Started on Web Research

Create questions, select a search engine, and start. The password is gogetter.

1. Select a topic.
2. Create questions about the topic. (See Chapter 5 for a memory jog.)
3. Select sources that are likely to provide answers to those questions.

sidebar/hint warning

4. Use those sources. Remember to keep track of

cross-reference

   a. Exactly where you have been (e.g., the exact http address of the documents you download)
   b. The titles and authors of the material you use
   c. The keywords you have used and an evaluation of their effectiveness
5. To gain access to the network
   a. Turn on your computer
   b. Double-click on the Webscape icon
   c. Enter this password at the $prompt:

reader types this ———— gogetter <r>.

© Cengage Learning

Figure 14. *Examples of Independent-Focus Text*

You can use color to indicate different levels in your document's hierarchy. Common functions that color can indicate are

- Marginal material
- Running information—in headers and footers, or in the head system

Figure 15 (p. 176) shows a page that has each of these items. Notice that the difference is achieved by differences in value, shape, and location.

Use Color to Indicate the Point of a Visual Aid Use color in visual aids to draw readers' attention to specific items (see Figure 16, p. 177). For instance,

- To highlight a single line in a table
- To highlight the data line in a line graph
- To focus attention on a particular bar or set of bars in a bar graph
- To differentiate callouts and leaders from the actual visual

Be aware that many of the graphics programs provide you with color but that the color is not chosen to make this visual aid more sense.

Summary Guidelines for Using Color

Follow these basic guidelines in your handling of color in your documents:

- Be consistent. For each type of item, use the same hue, value, shape, and location.
- Correctly use contrast. Follow the color wheel to select combinations of colors that create high visibility. Remember that black and white create the highest contrast.

(Continued)

header   **The Creative Concept**   **1**

level 1 head   ## Finding Your Gist Through Analogy

Here you will present a lot of report text. This text is very important, and readers should focus on this. After all, it is the main point of what they have to read, and you have spent a lot of time researching all the information and figuring out what it means and how you can present it to the readers so that they feel comfortable with it.

level 2 head

Note: If you want to add notes in the margin, you could try them in some value other than the one the heads are in. Notice the different shade and thinner shape of this text.

marginal comment

The Passenger Train Analogy.   You have finally found the gist of the material and have a wonderful creative concept—a fine metaphor that you have discovered only after trying six or seven of them. The one you are about to use is the Passenger Train analogy. You will explain the creation of websites to your readers by leading them through the train and pointing out where they have their regular seats, where they can go to sit to enjoy an unrestricted view of the countryside in the observation car, where their sleeping berth is, and, most important of all, where the dining car is.

You will also have a "before section" in which you explain how to decide which train to take, how to buy a ticket, and how to use the station effectively.

The Rejected Analogies.   You know now that you have decided against the menu metaphor in which the Web creators would be simply pulling things off the page and more or less plopping them onto their plate and the thunderstorm analogy in which they would build and swirl for a long time, perhaps with the surrounding area very still until finally they unleashed their website in a fury of lightning and thunder.

© Cengage Learning

Figure 15. *Examples of Color Function*

Colored text must change size to contrast with surrounding black text.

- Correctly use feeling and association. Use warm colors for action items, especially warnings. Use cool colors for reflection items.
- Generally use only one hue with varying tints and shades. Use two or more colors after you have practiced with color and have had your creations critiqued by readers and users.
- Help color-blind readers by using different brightnesses of the same color. Remember that location on the page will help color-blind readers also. (Even if they cannot "read" the red, they will know it is a marginal comment because of where it is located.)

TABLE 1
Frequency of Analogies Used in Technical Articles

	Train	Thunderstorm	Food Menu
Packaging	30	40	30
Hospitality	20	15	65
Automotive	10	75	15

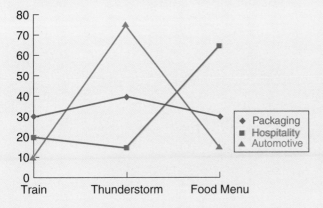

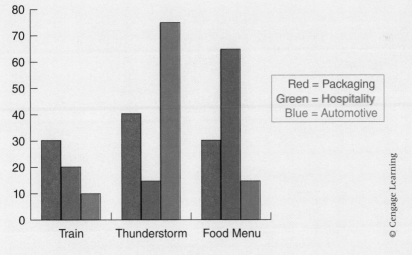

Figure 16. *Use of Color in Visual Aids*

© Cengage Learning

Using Visual Aids

CHAPTER CONTENTS

CHAPTER 7 **IN A NUTSHELL**

Visual aids help users by summarizing data and by showing patterns.

Common types of visual aids. *Tables* have rows and columns of figures. Place the items to compare (corporate sales regions) down the left side and the ways to compare them (by monthly sales) along the top. The data fill in all the appropriate spaces.

Line graphs illustrate a trend. Place the items to compare along the bottom axis (days of the week) and the pattern to illustrate (the price of the stock) along the left axis. The line shows the fluctuation.

Bar graphs illustrate a moment in time. Place the items to compare along the bottom axis (four cities) and the terms to compare (population figures) along the left axis. The bars show the difference immediately.

Pie charts represent parts of a whole. They work best when they compare magnitude.

Illustrations, either photos or drawings, show a sequence or a pattern—the correct orientation for inserting a disk into a computer, for example.

Guidelines for using visual aids.

▶ Develop a *visual logic*—place visuals in the same position on the page, make them about the same size, treat captions and rules (black lines) the same way. Be consistent.

▶ Create neat visuals to enhance your clarity and credibility.

▶ Tell readers what to notice and explain pertinent aspects, such as the source or significance of the data.

▶ Present visual aids below or next to the appropriate text.

V isual aids are an essential part of technical writing. Graphics programs for personal computers allow writers to create and refine visual aids within reports. This chapter presents an overview of visual aids. It explains the concept of "visual thinking," and then presents general information on using, creating, and discussing visual aids, as well as specific information on all the common types of visual aids.

Visual Thinking

Visual thinking is a skill just like writing. You can learn to conceptualize how visual aids will convey your point to your users (in this discussion the word *users* replaces *readers* because almost all communication now occurs digitally). Visual thinking "is the intuitive and intellectual process of visual idea generation and problem solving" (Brumberger 380). Visual thinking is a process that relies on the elements of visual languages, such as "images, shapes, patterns, textures, symbols, colors" (Brumberger 381). You need to practice this kind of thinking as you create documents. Brumberger suggests that communicators must think of visuals as solutions to problems inherent in the documents they are creating. In other words, it is not enough to only work through how you will design the text (as discussed in the previous chapter); you must also think through how visuals of any type will be part of, perhaps even the major part of, the way in which you deliver your ideas to your audience.

The Uses of Visual Aids

Visual aids have a simple purpose. According to noted theorist Edward Tufte, visual aids "reveal data" (13). This key concept controls all other considerations in using visual aids. You will communicate effectively if your visual aids "draw the reader's attention to the sense and substance of the data, not to something else" (91). Technical writers use visual aids for four purposes:

▶ To summarize data
▶ To give users an opportunity to explore data
▶ To provide a different entry point into the discussion
▶ To engage user expectations

To summarize data means to present information in concise form. Figure 7.1, a graph of a stock's price for one week, presents the day-end price of the stock for each day. A reader can tell at a glance how the stock fared on any day that week.

To give users an opportunity to explore data means to allow them to investigate on their own. Readers can focus on any aspects that are relevant to their needs. For instance, they might focus on the fact that the stock rose at the

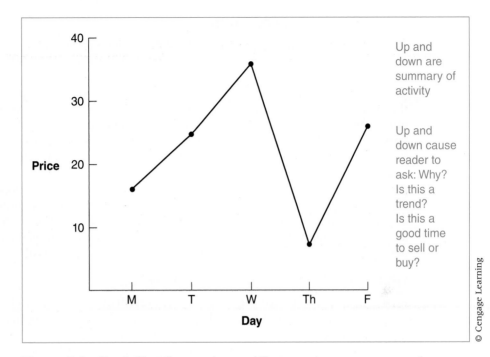

Figure 7.1 Graph That Summarizes and Engages

beginning and again at the end of the week, or that the one-day rebound on Friday equaled the two-day climb on Monday and Tuesday.

To provide a different entry point into the discussion means to orient readers to the topic even before they begin to read the text. Studying the graph of a stock's price could introduce the reader to the concept of price fluctuation or could provide a framework of dollar ranges and fluctuation patterns.

To engage user expectations is to cause users to develop questions about the topic. Simply glancing at the line that traces the stock's fluctuation in price would immediately raise questions about causes, market trends, and even the timeliness of buying or selling.

Creating and Discussing Visual Aids

How to Create Visual Aids

The best way to create a visual aid is to follow the basic communication process: plan, draft, and finish, always with an eye to ethical presentation (see box "Ethics and Visual Effects," pp. 195–196).

Plan the Visual Aid Carefully

Make the visual aid an opportunity both to present your data and to engage your readers. Your overall goal is to help your readers as they research the information they need and help them to make sense of it once they find it (Schriver). Consider your audience's knowledge and what they will do with the information. Are they experts who need to make a decision? Consider your goal in presenting the information. Is it to summarize data? To offer an opportunity to explore data? To provide a visual entry point to the discussion? To engage reader expectations? Your visual aid should have just one main point. Take into account the fact that graphs have an emotional impact. For example, in graphs about income, lines that slope up to the right cause pleased reactions, whereas those that slope down cause anxiety. Consider also any constraints on your process. How much time do you need to make a clear visual aid? Consider the layout of the visual aid. In tables, which items should be in rows and which in columns? In graphs, which items should appear on the horizontal axis and which on the vertical?

Draft the Visual Aid Just as You Draft Text

Revise until you produce the version that presents the data most effectively. For graphs, select wording, tick marks, and data line characteristics (solid, broken, dots). For tables, select column and row heads, enter the data, create a format for the caption and any rules, and add necessary notes. For charts and illustrations, select symbols, overall dimensions, a font for words, and a width for rules. Reread the visual aid to find and change unclear elements, in either content or form.

Finish by Making the Visual Aid Pleasant to View

Treat all items consistently. Reduce clutter as much as possible by eliminating unnecessary lines and words.

How to Discuss Visual Aids

Carefully guide the reader's attention to the aspect of the visual aid you want to discuss. Your goal is to enrich the reader's understanding of the topic (Schriver). For instance, you could choose to explain elementary, intermediate, or overall information (Killingsworth and Gilbertson). *Elementary* information is one fact: On Wednesday, the stock's price rose. *Intermediate* information is a trend in one category: As the week progressed, the stock price fluctuated. *Overall* information is a trend that relates several categories: After the price dropped, investors rushed to buy at a "low."

   In addition, you must explain background, methodology, and significance. *Background* includes who ordered or conducted the study of the stock, their reasons for doing so, and the problem they wanted to investigate. *Methodology* is how the data were collected. *Significance* is the impact of the data for some other concern—for instance, investor confidence. See pages 186 and 189 for examples of discussions of visual aids.

How to Reference Visual Aids

Refer to the visual aid by number. If it is several pages away, include the page number in your reference. You can make the references textual or parenthetical.

Textual Reference

A *textual* reference is simply a statement in the text itself, often a subordinate clause, which calls attention to the visual aid.

> . . . as seen in Table 1 (p. 10).
> If you look at Figure 4, ...
> The data in Table 1 show . . . .

Parenthetical Reference

A *parenthetical* reference names the visual aid in parentheses in the sentence. A complete reference is used more in reports, and an abbreviated reference in sets of instructions. In reports, use *see* and spell out *Table*. Although *Figure* or *Fig.* are both used, *Figure* is preferable in formal writing.

> The profits for the second quarter (see Figure 1) are . . . .
> A cost analysis reveals that we must reconsider our plans for purchasing new printers (see Table 1).

In instructions, you do not need to use *see*, and you may refer to figures as *Fig.*

> Insert the disk into slot A (Fig. 1).
> Set the CPM readout (Fig. 2) before you go on to the next step.

Do not capitalize *see* unless the parenthetical reference stands alone as a separate sentence. In that case, also place the period inside the parentheses.

> All of these data were described earlier. (See Tables 1 and 2.)

Guidelines for Effective Visual Aids

The following five guidelines (Felker et al.; MacDonald-Ross; Schriver) will help you develop effective visual aids. Later sections of the chapter explain which types of visual aids to use and when to use them.

1. Develop visual aids as you plan a document. Because they are so effective, you should put their power to work as early as possible in your project. Many authors construct visuals first and then start to write.

2. As you draft, make sure each visual aid conveys only one point. If you include too many data, readers cannot grasp the meaning easily. (Note, however, that tables often make several points successfully.)

3. Position visual aids within the draft at logical and convenient places, generally as close to their mention in the text as possible.

4. Revise to reduce clutter. Eliminate all words, lines, and design features (e.g., the needless inclusion of three dimensions) that do not convey data.

5. Construct high-quality visual aids, using clear lines, words, numbers, and organization. Research shows that the quality of the finished visual aid is the most important factor in its effectiveness (Felker et al.; MacDonald-Ross).

Using Tables

A table is a collection of information expressed in numbers or words and presented in columns and rows. It shows the data that result from the interaction of an independent and a dependent variable. An *independent variable* is the topic itself. The *dependent variable* is the type of information you discover about the topic (White, *Graphic*). In a table of weather conditions, the independent variable, or topic, is the months. The dependent variables are the factors that describe weather in any month: average temperature, average precipitation, and whatever else you might want to compare. The data—and the point of the table—are the facts that appear for each month.

When to Use a Table

Because tables present the results of research in complete detail, they generally contain a large amount of information. For this reason, professional and expert audiences grasp tables more quickly than do nonexperts. When your audience knows the topic well, use tables to do the following (Felker et al.):

▶ To present all the numerical data so that the audience can see the context of the relationships you point out
▶ To compare many numbers or features (and eliminate the need for lengthy prose explanations)

In the text, you should add any explanation that the audience needs to understand the data in the table.

Parts and Guidelines

Tables have conventional parts: a caption that contains the number and title, rules, column heads, data, and notes, as shown in Figure 7.2. The following guidelines will help you use these parts correctly (based in part on *Publication*).

1. Number tables consecutively throughout a report with Arabic numerals in the order of their appearance. Put the number and title above the table. Use the "double-number" method (e.g., "Table 6.3") only in long reports that contain chapters.

TABLE 1   Number

Winter Weather Conditions in Minnesota   Title

Month	Average Temp (°F)		Record Temp (°F)		Average Snow (in.)[a]	Record Snow (in.)	
	High	Low	High	Low		High	Low
January	20	3	59	241	9.0	46	0.0
February	25	9	64	240	7.7	26	Tr
March	38	23	83	232	9.6	40	.02

Spanner heads

Column heads

Data

Rule

[a]Snowfall data began in 1859.   Notes
Note: February includes calculations based on 28 and 29 days.

Source: Based on Minnesota Weatherguide Environment Calendar.   Source
The Freshwater Society. Excelsior, MN: 2012

© Cengage Learning

Figure 7.2 Elements of a Table

2. Use the table title to identify the main point of the table. Write brief but informative titles. Do not place punctuation after the title.

3. Use horizontal rules to separate parts of the table. Place a rule above and below the column heads and below the last row of data. Seldom use vertical rules to separate columns; use white space instead. If the report is more informal, use fewer or no rules.

4. Use a *spanner* head to characterize the column headings below it. Spanners eliminate repetition in column heads.

5. Arrange the data into columns and rows. Put the topics you want to compare (the independent variables) down the left side of the table in the *stub* column. Put the factors of comparison (the dependent variables) across the top in the column headings. Remember that columns are easier to compare than rows.

6. Place explanatory comments below the bottom rule. Introduce these comments with the word *Note.* Use specific notes to clarify portions of a table. Indicate them by raised (superscript) lowercase letters within the table and at the beginning of each note.

7. Cite the *source* of the data unless the data were obviously collected specifically for the paper. List the sources you used, whether primary or secondary.

The average caffeine content in each location is shown in Table 3. Results are reported for 1 standard cup (5 fl. oz.) and for a standard "medium" size (16 fl. oz.).

Samples from both Brew Devils and the Library both contained more caffeine than samples from Jarvis ($p < 0.0001$ and $p = 0.047$, respectively; Table 3). There was no difference between caffeine at Brew Devils and the Library ($p = 0.64$).

Table 3: Average Caffeine Content in Specialty Coffee Samples

Location	Average Caffeine Per 5 fl. oz. cup	Average Caffeine Per 16 fl. oz. cup
Brew Devils	101 ± 5.8 mg*	324 ± 19 mg
Swanson Library Express Cart	98 ± 11.1 mg**	315 ± 35 mg
HC2 Jarvis Express	80 ± 3.3 mg	255 ± 11 mg
	$p < 0.0001$ compared to Jarvis	**$p = 0.047$ compared to Jarvis*

© Cengage Learning 2014

Figure 7.3   Sample Text and Table

A Sample Table and Text

Table 3 in Figure 7.3 appeared in a lab report. Although the data are clear, the reader needs more help in order to understand the point being made. The explanatory text tells the reader what to notice and explains the significance of the numbers.

Using Line Graphs

A *line graph* shows the relationship of two variables by a line connecting points inside an *X* (horizontal) and a *Y* (vertical) axis. These graphs usually show trends over time, such as profits or losses from year to year. The line connects the points, and its ups and downs illustrate the changes—often dramatically. On the horizontal axis, plot the independent variable, the topic whose effects you are recording, such as months of a year. On the vertical axis, record the values of the dependent variable, the factor that changes when the independent variable changes, such as sales. The line represents the record of change—the fluctuation in sales (Figure 7.4).

When to Use a Line Graph

Line graphs depict trends or relationships. They clarify data that would be difficult to grasp quickly in a table. Research shows that expert readers grasp line graphs more easily than nonexperts (Felker et al.). Use a line graph

FIGURE 1
Five-Month Sales Data

© Cengage Learning

Figure 7.4   Dependent and Independent Variables

▶ To show that a trend exists (see Figure 7.4)
▶ To show that a relationship exists, say, of pollutant penetration to filter size
▶ To give an overview or a general conclusion, rather than fine points
▶ To initiate or supplement a discussion of cause or significance (Figure 7.4 alerts readers to ask why April is unusual)

Add explanatory text that helps the audience grasp the implications of the graph.

Parts and Guidelines

Line graphs have conventional parts: a caption that contains the number and title (with a source note when necessary), axis rules, tick marks and tick identifiers, axis labels, a data line, and a legend. These parts are illustrated in Figure 7.5, page 188. The following guidelines will help you treat these parts correctly.

1. Number figures consecutively throughout the report, using Arabic numerals. Use double-numbering (e.g., "Figure 6.6") only if you have numbered chapters.

2. Use a brief, clear title to specify the content of the graph. Do not punctuate after the title.

3. Put the caption below the figure. (Many computer programs automatically place the caption above the figure. This method is acceptable, especially for informal reports. Choose one placement or the other and use it consistently throughout the document.)

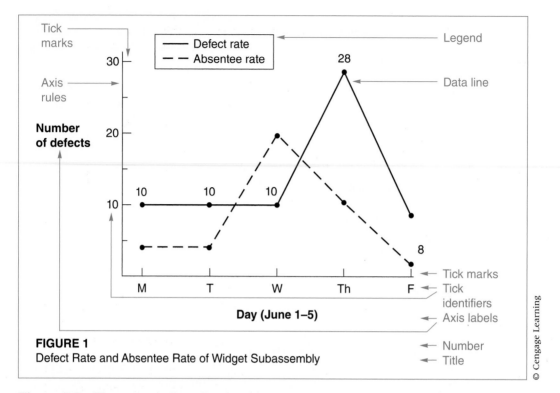

FIGURE 1
Defect Rate and Absentee Rate of Widget Subassembly

Figure 7.5   Elements of a Line Graph

dependent

4. Place the independent variable on the horizontal axis; place the dependent variable on the vertical axis.

5. Space tick marks equally along the axis rule. Use varying thicknesses to indicate subordination. (Note that in Figure 7.5 the 15 and 25 tick marks are smaller than the 20 and 30 marks.)

6. Provide clear axis labels. In general, spell out words and write out numbers from left to right. Use abbreviations only if they are common.

7. Present a data line with definite marks that indicate the intersection of the two axes (such as the dot in Figure 7.5 where Monday meets 10). Add explanatory numbers or words inside the graph.

8. If the graph has two or more lines, make them visually distinct and identify them with labels or in a legend, or key.

Sample Graphs and Text

The following line graph (Figure 7.6; Tans and Keeling) is taken from a National Oceanic and Atmospheric Administration report, which illustrates the monthly mean of carbon dioxide at the Mauna Observatory in Hawaii.

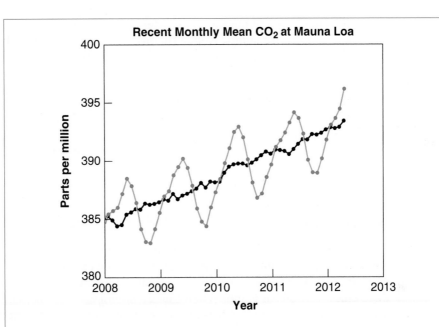

The graph shows recent monthly mean carbon dioxide measured at Mauna Loa Observatory, Hawaii.

The last four complete years of the Mauna Loa CO_2 record plus the current year are shown. Data are reported as a dry air mole fraction defined as the number of molecules of carbon dioxide divided by the number of all molecules in air, including CO_2 itself, after water vapor has been removed. The mole fraction is expressed as parts per million (ppm). Example: 0.000400 is expressed as 400 ppm.

In the above figure, the dashed **red line** with diamond symbols represents the monthly mean values, centered on the middle of each month. The **black line** with the square symbols represents the same, after correction for the average seasonal cycle. The latter is determined as a moving average of SEVEN adjacent seasonal cycles centered on the month to be corrected, except for the first and last THREE and one-half years of the record, where the seasonal cycle has been averaged over the first and last SEVEN years, respectively.

The last year data are still preliminary, pending recalibrations of reference gases and other quality-control checks. The Mauna Loa data are being obtained at an altitude of 3400 m in the northern subtropics, and may not be the same as the **globally averaged CO_2 concentration at the surface**.

Figure 7.6   Effective Professional Use of Line Graphs

The paragraph describing the figure uses several strategies that are effective in relating graphs to readers. Notice that as you read the text and review the graph, you can grasp the point even if you know nothing about molecules per million. The basic strategies are to

▶ Provide necessary background: "Data are reported as . . . ."
▶ Provide a topic sentence that announces the main point of the paper: "The graph shows recently monthly . . . ."
▶ Indicate what to notice: The red and black lines are both explained.
▶ Significance is noted: that the data is a moving average and that the last year of data is preliminary.

Using Bar Graphs

A *bar graph* uses rectangles to indicate the relative size of several variables. Bar graphs contrast variables or show magnitude. They can be either horizontal or vertical. Horizontal bar graphs compare similar units, such as the populations of three cities. Vertical bar graphs (often called *column graphs*) are better for showing discrete values over time, such as profits or production at certain intervals.

In bar graphs, the independent variable is named along the base line (Figure 7.7). The dependent variable runs parallel to the bars. The bars show the data. In a graph comparing the defect rates of three manufacturing lines, the lines are the independent variable and are named along the base line. The defect rate is the dependent variable, labeled above the line parallel to the bars. The bars represent the data on defects.

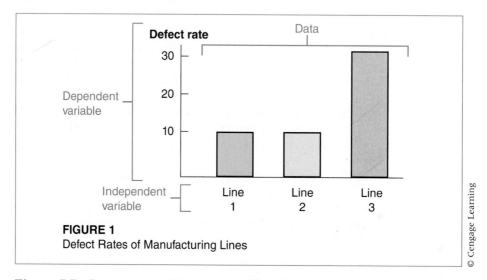

FIGURE 1
Defect Rates of Manufacturing Lines

© Cengage Learning

Figure 7.7   Dependent and Independent Variables

Parts and Guidelines

Bar graphs have conventional parts: a caption that contains the number and title (with source line when necessary), axis rules, tick marks and tick identifiers, axis labels, the bars, and the legend (Figure 7.8). The following guidelines (based on Tufte) will help you treat these parts effectively. (These guidelines are for vertical bar graphs; rearrange them for horizontal bar graphs.)

1. As with other visual aids, provide an Arabic numeral and a title that names the contents of the graph. This caption material appears below the figure in formal reports but is often found above in informal reports.

2. Place the names of the items you are comparing (the independent variable) on the horizontal axis; place the units of comparison (the dependent variable) on the vertical axis.

3. Space tick marks equally along the axis rule. Use varying thicknesses to indicate subordination.

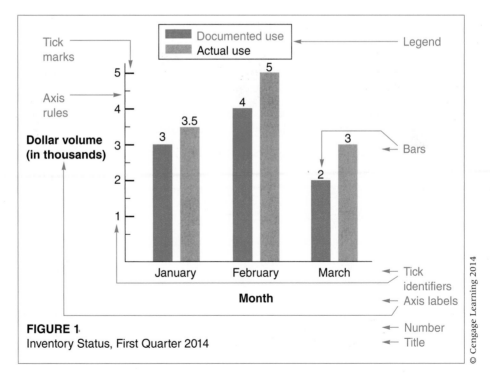

© Cengage Learning 2014

Figure 7.8   Elements of a Bar Graph

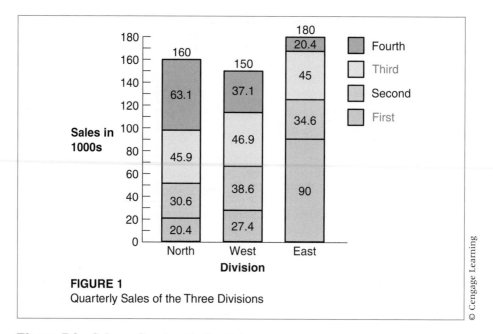

FIGURE 1
Quarterly Sales of the Three Divisions

Figure 7.9   Column Graph with Shading

4. Provide clear axis labels. Spell out words and write out numbers.

5. Make the spaces between the bars one-half the width of the bars. (You may have to override the default in your computer program.)

6. Use a legend or callouts to identify the meaning of the bars' markings.

7. Avoid elaborate cross-hatching and striping, which create a hard-to-read op art effect.

8. Use explanatory phrases at the end of bars or next to them.

9. Subdivide bars to show additional comparisons (Figure 7.9).

When to Use a Bar Graph

Bar graphs compare the relative sizes of discrete items, usually at the same point in time. Like line graphs, they clarify data that would be difficult to extract from a table or a lengthy prose paragraph. Nonexpert readers find bar graphs easier to grasp than tables. Use a bar graph

 ▶ To compare sizes
 ▶ To give an overview or a general conclusion

▶ To initiate or supplement a discussion of cause or significance (e.g., the bar graph of the defect rates of three widget lines [see Figure 7.7, p. 190] prompts a reader to ask why the rate was high on line 3.)

Use accompanying text to help readers grasp the implications of the graph.

Using Pie Charts

A *pie chart* uses segments of a circle to indicate percentages of a total. The whole circle represents 100 percent, the segments of the circle represent each item's percentage of the total, and the callouts identify the segments in the graph. Data words or symbols provide detailed information.

When to Use a Pie Chart

Pie charts work best when they are used to compare magnitudes that differ widely (see Figure 7.11, p. 194). Because pie chart segments are not very precise, most people cannot distinguish between close values, such as 17 percent and 20 percent. Nonexpert audiences find pie charts easier to use than tables (Felker et al.). Use pie charts

▶ To compare components to one another
▶ To compare components to the whole
▶ To show gross differences, not fine distinctions

Parts and Guidelines

Like other graphs, pie charts have conventional elements: a caption, the circle, the segments ("pie slices"), and callouts (Figure 7.10). The following guidelines (based in part on *Publication*) will help you treat these parts effectively.

1. The caption may appear above or below the chart. Use Arabic numerals and a title that names the contents of the graph. Informal charts often have only a title.

2. Start at "12 o'clock," and run the segments in sequence, clockwise, from largest to smallest. (Sometimes you cannot satisfy all these requirements. If you are comparing the grades in a chemistry class, the A segment is usually smaller than the C segment. However, you would logically start the A segment at 12 o'clock.)

3. Identify segments with callouts or legends. *Callouts* are phrases that name each segment (see Figure 7.10, p. 194). Use callouts if space permits, and arrange them around the circumference of the circle. A *legend* is a small sample of each segment's markings plus a brief identifying phrase, as shown in Figure 7.11 (p. 194). Cluster all the legend items in one area of the visual.

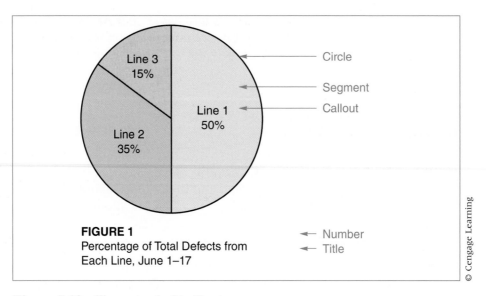

Figure 7.10 Elements of a Pie Chart

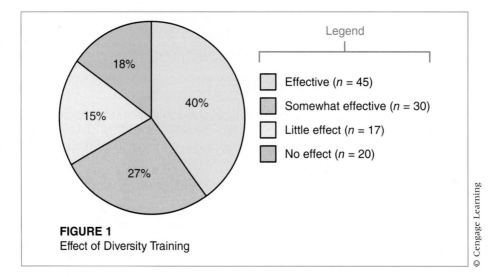

Figure 7.11 Pie Chart with Legend

4. In general, place percentage figures inside the segments.

5. For emphasis, shade important segments, or present only a few important segments rather than the whole circle.

6. In general, divide pie charts into no more than five segments. Readers have difficulty differentiating the sizes of small segments. Also, a chart with many segments, and thus many callouts, looks chaotic.

Ethics and Visual Effects

Deciding on the visual components of a document is complex, entailing everything from font and page properties to the use of images and their download capabilities for Web use. Clarity and order are paramount to making your point clear to your audience. Hiding or blurring the facts by obscuring the truth with lots of visuals is not only unfair to your audience, it is unethical as well. Make your visual effects user-centered. Carefully considering visual effects means connecting with your audience and understanding their needs, as well as creating a document that is aesthetically pleasing, functional, and ethically sound.

Nancy Allen has suggested several ethical guidelines for writers to consider when incorporating visual material:

Selection: Is all the information included that viewers need to make a decision?

Emphasis: What subject is focused upon? Is that the subject one dealt with in the document?

Framing: What items are being related or grouped by the way the graphic is framed? What purpose is served by relating these items and presenting them in this manner?

Accuracy: Is factual information accurate? Does the overall impression support or distort accurate perception? (102)

Jan White in *Using Charts and Graphs* illustrates ways in which perception can be distorted, thus causing an unethical presentation of data.

1. Changing the width of the units on the X axis alters the viewer's emotional perception of the data. The graphs in Figure 7.12 plot exactly the same data. Using the middle graph could make the situation seem more urgent than it is.

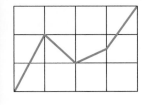 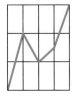 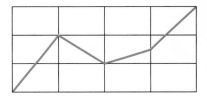

a. Normal　　　　　b. Dramatic rise　　　　　c. Gradual rise

Figure 7.12　Three Graphs That Plot the Same Data

Source: Jan White, Using Charts and Graphs (New York: Bowker, 1984). Reprinted by permission of R. R. Bowker LLC.

(Continued)

2. The nearer a highlighted feature appears, the more impact it has on one's consciousness. The pie charts in Figure 7.13 report the same data. Note that the wedge on the left appears largest and that the wedge on the right is forced into the viewer's consciousness. The middle wedge appears unimportant because it is far away.

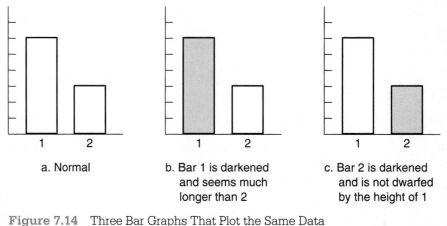

Figure 7.13   Three Pie Charts That Plot the Same Data

Source: Jan White, Using Charts and Graphs (New York: Bowker, 1984). Reprinted by permission of R. R. Bowker LLC.

3. Darker elements seem more important than lighter elements (Figure 7.14). If the real subject in the middle graph is the data represented by the white bar, then the graph is presenting the data unethically.

a. Normal

b. Bar 1 is darkened and seems much longer than 2

c. Bar 2 is darkened and is not dwarfed by the height of 1

Figure 7.14   Three Bar Graphs That Plot the Same Data

Source: Jan White, Using Charts and Graphs (New York: Bowker, 1984). Reprinted by permission of R. R. Bowker LLC.

Globalization and Visual Aids

Visual aids can make your documents easier for foreign audiences to understand and to translate (Porter). In countries where literacy rates are low, pictures, images, and icons are often the best way to convey the intended message (Visual). In addition, images can convey the entire meaning of the document, eliminating the need for words. Consider the Swedish corporation IKEA, which sells household furnishings worldwide. Many of their products, such as tables, have to be assembled by the purchaser. Rather than having an elaborate document in many languages, IKEA uses only drawings with measurements, item counts, process drawings, and exploded views.

Visual aids require careful research on the culture for which you are writing and must be chosen with sensitivity to cultural norms. Certain symbols are broadly understood by Americans, and so we assume that they are accepted internationally. However, these shorthand devices can be meaningless, inappropriate, or even offensive in other cultures. For example, the computer mailbox icon symbolizing e-mail may be confusing in parts of the world, where mailboxes generally look quite different. An envelope is a better way to signify mail or e-mail (Thrush). Pictures of body parts or hand gestures are best avoided, as they often have very different, sometimes even obscene, meanings.

Color is an interesting dimension to add to visual aids, but requires considerable cultural sensitivity.

What seems a normal use of color in one culture is not normal at all in another. For instance, in the United States to "see red" is an indication of anger, but in Russia red is related to the idea of beauty. In countries such as China and India, red is the color of celebration. In the United States, to "feel blue" is to be sad, but in France blue is connected with fear. In the United States, when one is jealous, one is "green-eyed," but in China one has the "red-eye disease" (Bortin).

While white is a color of purity in the United States, in many Asian countries it symbolizes death. Purple is often a polarizing and unsafe color to use.

In Ireland, orange should be avoided, because that color is associated with Protestant England, as opposed to Catholic Ireland.

Other areas of concern include flags and maps. The appropriate orientation of a flag is important to viewers from that country but sometimes difficult to discern for non-natives. A flag presented upside down, for instance, can be insulting. Maps seem like harmless images to Americans, whose borders are very well defined. However in some countries national boundaries, even the names of land areas, are in flux or in contention. Showing the "other side's" map can cause difficulties in the way the audience perceives you ("Writing").

Graphics and text combined must meet audience expectations. Or maybe the purpose of the message. If the purpose is to provoke or create interest, the choice of text and graphics may be placed along the borderline of what is acceptable to the audience. However if the purpose is to create credibility, trustworthiness, and ease among readers, you must actively focus on their accepted cultural norms.

Using Charts

Chart is the catchall name for many kinds of visual aids. Charts represent the organization of something: either something dynamic, such as a process, or something static, such as a corporation. They include such varied types as troubleshooting tables, schematics of electrical systems, diagrams of the sequences of an operation, flow charts, decision charts, and layouts. Use the same techniques to title and number these as you use for graphs. (Some software companies, including Microsoft, use *chart* to mean *graph,* and you will find these terms used interchangeably.)

Troubleshooting Tables

Troubleshooting tables in manuals identify a problem and give its probable cause and cure. Place the problem at the left and the appropriate action at the right. You can also add a column for causes (Figure 7.15). Almost all manuals include these tables.

TABLE 1
Troubleshooting Table

You Notice	This May Mean	Caused by	You Should
Containers do not center with nozzle	Infeed Starwheel is out of time	Misadjustment or loose mounting bolts	Readjust & Retighten
	Machine speed is too fast	Speed not reset during change-over	Reset speed (see "Speed Adjust," p. 5)
	Wrong change parts being used		Install correct change parts (see "Change Parts Data Sheet," p. 27)
Machine vibrates	Lack of maintenance	Tight roller chains (chains that drive the conveyor) or the two chains on the end of the Spiral Screw Feed	Loosen chains
		Roller chains running dry	Lubricate chains
		Cross-beam bearing dry	Lubricate (see Figure 5–148)

Figure 7.15   Troubleshooting Table
Source: RFPC Rotary Filler Manual by Cozzoli Machine Co.

Flow Charts

Flow charts show a time sequence or a decision sequence. Arrows indicate the direction of the action, and symbols represent steps or particular points in the action. In many cases, especially in computer programming, the symbols have special shapes for certain activities. A rectangle signals an action to perform, a diamond signals a decision point, an oval signals the first or last action. The ovals are often eliminated in very short charts. A simple flow chart (Figure 7.16) merely gives the broad outline of a sequence of actions. A decision tree (in many instances also just called a flow chart) uses graphics to explain whether or not to perform a certain action in a certain situation. At each point, the reader must decide *yes* or *no* and then follow the appropriate path until the final goal is reached (Figure 7.17).

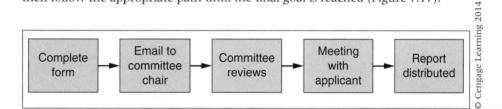

Figure 7.16   Flow Chart

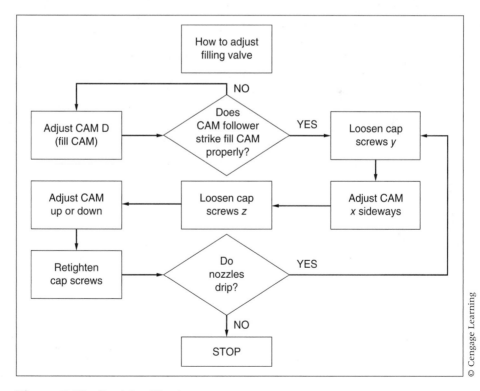

Figure 7.17   Decision Chart

Gantt Charts

A *Gantt chart* (named after its inventor) represents the schedule of a project. Along the horizontal axis are units of time; along the vertical axis are subprocesses of the total project (Figure 7.18). The lines indicate the starting and stopping points of each subprocess.

Layouts

A *layout* is a map of an area seen from the top. As Figure 7.19 shows, layouts can easily show before and after arrangements. Draw simple lines, and use callouts and arrows precisely. Callouts may appear inside the figure.

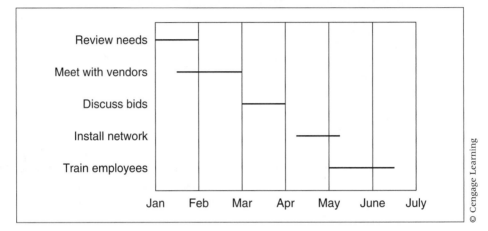

Figure 7.18 Gantt Chart

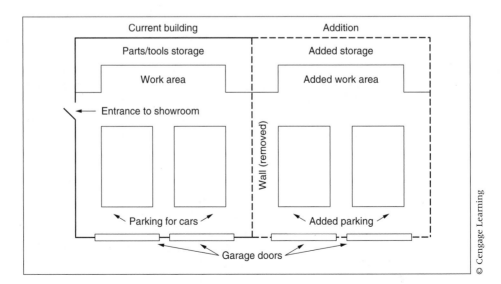

Figure 7.19 Sample Layout

Using Illustrations

Illustrations, usually photographs or drawings of objects, are often used in sets of instructions and manuals.

Guidelines

There are two basic guidelines for using illustrations.

1. Use high-quality illustrations. Make sure they are clear, large enough to be effective, and set off by plenty of white space.

2. Keep the illustrations as simple as possible. Show only items essential to your discussion.

   Use an illustration (Felker et al.)

   ▶ To help explain points in the text
   ▶ To help readers remember a topic
   ▶ To avoid lengthy discussions (A picture of a complex part is generally more helpful than a lengthy description.)
   ▶ To "give the reader permission" (A visual of a computer screen duplicates what is obviously visible before the user, but gives the user permission to believe his or her perception. It reassures the reader.)

Photographs

A good photograph offers several advantages: It is memorable and easy to refer to, it duplicates the item discussed (so audiences can be sure they are looking at what is being discussed), and it shows the relationships among various parts. The disadvantages are that it reduces a three-dimensional reality to two dimensions and that it shows everything, thus emphasizing nothing. Use photographs to provide a general introduction or to orient a reader to the object (Killingsworth and Gilbertson). In manuals, for instance, writers often present a photograph of the object on the first page. Figure 7.20 (p. 202) shows a photograph that clearly indicates not only the part but its relationship to other parts on the machine.

Drawings

Drawings, whether made by computer or by hand, can clearly represent an item and its relationship to other items. Use drawings to eliminate unnecessary details so that your reader can focus on what is important. Two commonly used types of drawings are the exploded view and the detail drawing.

Figure 7.20 Photograph Identifies Part

Exploded View

As the term implies, an *exploded view* shows the parts disconnected but arranged in the order in which they fit together (Figure 7.21). Use exploded drawings to show the internal parts of a small and intricate object or to explain how it is assembled. Manuals and sets of instructions often use exploded drawings with named or numbered parts.

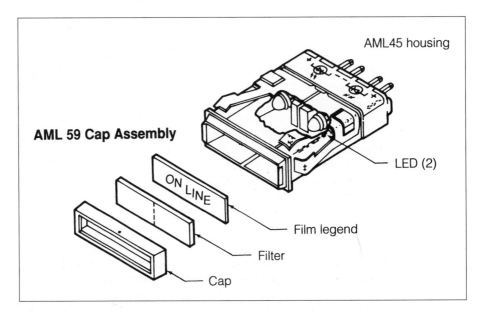

Figure 7.21 Exploded View

Source: John R. Mancosky, Manual Controls: Independent Study Workbook. Used by permission of Microswitch and John R. Mancosky

Detail Drawings

Detail drawings are renditions of particular parts or assemblies. Drawings have two common uses in manuals and sets of instructions.

▶ They function much as an uncluttered, well-focused, cropped photograph, showing just the items that the writer wishes.
▶ They show cross-sections; that is, they can cut the entire assembled object in half, both exterior and interior. (In technical terms, the object is cut at right angles to its axis.) A cross-sectional view shows the size and the relationship of all the parts. Two views of the same object—front and side views, for example—are often juxtaposed to give the reader an additional perspective on the object (Figure 7.22).

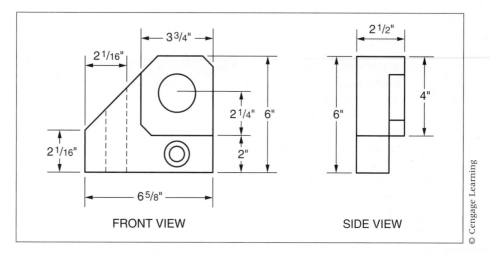

© Cengage Learning

Figure 7.22   Detail Drawing

Worksheet for Visual Aids

☐ **Name the audience for the visual aid.**

☐ **What should your visual aid do?**

☐ **Summarize data? Present an opportunity to explore data? Provide a visual entry point? Engage expectations?**

☐ **Choose a format: Where will you place the caption—above or below?**

(Continued)

☐ Decide on a way to treat this visual's conventional parts. Follow the guideline lists in this chapter.

☐ How large will the visual aid be? Do you have room for it in your document?

☐ How much time do you have? Will you be able to construct a high-quality visual in that time?

☐ Select a method for referring to the visual in your text.

☐ Create a draft of the visual aid. Review it to determine whether you have treated its parts consistently.

☐ Eliminate unnecessary clutter.

Create Helpful Visuals

Suppose readers had to know the exact location of the emergency stop button in order to operate a machine safely. To help them find the button quickly, you decide to include a visual aid. The two examples in Figure 7.23 indicate an imprecise way and a precise way of doing so.

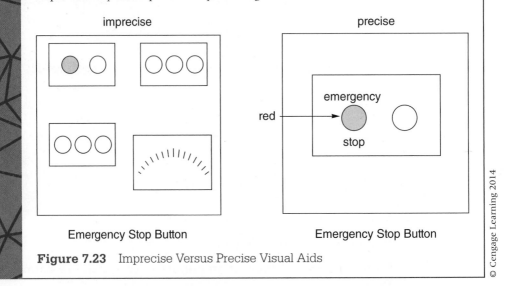

Figure 7.23 Imprecise Versus Precise Visual Aids

© Cengage Learning 2014

Exercises

▶ You Create

1. Collect data from the entire class on an easy-to-research topic, such as type of major, year in school, years with the company (e.g., 1, 2, 3–5, 6–10, 10+), or population of hometown or birthplace (e.g., 100K+, 50–100K, 5–50K). Collect data from the entire class. Then create two visual aids: a bar graph and a pie chart. Make them as neat and complete as possible, including captions and callouts. Write a brief report that states a conclusion you can read by looking at the visual or that tells an audience what to notice in the visual.

   Alternates: (1) Create the visuals in groups of three to four. (2) Create the visuals on a computer. (3) Write a report explaining the process you used to create the visuals. (4) Write an IMRD (see Chapter 10) to explain your project.

2. Use one of the visuals created in Exercise 1 as the basis for a brief story for the student newspaper or company newsletter, explaining the interesting diversity of your technical writing class.

3. Use a line graph to portray a trend (e.g., sales, absentees, accepted suggestions, defect rate) to a supervisor. Write a brief note in which you name the trend, indicate its significance, and offer your idea of its cause.

4. Use the following data to create a table (Jaehn). Monthly from July to December, the following machines had the following percentages of rejection for mechanical quality defects: #41—4.0, 3.1, 3.9, 4.3, 3.5, 3.2; #42—3.0, 3.3, 2.4, 3.2, 3.7, 3.1; #43—3.4, 3.7, 4.1, 4.5, 4.4, 4.8. Include averages for each month and for each machine. Additional: Write a report to a supervisor explaining the importance of the trend you see in the data.

5. If you have access to a computer graphics program, make three different graphs of the data in Exercise 4. In a brief paragraph, explain the type of reader and the situation for which each graph would be appropriate.

6. Divide the class into groups numbered 1–3. Have individuals in each section convert the numbers in either of the following paragraphs into visual aids. Section 1 should make a line graph, section 2 a bar graph, and section 3 a pie chart. Have one person from each section put that group's visual on the board. Discuss their effectiveness. Here are the figures:

   Respondents to a survey were asked whether they would pay more for a tamper-evident package: 8.2% said they would pay up to $.15 more; 25.8%

were unwilling to pay more; 51.6% would pay $.05 more; and 14.4% would pay $.10 more.

Industrial designers report that they use various tools to reduce the time they spend on design. Those tools are faster computers (57%), 3D solid modeling (32%), 2D CAD software (26%), Computerized FEA (24%), Rapid Prototyping (15%), other (3%) (based on Colucci).

7. Convert the following paragraph into a table. Then rewrite the paragraph for your manager, proposing that the company start a recycling program. Refer to specific parts of the table in your paragraph. Alternate: Recommend that the company not start a recycling program. Refer to specific parts of the table.

> The company can recycle 100 pounds of aluminum per week. The current rate for aluminum is 25¢ per pound. This rate would earn the company $25.00 per week and $1200.00 per year. The company can also recycle 200 pounds of paper per week. The current rate for paper is 5¢ per pound. Recycling paper would earn the company $10.00 per week and $520.00 per year. The total earnings from recycling are $1820.00 per year.

8. Read over the next four paragraphs from a report. Then construct a visual aid that supports the writer's conclusion.

> The pipe cutter must be small enough so it can be transported in our truck. The Grip-Tite model takes up 2 cubic feet, and it comes with a stand so that it can be folded out when in use and then folded up when not in use. The Mentzer model takes up 5 cubic feet and doesn't come with a stand.
>
> The pipe cutter must be able to run off 110-volt electricity. Because 220-volt electricity is not easily accessible on the job site, 110-volt must be used. The Grip-Tite model is capable of running off 220-volt, 110-volt, and DC current. The Mentzer cutter is capable only of running off 220-volt.
>
> The cutter should be able to switch threaders to accommodate ⅜" pipe up to 3" pipe. The Grip-Tite model has two threaders. One can be adjusted from ¼" up to 2", and the other can be adjusted from 2" up to 3". The Mentzer model has only one threader that threads ¾" pipe.
>
> The cost of the pipe cutter should not exceed $2000. The Grip-Tite model costs $1750, whereas the Mentzer model costs $2250.

▶ You Revise

9. Redo this table and the paragraph that explains it.

> I have estimated construction costs at a price of $50.00 per square foot as quoted by Lamb and Associates Construction Company (see Table 3). This will include construction of perimeter walls, carpeting, decor, and lighting. The cost of this will be $37,500.

TABLE 3
Renovation Costs   *Main point*

Construction:			$37,500.00
carpeting	15,000		
décor	8,000		
lighting	6,000		
perimeter walls	8,500		
Loose Fixtures:			
#5021 (7 ball Chrome)	15 @ $5.95 each	$ 89.25	
#4596 (4 straight arm)	4 @ $78.00 each	$312.00	
#4597 (2 straight 2 slant)	3 @ $81.00 each	$243.00	
			$644.25
			$38,144.25
Personnel:			
2 full-time stock men 2/8 hour days @ $4.50/hr.			$144.00
Total Renovation Costs:			$38,288.25

▶ You Analyze

10. Discuss the differences between these two table designs. Which one is more effective? Why? Alternate: Create your own table for these data.

TABLE 1
Comparison of Three Trucks

Brand	Purchase Price	3-Year Maintenance Cost	Warranty	Warranty
Big Guy	$11,999	$ 700	2 years	24,000 miles
Friend	13,200	1,000	3 years	30,000 miles
Haul	13,700	850	5 years	50,000 miles

TABLE 1
Comparison of Three Trucks

Brand	Purchase Price in Dollars	3-Year Maintenance Cost	Warranty	
			Years	Miles
Big Guy	$11,999	$ 700	2	24K
Friend	13,200	1000	3	30K
Haul	13,700	850	5	50K

❱ Group

11. Divide into groups of three or four, by major if possible. Select a process you are familiar with from your major or from your campus life. Possibilities include constructing a balance sheet, leveling a tripod, focusing a microscope, constructing an isometric projection, threading a film projector, finding a periodical in the library, and making a business plan. As a group, construct a flow chart of the process. For the next class meeting, each person should write a paragraph explaining the process by referring to the chart. Compare paragraphs within your group; then discuss the results with the class.

Writing Assignments

1. Write a report to your manager to alert him or her to a problem you have discovered. Include a visual aid whose data represent the problem (such as a line graph that shows a "suspect" defect rate).

2. Divide into groups of three. From some external source, acquire data on some general topic such as population, budget, production, or volume of sales. Sources could be your Chamber of Commerce, a government agency, an office in your college, or your corporation's human resources office. As a group, decide on a significant trend that the data show. Write a report to the appropriate authority, informing him or her of this trend and suggesting its significance. Use a visual aid to convey your point.

3. As a class, agree on a situation in which you will present information to alert someone to a trend or problem. Then divide into several groups. Each group should select one of the four reasons for using a visual aid (pp. 180–181). Write a report and create the visual aid that illustrates the reason your group chose it (e.g., to provide a different entry point). Make copies for each of the other groups. When you are finished, circulate your reports and discuss them.

4. Use a flow chart and an accompanying report to explain a problem with a process—for example, a bottleneck or a step where a document must go to two places simultaneously. Your instructor may ask you to go on to Writing Assignment 5.

5. Use Writing Assignment 4 as a basis to suggest a solution to a problem. Create a new flow chart that illustrates how your solution solves the problem. Discuss the solution in a report to your manager requesting permission to put it into effect.

6. Write a learning report for the writing assignment you just completed. See Chapter 5, Writing Assignment 8, page 137, for details of the assignment.

Web Exercise

Write a brief report or an IMRD (see Chapter 10) to your classmates in which you analyze the use of visual aids in websites. Your goal is to give hints on how to incorporate visuals effectively into their Web documents. Do the visual aids exemplify one of the four uses of visual aids explained in this chapter? Explain why the visual aid is effective in the Web document.

Works Cited

Adkins, Jill. *Rotary Piston Filler: Eight Head.* Menomonie, WI: MRM/Elgin, 1985. Print.

Allen, Nancy. "Ethics and Visual Rhetoric: Seeing's Not Believing Anymore." *Technical Communication Quarterly* 5 (1996): 87–105. Print.

Bortin, Meg. "In a Word: When Colors Take on Different Cultural Hues." *New York Times Archive.* 28 Sept. 2002. Web. 20 May 2012. <http://www. nytimes. com/2002/09/28/news/28iht-rinaword_ed3_.html>.

Brumberger, Eva. "Making the Strange Familiar: A Pedagogical Exploration of Visual Thinking." *Journal of Business and Technical Communication* 21 (October 2007): 376–401. Print.

Colucci, D. "How to Design in Warp Speed." *Design News* (1996): 64–76. Print.

Felker, Daniel B., Francis Pickering, Veda R. Charrow, V. Melissa Holland, and James C. Redish. *Guidelines for Document Designers.* Washington, DC: American Institutes for Research, 1981. Print. This volume was used extensively in the preparation of this chapter.

Jaehn, Alfred H. "How to Effectively Communicate with Data Tables." *Tappi Journal* 70 (1987): 183–184. Print.

Killingsworth, M. Jimmie, and Michael Gilbertson. "How Can Text and Graphics Be Integrated Effectively?" *Solving Problems in Technical Writing.* Ed. Lynn Beene and Peter White. New York: Oxford University Press, 1988. 130–149. Print.

MacDonald-Ross, M. "Graphics in Texts." *Review of Research in Education.* Vol. 5. Ed. L. S. Shulman. Itasca, IL: F. E. Peacock, 1978. Print.

Mancosky, John R. *Manual Controls: Independent Study Workbook.* Freeport, IL: Microswitch, 1991. Print.

Porter, Alan J. "The Global Language: Using Symbols and Icons When Delivering Technical Content." *intercom.* December 2010. Web. 12 May 2012. <http://intdev.stc.org/2011/01/the-global-language-using-symbols-and-icons-when-delivering-technical-content/>.

Publication Manual of the American Psychological Association. 4th ed. Washington, DC: APA, 1994. Print.

Schriver, Karen A. *Dynamics in Document Design: Creating Texts for Readers.* New York: Wiley, 1997. Print.

Tans, Pieter, and Ralph Keeling. "Trends in Atmospheric Carbon Dioxide." 2012. Web. 1 June 2012. <http://www.esrl.noaa.gov/gmd/ccgg/trends/>.

Thrush, Emily. "Writing for an International Audience." Suite101.com. 2000. Web. 18 May 2012. <http://archive.suite101.com/article.cfm/communications_skills/32233>.

Tufte, Edward R. *The Visual Display of Quantitative Information.* Cheshire, CT: Graphics Press, 1983. Print.

"Visual Communication in Global Health Online Course: Module 4 Creating Effective Visual Literacy Tools."Unite for Sight.2000-2011. Web. September 2012. <http://www.uniteforsight.org/visual-literacy/module4>.

White, Jan. *Graphic Design for the Electronic Age.* New York: Watson-Guptill, 1988. Print.

———. *Using Charts and Graphs.* New York: Bowker, 1984. Print.

"Writing for an international audience." IBM. 2012. Web. 2 June 2012. <http://www-01.ibm.com/software/globalization/topics/writing/culture.html>.

8

Describing

CHAPTER CONTENTS

CHAPTER 8 **IN A NUTSHELL**

Description orients readers to objects and processes. Your goal is to make the readers feel in control of the subject.

▸ Show readers how this subject fits into a larger context that is important to them.

▸ Tell them what you are going to say; write this kind of document in a top-down manner.

▸ Choose clear heads; write in manageable chunks; define terms sensibly; use visual aids that helpfully communicate.

▸ Choose a tone that enables readers to see you as a guide.

Mechanism descriptions. In the *introduction*

▸ Define the mechanism and tell its purpose.

▸ Provide an overall description.

▸ List the main parts.

For each *main part*

▸ Define the mechanism and describe it in terms of size, shape, material, location, color.

▸ List and describe any subparts.

Process descriptions. In the *introduction* you need to

▸ Tell the goal and significance of the process.

▸ Explain principles of operation.

▸ List the major sequences.

For each sequence or step

▸ Tell the end goal of the process.

▸ Describe the action in terms of qualities and quantities.

Description is widely used in technical writing. Many reports require that you describe something—a machine, process, or system. Sometimes you will describe in intricate detail, other times in broad outline. This chapter shows you how to describe a mechanism, an operation, and a process focused on a person in action.

Planning the Mechanism Description

The goal of a mechanism description is to make the readers confident that they have all the information they need about the mechanism. Obviously, you can't describe every part in minute detail, so you select various key parts and their functions. When you plan a description of a mechanism, consider the audience, select an organizational principle, choose visual aids, and follow or adopt the usual form for writing descriptions.

Consider the Audience

To make the audience feel confident, consider their knowledge level and why they need the information. Basically, the principle is to give them the physical details that they need to act. The details you choose and the amount of definition you provide reflect your understanding of their knowledge and need. Here is a brief, simple description for an audience that must make a decision about a topic easily understood by most people. The author can safely assume that length and width terms need no definition.

> The truck box size is an important factor because we frequently transport 4 ft by 8 ft sheets of wood. The box size of the Hauler at the floor is 3.5 ft by 6 ft. The box size of the X-200 at the floor is 4 ft by 8 ft. This factor means we should purchase the X-200.

Here, however, is another brief description, also the basis for a decision, directed at an audience that has a specialized technical knowledge. The writer here assumes that the audience understands terms like "dpi resolution." If the writer assumed the audience did not understand the term, he or she would have to enlarge the discussion to define all the terms.

> The ABC scanner has 4800 dpi resolution, 48-bit color scanning, and a built-in transparency unit. The XYZ scanner has 6400 dpi resolution, 48-bit color, and a built-in transparency unit. The higher resolution capabilities make XYZ the preferable purchase to fill our needs.

Select an Organizational Principle

You can choose from several organizational principles. For instance, you can describe an object from

▶ Left to right (or right to left).
▶ Top to bottom (or bottom to top).
▶ Outside to inside (or inside to outside).
▶ Most important to least important (or least important to most important).

Base this decision on your audience's need. For a general introduction, a simple sequence like top to bottom is best. For future action, say, to decide whether or not to accept a recommendation, use most to least important.

An easy way to check the effectiveness of your principle of organization is to look for *backtracking*. Your description should move steadily forward, starting with basic definitions or concepts that the audience needs to understand later statements. If your description is full of sections in which you have to stop and backtrack to define terms or concepts, your sequence is probably inappropriate.

Choose Visual Aids

Use visual aids to enhance your description of a mechanism. As the figure of a paper micrometer (Figure 1, p. 214) demonstrates, overviews show all the parts in relationship. Details focus readers on specific aspects. Often a visual aid of a detail can dramatically shorten a text discussion. Consider the followiing brief discussion of a problem with broken piping joints. It would be much longer and difficult to comprehend if it were all text.

Figure 1a shows a typical cross-section view of a copper-to-copper tube joint soldered together. Notice how the solder covers up the entire opening between the two tubes. Figure 1b shows the break in the solder due to a change in temperature.

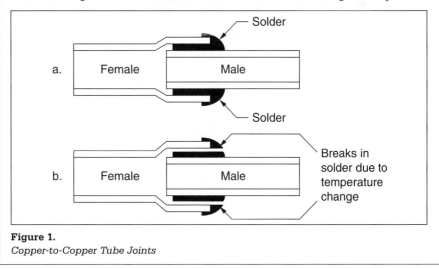

Figure 1.
Copper-to-Copper Tube Joints

© Cengage Learning

Often your visual aid focuses the text. In effect, your words describe the visual aid.

Follow the Usual Form for Descriptions

Generally, descriptions do not stand alone but are part of a larger document. However, they still have an introduction and body sections; conclusions are optional. Use conclusions only if you need to point out significance. Make the introduction brief, stating either your goal for the reader or the purpose of the mechanism. To describe a part, point out whatever is necessary about relevant physical details—size, shape, material, weight, relationship to other parts, or method of connection to other parts. If necessary, use analogies and statements of significance to help your reader understand the part.

Writing the Mechanism Description

A stand-alone mechanism description has a brief introduction, a description of each part, and an optional conclusion.

Introduction

The introduction gives the reader a framework for understanding the mechanism. In the introduction, define the mechanism, state its purpose, present an overall description, and preview the main parts.

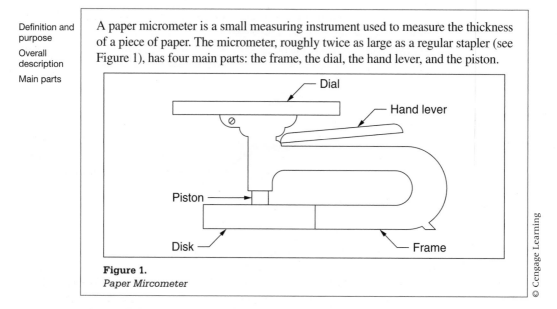

Definition and purpose
Overall description
Main parts

A paper micrometer is a small measuring instrument used to measure the thickness of a piece of paper. The micrometer, roughly twice as large as a regular stapler (see Figure 1), has four main parts: the frame, the dial, the hand lever, and the piston.

Figure 1.
Paper Mircometer

© Cengage Learning

Figure 8.1   Sample Introduction of Mechanism Description

TIP ▏▏▏▏

Mechanism Description

A quick way to plan a mechanism description is to use this outline:

I. Introduction
   A. Definition and Purpose
   B. Overall Description (size, weight, shape, material)
   C. Main Parts

II. Description
   A. Main Part A (definition followed by detailed description of size, shape, material, location, method of attachment)
   B. Main Part B (definition followed by overall description, and then identification of subparts)
      1. Subpart X (definition followed by detailed description of size, shape, material, location, method of attachment)
      2. Subpart Y (same as for X)
   C. Other Main Parts
   D. Etc.

Body: Description of Mechanism

The *body* of the description contains the details. Identify each main part with a heading and then describe it. In a complex description like the following one, begin the paragraph with a definition, then add details. Use coordinate structure (see pp. 71–72) for each section. If you put size first in one section, do so in all of them.

Definitions	**THE FRAME**
	The frame of the paper micrometer is a cast piece of steel that provides a surface to which all the other parts are attached. The frame, painted gray, looks
Color and analogy	like the letter *C* with a large flat disk on the bottom and a round calibrated
Size and analogy	dial on top. The disk is 4½ inches in diameter and resembles a flat hockey puck. The frame is 5⅛ inches high and 7½ inches long. Excluding the bottom disk, the frame is approximately 1¼ inches wide. The micrometer weighs
Weight	8 pounds.
	THE DIAL
Definition and analogy	The dial shows the thickness of the paper. The dial looks like a watch dial except that it has only one moving hand. The frame around the dial is made of chrome-plated

(Continued)

metal. A piece of glass protects the face of the dial in the same way that the glass crystal on a watch protects the face and hands. The dial, 6 inches in diameter and ⅞ inch thick, is calibrated in .001-inch marks, and the face of the dial is numbered every .010 inch. The hand is made from a thin, stiff metal rod, pointed on the end.

Analogy size

Appearance

THE HAND LEVER

The hand lever, shaped like a handle on a pair of pliers, raises and lowers the piston. It is made of chrome-plated steel and attaches to the frame near the base of the dial. The hand lever is 4 inches long, ½ inch wide, and ¼ inch thick. When the hand lever is depressed, the piston moves up, and the hand on the dial rotates. When the hand lever is released and a piece of paper is positioned under the piston, the dial shows the thickness of the paper.

Analogy and definition

Relationship to other parts

Effect

THE PISTON

The piston moves up and down when the operator depresses and releases the hand lever. This action causes the paper's thickness to register on the dial. The piston is ⅜ inch in diameter, flat on the bottom, and made of metal without a finish. The piston slides in a hole in the frame. The piston can measure the thickness of paper up to .300 inch.

Definition

Function

Size

Relationship to other parts

Figure 8.2 Body of Mechanism Description

Other Patterns for Mechanism Descriptions

Two other patterns are useful for describing mechanisms: the function method and the generalized method.

The Function Method

One common way to describe a machine is to name its main parts and then give only a brief discussion of the function of each part. The *function method* is used extensively in manuals. The following paragraph is an example of a function paragraph:

FUNCTION BUTTONS

The four function buttons, located under the liquid crystal display, work in conjunction with the function switches. The four switches are hertz (Hz), decibels (dB), continuity (c), and relative (REL).

List of subparts

The hertz function allows you to measure the frequency of the input signal. Press the button a second time to disable. The decibel function allows you to measure

Function and size of subpart 1

Function and size of subpart 2

the intensity of the input signal, which is valuable for measuring audio signals. It functions the same way as the hertz button.

Function and size of subpart 3

The continuity function allows you to turn on a visible bar on the display, turn on an audible continuity signal, or disable both of them. The relative function enables you to store a value as a reference value. For example, say you have a value of 1.00 volt stored; every signal that you measure with this value will have 1.00 volt subtracted from it.

Function and size of subpart 4

© Cengage Learning

Figure 8.3 Sample Function Method Description

The Generalized Method

The *generalized method* does not focus on a part-by-part description; instead, the writer conveys many facts about the machine. This method of describing is commonly found in technical journals and reports. With the generalized method, writers use the following outline (Jordan):

1. General detail

2. Physical description

3. Details of function

4. Other details

General detail consists of a definition and a basic statement of the operational principle. *Physical description* explains such items as shape, size, appearance, and characteristics (weight, hardness, chemical properties, methods of assembly or construction). *Details of function* explain these features of the mechanism:

▶ How it works, or its operational principle
▶ Its applications
▶ How well and how efficiently it works
▶ Special constraints, such as conditions in the environment
▶ How it is controlled
▶ How long it performs before it needs service

Other details include information about background, information about marketing, and general information, such as who makes it.

Here are two sample general descriptions:

Low weight, portability and ease of use are not words and phrases used to describe the average mid-size tripod, but the Tamrac ZipShot is designed to satisfy all three requirements. Unlike conventional 'telescoping' leg tripods, the ZipShot is constructed of thin, circular aluminum legs that are segmented in four places and strung together via a hefty elastic band. When the two maroon safety bands are released, the tripod's legs will fall, snapping into place in an instant. The Tamrac ZipShot is an odd piece of equipment by traditional standards, but at a street price of around $50, it might be the ideal solution for those who like to pack lightly.

Undoubtedly, the ZipShot's key feature is its simplicity. In just over five seconds the tripod can be fully deployed, awaiting a camera to be mounted to its omni-directional ball head. Tamrac offers a Quick Release accessory kit for an additional $10 that comes with attachments for a point-and-shoot and a DSLR, enabling photographers to snap a camera into place within seconds. With the Quick Release system, I was able to setup the ZipShot with a point-and-shoot mounted in under 10 seconds.

As for bulk and dimensions, the Tamrac ZipShot weighs 11 oz. and measures 15" in length when folded down for storage. The tripod can easily be tossed into a backpack. When the ZipShot is fully deployed, it offers a shooting height of 44" (1.1m). This is a decent height to shoot from, but keep in mind that the legs do not telescope, so it's fixed. The ZipShot can also be used as a monopod by bundling the legs together with the maroon safety straps, which increases total elevation by an additional four inches.

Information courtesy of the Internet Movie Database (IMDb).

http://www.dpreview.com/articles/6145161331/accessory-review-tamrac-zipshot-tripod. Used with Permission.

Planning the Process Description

Technical writers often describe processes such as methods of testing or evaluating, methods of installing, flow of material through a plant, the schedule for implementing a proposal, and the method for calculating depreciation. Manuals and reports contain many examples of process descriptions. Policy documents often include concise statements of what happens in each step of a process, such as admitting a patient, or filing a grievance.

As with a mechanism description, the writer must consider the audience, select an organizational principle, choose visual aids, and follow the usual form for writing descriptions.

Consider the Audience

Your goal is to make your audience confident that they have all the information they need about the process. The knowledge level of audiences and their potential use of the document will vary. Their knowledge level can range from advanced to beginner. Uses will vary; often they use the description to make a decision. For instance, a plant engineer might propose a change in material flow in a plant because a certain step is inefficient, causing a bottleneck. To get the change approved, he or she would have to describe the old and new processes to a manager, who would use that description to decide whether to implement the new process.

Process descriptions also explain theory, thus answering the audience's need for a background understanding. The writer can describe how a sequence of actions has a cause–effect relationship, thus allowing the reader to understand where the trouble might be in a machine or what the significance of an action might be. Here is a brief process description that allows a reader to analyze his or her own leaking faucet:

When the hot- and cold-water handles of the stem faucets are turned on, the rotating stems ride upward on their threads. As they rise, the stems draw the washers away from the brass rings, called faucet seats, at the tops of the water supply lines, allowing water to flow. When both hot and cold water flow through faucets, they mix in the faucet body and run from the spout as warm water. When the handles are turned to the off position, the stems ride downward on their threads. The washers press against the faucet seats, shutting off the flow of water (*How* 71).

Process descriptions can also present methodology, the steps a person took to complete a project or solve a problem. These statements often reassure the reader by showing that the writer has done the project the "right" way. Here is a brief methodology statement.

I started the project by sending out for price quotes on the 21 modular containers. I chose three vendors from the Phoenix area: Box Company, Johnson Packaging, and Packages R Us. I chose the first two based on their past services for the JCN-Tucson plant. I chose Packages R Us because they are the national vendor for the Modular program for the entire JCN Corporation. From these vendors I asked for two price quotes: one based on a just-in-time (JIT) inventory system, and the other based on the existing inventory system.

Select an Organizational Principle

The organizational principle for processes is *chronological:* Start with the first action or step, and continue in order until the last. Also consider whether you need to use cause-effect in the arrangement. Many processes have obvious sequences of steps, but others require careful examination to determine the most logical sequence. If you were describing the fashion cycle, you could easily determine its four parts (introduction, rise, peak, and decline). If you had to describe the complex flow of material through a plant, however, you would want to base your sequence of steps on your audience's knowledge level and intended use of the description. You might treat "receiving" as just one step, or you might break it into "unloading," "sampling," and "accepting." Your decision depends on how much your audience needs to know.

Choose Visual Aids

Choose a visual aid that orients your reader to the process, either to see the entire process at a glance or to see the working of one step. If your subject is a machine in operation, visuals of the machine in different positions will clarify the process. If you are describing a process that involves people, a flow chart can quickly clarify a sequence.

Follow the Usual Form for Writing Descriptions

The process description takes the same form as the mechanism description: a brief introduction, which gives an overview, and the body, which treats each step in detail, usually one step to a section. Make the introduction brief, either a statement of your goal or the purpose of the process. Use conclusions only if you need to point out significance. In each paragraph, first define the step (often in terms of its goal or end product), and then describe it. Use coordinate structure (see pp. 71–72) for each section. If you follow a definition of an end goal with a brief description of the machine and then the action in one section, do so in all sections.

Define a step's end goal or purpose, and then describe the actions that occur during that step. Point out qualities like "fast" and quantities like "60 times." Add statements of significance if you need to.

Writing the Process Description

The outline in the Tip on page 224 shows the usual form for a description of a process. Other examples appear in the Examples section (pp. 229–238). Include an introduction, a description of each step, and an optional conclusion.

Introduction

The introduction provides a context for the reader. Define the process, explain its principles of operation (if necessary), and preview the major sequences. The following introduction performs all three tasks:

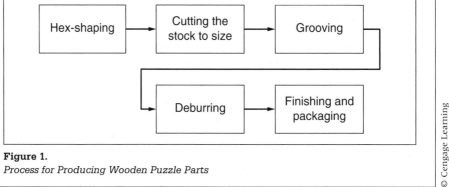

Title

Author

Background

Common examples

Preview of major sequences

Processes involved in producing wooden puzzle in automated manufacturing cell

By Brad Tanck, First Shift Floor Supervisor

This report explains the process of continuous production of wooden puzzles by an automated manufacturing cell. The wooden puzzle consists of nine hex-shaped parts arranged in a spherical configuration when fully assembled. The automated manufacturing consists of five processes: hex-shaping the stock, cutting the stock to size, grooving the parts, deburring the parts, and finishing and packaging the parts (see Figure 1).

Figure 1.
Process for Producing Wooden Puzzle Parts

© Cengage Learning

Figure 8.4   Introduction of Process Description

Body: Description of the Operation

In the body of the paper, write one paragraph for each step of the process. Each paragraph should begin with a general statement about the end goal or main activity. Then the remainder explains in more detail the action necessary to achieve that goal. In the following example, each paragraph starts with an overview, and all the paragraphs are constructed in the same pattern. The flow chart gives readers an overview before they begin to read.

End goal	**HEX-SHAPING THE STOCK** The hex-shaping process entails running 30-inch lengths of ¾-inch-square maple stock through a set of shapers to cut the stock into a hex shape. The maple
Action Action	stock lengths are hand-placed in a gravity hopper, where they drop down onto a moving conveyor system. From here, the conveyor transports the stock until it reaches a set of push rollers, one on each side of the stock. Friction from the push rollers forces the stock through a set of diamond-shaped shaper bits, which are offset at 60-degree angles to one another in order to shape the square stock into perfect hex stock. From here, the hex stock rides on a conveyor belt to the next station.
End goal	**CUTTING THE STOCK TO SIZE** The purpose of cutting the stock to size is to get nine parts, all three inches in length. When the hex stock reaches the cut-to-size station, it trips a limit switch.
Action	This switch causes a stop to eject and prevents the stock from further advancing. At this point, a pneumatic circular saw is activated, which cuts a three-inch part off the hex stock.
Action	When this process is completed, the stop is retracted and the stock moves onward, pushing the cut-to-size part down a chute to the next station and signaling the stop in preparation for cutting another part from the stock.
End goal	**GROOVING THE PARTS** In order for the puzzle parts to fit together, several different grooves must be cut into the parts at various angles. Of the nine parts that are processed,
Background	three parts have two grooves in them, three parts have one groove in them, and three parts have three grooves in them.
Action Background	After the part arrives at the grooving station, it is grasped by a set of pinch rollers. The pinch rollers feed the part into a hex-shaped aluminum collet mounted on an x-y table. This collet is specially designed with two slots cut out of the top of it, both at 70-degree angles, and one slot cut out of its bottom at a 120-degree

(Continued)

Action

angle. The table moves the part and collet through a set of two routers, one on top and one on the bottom, which cut the required grooves in the part through the slots in the collet.

Action

When this grooving process is completed, the pressure on the collet is released. The ram of a pneumatic cylinder pushes the part onto an outfeed chute, where it travels to the next station.

DEBURRING THE PARTS

End goal

Deburring the parts includes removing any burrs and shavings from the part and cleaning up its surface before it is finished.

Action

Action

When the part arrives at the deburring station, it lands in a gravity hopper at the head of the station. The orientation of the part is changed at this time. As the part comes through the hopper, it is handled side for side instead of end for end.

Action

After the part drops down onto the stationary feed table, pressure from a small pneumatic cylinder secures it to the back edge of the moving feed table. At this time, a larger secondary pneumatic cylinder is actuated. This cylinder pushes the attached moving feed table with the part through a set of abrasive wheels, one above the part and one below it. These wheels take the burrs off the part and clean it up for finishing.

Action

At the conclusion of its stroke, the larger cylinder pauses and the smaller cylinder retracts, releasing the part to the next station. The larger cylinder then extends in order to deburr the next part.

FINISHING AND PACKAGING THE PARTS

End goal

The finishing and packaging process includes spraying a lacquer finish on the part, drying the finish, and packaging the puzzle parts in a sealed plastic carton.

Action

Background

As the part drops from the deburring station, it lands on a moving conveyor system. The conveyor system's ¾-inch pins protruding from the belt keep the part from coming in contact with the belt itself. The conveyor system transports the part through a spray booth, where a lacquer finish is applied. From here the conveyor keeps moving through a drying booth, where high-intensity heat is blown on the part to dry it before packaging.

Action

Finally, the part drops off the end of the conveyor, activating a photo-electric switch, which counts the part, and a pick-and-place robot. The pick-and-place robot takes the part and places it in a plastic carton.

Action

When all nine puzzle parts are placed in the carton, a plastic wrap is placed over the finished product and it is sealed by an electronic heat seal device.

Figure 8.5 Process Description Body

Conclusion

Conclusions to brief descriptions of operation are optional. At times, writers follow the description with a discussion of the advantages and disadvantages of the process or with a brief summary. If you have written a relatively brief, well-constructed description, you do not need a summary.

Planning the Description of a Human System

A human system is a sequence of actions in which one or more people act in specified ways. If you describe in general what happens in such a system, you have a process description. But if you tell people how to act, you have a procedure or a set of instructions. The description of the product tester selection policy in the next section describes actions that the members of the group usually take on. A set of instructions, by contrast, would tell a person exactly what to do, in detail, for each step.

You plan for such a document as for a regular process description (see pp. 218–220). Like the process description shown earlier (pp. 220–222), this document contains an introduction, which defines the process and its major sequences, and the body, which describes the process in detail.

Writing the Description of a Human System

The following section shows the usual form for writing a description of a human system. Include an introduction and a body. The conclusion is optional.

Introduction

In the introduction, orient the reader to the process. The following introduction states the purpose of the document, explains why the reader needs the information, and lists the major steps in the process (eBay Account Registry).

TIP ||||

Process Description

To create a process description quickly, follow this outline:

I. Introduction
   A. Define process
   B. Explain principles of operation or give common examples
   C. Preview main steps in the process

II. Describe process
   A. Main Step One
     1. Define the step's goal
     2. Add necessary background material
     3. Present details of action
   B. Main steps
   C. Etc.

PRODUCT TESTER SELECTION POLICY

REGISTERING FOR AN eBay ACCOUNT
eBay has become one of the most popular ways to purchase and sell new and used items over the Internet. The wide variety of product categories means that there's something for everyone. Plus, it's easy to set up an account and get started. If you've ever considered buying and selling on eBay, these instructions will help you get started.

eBay ACCOUNT REGISTRATION PROCESS
This document explains the eBay registration process, so you can set up your account. Unless you follow these steps, you will not be allowed to buy or sell items on eBay. If you are interested in getting your eBay account up and running, you should read this document carefully. The three steps involved in the process are:

1. Navigate to the eBay Registration Page
2. Fill out the Registration Form
3. Verify Your Account

© Cengage Learning 2014

Figure 8.6   Introduction to Description of Human System

Body: Sequence of a Person's Activities

In the body, describe each step in sequence. Present as much detail as necessary about the quality and quantity of the actions (eBay Account Registration).

FILL OUT THE REGISTRATION FORM

To continue the registration process, fill out the required form. All fields must be filled out, or you will not be allowed to create an account. You will begin by providing your contact information including:

- name
- address
- zip code (which will auto populate your city and state)
- e-mail address
- telephone number

Next, you will create a user ID and password. You will be asked to create a user name which will be displayed when you are transacting with others. You'll also need to create a password and reenters the password in order to authenticate it. To help you if you forget your password, you will be provided with several "secret questions" to choose from and will be asked to provide an answer to the question. This safeguard protects your account by making sure only you can access your account if there is a problem with your password. You will also be asked to provide your date of birth, to verify that you are old enough to participate in transactions.

Please read the User Agreement and Privacy Policy, and then select "Submit."

VERIFY YOUR ACCOUNT

After you've submitted your registration, you will need to verify the creation of your account. To do this, you will start by checking your e-mail at the address you provided. eBay makes this easy by providing a big blue "Check your e-mail" button, which you will select. You will be taken to your e-mail account.

Next, you will open your e-mail inbox and click on the e-mail from eBay. On the screen you will see a message from eBay and a blue "Confirm" button. To confirm your eBay registration, and activate your account, select the "Confirm" button.

After verification, you will be taken back to eBay, where you can choose to learn more about eBay by following the "New to eBay?" link. Otherwise you can begin putting your items up for sale by clicking the "Start Selling" link, or manage your buying and selling by clicking "Visit My eBay."

© Cengage Learning 2014

Figure 8.7 Body of Human System Description

Conclusion (Optional)

A conclusion is optional. If you choose to include one, you might discuss a number of topics, depending on the audience's needs, including the advantages and disadvantages of the process.

Worksheet for Planning a Description

☐ **Name the audience for this description.**

☐ **Estimate the level of their knowledge about the concepts on which this description is based and about the topic itself.**

☐ **Name your goal for your readers.**

Should readers know the parts or steps in detail or in broad outline?
Should they focus on the components of each step or part, or on the effect or significance of each step or part?
Should they focus just on the machine or process, or grasp the broader context of the topic (such as who uses it, where and how it is regulated, who makes it, and its applications and advantages)?

☐ **Select an approach.**

What will you do first in each paragraph?
In what sequence will you present the explanatory detail?

☐ **Name the audience for this description.**

☐ **Estimate the level of their knowledge about the concepts on which this description is based and about the topic itself.**

☐ **Name your goal for your readers.**

Should readers know the parts or steps in detail or in broad outline?
Should they focus on the components of each step or part, or on the effect or significance of each step or part?
Should they focus just on the machine or process, or grasp the broader context of the topic (such as who uses it, where and how it is regulated, who makes it, and its applications and advantages)?

☐ **Select an approach.**

What will you do first in each paragraph?
In what sequence will you present the explanatory detail?

☐ **Plan a visual aid.**

What is your goal with the visual aid? To provide a realistic introduction? To give an overview? To be the focus of the text? To supplement the text?

Will you have one visual aid for each step or part, or will you use just one visual and refer to it often?

☐ **Choose the type of visual aid. Use a visual that will help your reader grasp the topic.**

☐ **Decide on the visual aid's size (not too large or too small) and placement (for example, after the introduction).**

☐ **Construct a rough visual now. Finish it later.**

☐ **Devise a style sheet. Decide how you will handle heads, margins, paragraphing, and visual aid captions.**

☐ **To write a description of mechanisms**

Name each part.
Name each subpart.
Define each part and subpart.
List details of size, weight, method of attachment, and so forth.
Tell its function.

☐ **To write a description of processes**

Name each step.
Name each substep.
Tell its end goal.
List details of quality and quantity of the action.
Tell significance of action.

Worksheet for Evaluating a Description

1. For each part of the report—introduction, body, visual aid—answer the following questions:

   • Is it appropriate to the goal for the audience?
   • Is it consistent with all the other parts and its own subparts?
   • Is it clear to the audience and faithful to the reality?

2. Do the visual aid and the text helpfully, actively interact with each other?

3. Does the report build a mental model that an audience can use for future action?

(Continued)

4. Does the writer convince you that he or she is believable? (Consider two dimensions: statement of background and method of presentation.)

5. Check these items for a **mechanism description.**

Introduction

- Purpose statement is present and clear.
- Cause of your writing is present and clear.
- Preview of the paper is present.
- Topic is named and defined.
- Each item in list is a thing (not an action).

Body

- Each new section has effective use of keyword.
- Either first or second sentence defines each part.
- Any necessary background is given.
- Subparts are listed early in the paragraph.
- Each part has enough details (size, shape, material, location, method of attainment).

Format

- Heads have consistent format.
- Visual aid is clearly drawn.
- Visual aid has clear, correct caption at the bottom.
- Visual aid has callouts.
- Callouts are keywords in text.

Style

- Sentences use active voice if possible.
- All technical and jargon terms are defined.
- No spelling or grammar mistakes remain.

6. Check these items for a **process description.**

Introduction

- Purpose statement is present and clear.
- Cause of your writing is present and clear.
- Preview of the paper is present.
- Topic is named and defined.
- Each item in list is an action (not a thing).

Body

- Each new section has effective use of keyword.
- Either first or second sentence defines each step (check "no" if any of the steps are not defined).
- Any necessary background is given.
- End result or overall goal of each step is explained.
- Each step has enough details (specific actions, substeps, quantity and quality of actions)

Format

- Heads have consistent format.
- Visual aid is clearly drawn.
- Visual aid has clear, correct caption at the bottom.
- Visual aid has callouts.
- Callouts are keywords in text.

Style

- Sentences use active voice if possible.
- All technical and jargon terms are defined.
- No spelling or grammar mistakes remain.

Examples

Examples 8.1 to 8.6 describe a mechanism and four processes.

Example 8.1

Two Samples of Generalized Description

A business process is a set of logically related business activities that combine to deliver something of value (e.g. products, goods, services or information) to a customer.

A typical high-level business process, such as "Develop market" or "Sell to customer," describes the means by which the organization provides value to its customers, without regard to the individual functional departments (e.g., the accounting department) that might be involved.

As a result, business processes represent an alternative—and in many ways more powerful—way of looking at an organization and what it does than the traditional departmental or functional view.

Business processes can be seen individually, as discrete steps in a business cycle, or collectively as the set of activities that create the value chain of an organization and associate that value chain with the requirements of the customer.

It is important to recognize that the "customer" of a business process can be several different things, according to the process's position in the business cycle. For example, the customer of one process could be the next process in the cycle (in which case the output from one process is input to the next, "customer" process). Equally, the customer can be the end purchaser of a product (Cousins and Stewart).

Example 8.2

Description of
a Mechanism

SKINFOLD CALIPER

The following information explains the skinfold caliper and its individual parts. The skinfold caliper (see Figure 1) is an instrument used to measure a double layer of skin and subcutaneous fat (fat below the skin) at a specific body site. The measurement that results is an indirect estimate of body fatness or calorie stores. The instrument is approximately ten inches long, is made of stainless steel, and is easily held in one hand. The skinfold caliper consists of the following parts: caliper jaws, press and handle, and gauge.

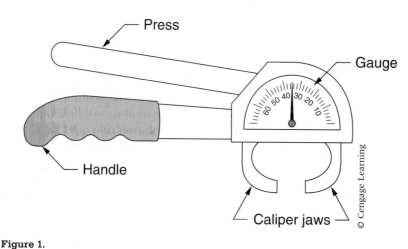

Figure 1.
Skinfold Caliper

CALIPER JAWS

The caliper jaws consist of two curved prongs. Each prong is approximately ¼ inch long. The prongs project out from the half-moon-shaped gauge housing. They are placed over the skinfold when the measurement is taken. They clasp the portion of the skinfold to be measured.

PRESS AND HANDLE

The press is the lever that controls the caliper jaws. Engaging the press opens the caliper jaws so they can slip over the skinfold. Releasing the press closes the jaws on the skinfold, allowing the actual measurement. The press is 4.5 inches long and 0.5 inch thick. It is manipulated by the thumb while the fingers grip the caliper handle. The caliper handle is 6 inches long and 0.5 inch thick. The outside edge of the handle has three indentations, which make the caliper easier to grip.

GAUGE

The gauge records the skinfold measurement. It is white, half-moon shaped, with 65 evenly spaced black markings and a pointer. Each marking represents 1 centimeter. The pointer projects from the middle of the straight edge of the half-moon-shaped gauge to the black markings. When the jaws tighten, the pointer swings to the marking that is the skinfold thickness.

Example 8.3

Description of a Process

PROCESS DESCRIPTION: IDENTIFICATION OF UNKNOWN CHEMICALS

This report is to familiarize you with one of the most common procedures done in the research laboratory at ACME Pharmaceuticals, Inc. Mr. Dupin, have you ever come across a container in your household cleaning cupboard and, because the label had fallen off, had absolutely no idea of what it was? Well, even though we in the lab do not solicit door to door for work, this whole concept of taking an unknown compound and identifying it is one of the most important aspects of our research lab. We deal almost exclusively with medicinal agents, but are often called on by the local police department to identify their seized unknown chemicals and by the hospital to work on unidentifiable agents found in blood and tissue samples. Figure 1 lists the six steps we use to identify an unknown chemical.

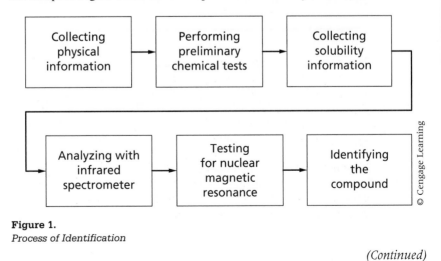

© Cengage Learning

Figure 1.
Process of Identification

(Continued)

COLLECTING PHYSICAL INFORMATION

Collecting physical information about an unknown compound gives general information pertaining to the overall chemical. Taking note of the physical state, color, smell, melting point, and boiling point of the compound gives you important basic information that deals with the most fundamental chemical properties of whatever it is you are working with.

PERFORMING PRELIMINARY CHEMICAL TESTS

Running universal preliminary chemical tests on the unknown identifies what major compound classification it falls under. All organic compounds can be classified into approximately 15 different categories, and running some very basic tests helps us to place the unknown into its general category. For example, if the ignition test produces black, sooty smoke, then it is evident that the general classification of the compound is an aromatic compound.

COLLECTING SOLUBILITY INFORMATION

Collecting information on the solubility of unknown compounds identifies the properties they exhibit when they are introduced to other compounds. By observing how an unknown compound reacts with a variety of other compounds, it is possible to gain insight into some of the compound's specific chemical structure. If, for example, the unknown is soluble in water, then it is clear that it is polar.

ANALYZING WITH INFRARED SPECTROMETER

Analysis of the infrared spectrometer test shows what functional groups are attached to the carbon "backbone" of the compound. By knowing the kinds of chemicals attached to the parent chemical, it is possible to begin sketching a picture of what the unknown is. These attached chemicals are mainly responsible for how the unknown reacted in the solubility tests.

TESTING FOR NUCLEAR MAGNETIC RESONANCE

Running the nuclear magnetic resonance test on an unknown compound gives the essential information about the parent carbon structure (this is the compound that is at the base of the molecule and is often called the *parent compound*). This test produces evidence relating to the number of carbons present, how they are bonded to one another, and how they interact with the attached hydrogens. This information is especially important because it shows what is at the core of what may be a very big structure.

IDENTIFYING THE COMPOUND

Identification of the unknown is now possible. All the information from the tests run by the researcher can be compiled, and in most cases, the unknown compound can be identified. One of the exceptions to this process is that, even with all this data, more extensive tests need to be completed before the compound is identifiable. The other exception is that the compound is common and could be identified after doing the preliminary chemical tests.

The process by which unknown compounds are identified begins by collecting very general information and continues until the information that is collected is very specific. These tests are run daily and are a vital function of the work done in the lab.

Example 8.4

Description of a Process
Source: From *Ultimate Chocolate* by Patricia Lousada. Copyright © 1997 Dorling Kindersley. Reprinted by permission of Dorling Kindersley.

MAKING CHOCOLATE

Chocolate, like coffee, originates in a bean, but one that grows on a tree, not a bush. The exotic cocoa tree produces a blizzard of pink and white flowers, green unripe fruit, and bright golden cocoa pods, all at the same time. Encased in the pod is the dark little cocoa bean. Put through an intricate production process, the bean is transformed into cocoa mass, which ultimately becomes that magically pleasurable ingredient, chocolate (Lousada 28).

THE COCOA TREE

The region between the twentieth parallels, with the exception of parts of Africa, is the home of the cocoa tree. The tree begins to bear fruit once it is four years old and has an active lifespan of at least sixty years. Its fruit grows directly out of the older wood of the trunk and main branches, reaching the size of a small football and ripening to a rich golden color. Inside the ripe pods, purplish-brown cocoa beans are surrounded by pale pink pulp. After the pods are cut from the tree, beans and pulp are left to ferment together. The beans turn a dull red and develop their characteristic flavor. After fermentation, the beans are dried in the sun, acquiring their final "chocolate" color. They are now ready for shipping to the manufacturing countries. Various bean types have been bred and experts take pride in their ability to distinguish chocolate made from Criollo or Trinitario beans from that made from Forastero.

PROCESSING THE BEANS

When the dried beans reach the processing plant, they are cleaned and checked for quality, and then roasted. Roasting, an important stage in the manufacturing process, develops the flavor of the beans and loosens the kernels from the hard outer shell. Each chocolate manufacturer has its own roasting secrets, which contribute significantly to the chocolate's flavor. After roasting, the beans have

(Continued)

a distinctive chocolaty smell. The next step is to crack the beans open, discarding their shells and husks, to obtain the kernels, called nibs. It is the processing of these small, brown nibs that gives us chocolate. The roasted nibs, which contain on average 54 percent cocoa butter, are ground into a dark, thick paste called cocoa mass or solids. When more pressure is applied to the cocoa mass, the resulting products are cocoa butter and a solid cocoa cake. From the cocoa cake, when it is crushed into cocoa crumbs and then finely ground, comes cocoa powder.

CONCHING

Chocolate is generally cocoa solids and sugar, with added cocoa butter (in the case of semisweet chocolate), or milk (in the case of milk chocolate), plus vanilla and other flavorings. Conching, in which the chocolate mixture is heated in huge vats and rotated with large paddles to blend it, is the final manufacturing process. Small additions of cocoa butter and lecithin, an emulsifier, are made to create the smooth, voluptuous qualities essential to the final product.

SWEETENING CHOCOLATE

Baking or bitter chocolate is simply cocoa solids and cocoa butter. To produce the great range of chocolates, from bittersweet to semisweet to sweet, more cocoa butter plus varying amounts of sugar, vanilla, and lecithin are added. The flavor and sweetness of a chocolate will be unique to its maker, with one brand's bittersweet tasting like another brand's semisweet. Changing your usual brand of chocolate can make a difference in the flavor of a favorite recipe. To make milk chocolate, milk solids replace some of the cocoa solids. White chocolate is not, in fact, a real chocolate since it is made without cocoa solids; brands containing all cocoa butter, rather than vegetable oil, are best (Lousada 28–29).

Example 8.5

Process
Description

CONCRETE, SLUMP, AND COMPRESSIVE STRENGTH

Concrete is one of the most universal construction materials in the world because its component raw materials are inorganic, noncombustible, highly versatile, and relatively low in cost in comparison to other materials. It is rated by its compressive strength after a 28-day curing period. Specified compressive strengths of concrete are produced by varying the proportions of cement, sand (a.k.a. fine aggregate), coarse aggregate, and water combined to make the concrete paste.

One of the most important factors affecting the strength of concrete is the water-to-cement ratio, expressed in pounds or gallons of water per sack of cement. The sections that follow will explain the nature of slump tests, discuss the testing process, and end with slump test indications.

NATURE OF SLUMP TESTING

Concrete must always be made with a workability, plasticity, and consistency suitable for job conditions. Workability is a measure of how easy or difficult it is to place, consolidate, and finish concrete. Plasticity determines concrete's ease of molding. Consistency is the ability of freshly mixed concrete to *flow*. In general, the higher the slump, the wetter the mixture; however, too much water in the mix may cause segregation of the mid-components (aggregates, etc.), producing nonuniform concrete.

Concrete for buildings is usually mixed in a transit mix truck and tested in the field (i.e., on the job site) for the water-to-cement ratio by means of a slump test. The slump test is a measure of concrete consistency. This test is performed by rodding concrete from the transit mix chute into a conical steel form. When the form is removed, the amount of "drop" or "slump" is measured. Figure 1 illustrates this method of determining concrete consistency.

TESTING PROCESS

A slump test is made to ensure the concrete conforms to specifications and has the flowability required for placing. The test is carried out as soon as a batch of concrete is mixed, and a standard slump cone and tamping rod are required to carry it out. The cone is made of sheet metal, 4 inches in diameter at the top, 8 inches in diameter at the bottom, and 12 inches high. The rod is a 5/8-inch bullet-nosed rod about 24 inches long.

Obtaining a Sample

Concrete is poured from the transit mix truck chute directly into the slump test cone as soon as the batch is mixed in order to allow for the most accurate, efficient test results. If testing is done at a later point, the concrete will give incorrect slump measurements due to partial setup (hardening) or "bleeding" caused by separation of components. The cone is filled in three equal layers, each tamped 25 times with the rod. This is done to compact the sample so excess air and voids caused by larger coarse aggregates are eliminated, and a more accurate slump measurement is given.

Form Removal

After the third layer is in place and has been tamped, the concrete is struck off level, the cone is lifted carefully and set down beside the slumped concrete, and the rod is laid across the top of the cone. If the cone is not carefully removed from the sample, the test results will be inaccurate because of a shift of the concrete.

(Continued)

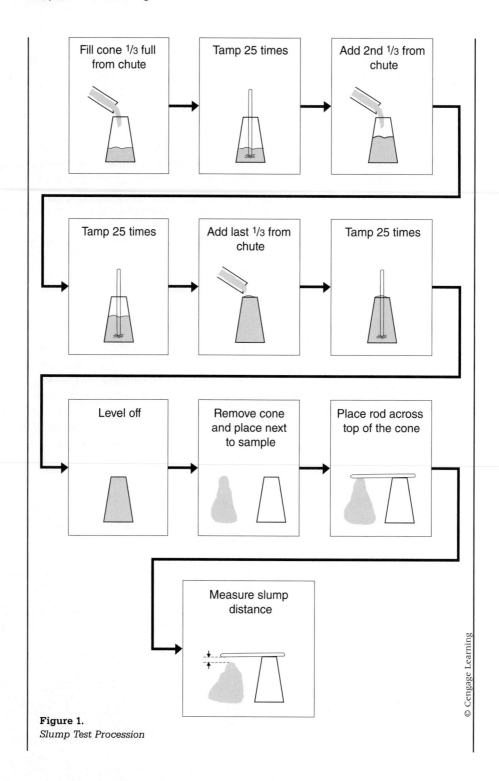

Figure 1.
Slump Test Procession

© Cengage Learning

Slump Measurement

The distance from the underside of the rod to the average height of the top of the concrete is measured and registered as the amount of slump in inches. Different slumps are needed for various types of concrete construction. Slump is usually indicated in the job specifications as a range, such as two to four inches, or as a maximum value not to be exceeded. Table 1 shows slump values for unspecified jobs according to general industry recommendations. When making batch adjustments, the slump can be increased by about one inch by adding ten pounds of water per cubic yard of cement.

TEST INDICATIONS

Slump is indicative of workability when assessing similar mixtures. However, it should not be used to compare mixtures of totally different proportions. When used with different batches of the same mixture, a change in slump indicates a change in consistency and in the characteristic of materials, mixture proportions, or watery content.

TABLE 1 Recommended Slumps for Various Types of Construction

	Slump (inches)	
Concrete Construction	**Maximum**	**Minimum**
Reinforced foundation walls and footings	3	1
Plain footings, caissons, and substructure walls	3	1
Beams and reinforced walls	4	1
Building columns	4	1
Pavements and slabs	3	1
Mass concrete	2	1

© Cengage Learning

Example 8.6

Human
System
Process
Description

A protocol is essentially a set of rules. The protocol below explains the procedure that a presenter will follow when conducting a "tuning," which is a staff development process. Only one step of the seven-step process is given here.
Tuning Protocols

2. Presentation—15 minutes

The presenter sets the context, describing the teaching/learning situation to be discussed and distributing materials related to the practice being described—for

(Continued)

example, collections of student work, audio or video tapes of students in the classroom, assessments, or lesson plans. Participants say nothing but take notes. The presenter then poses one or two key questions he or she wants the group to address. For example, the presenter brings a student's portfolio to the meeting and, after describing the assignment that led to the portfolio, asks: What habits of mind does this portfolio convey? How can I use portfolios to push the student's thinking deeper? (Brown)

Exercises

▶ You Create

1. In class, develop a brief mechanism description by brainstorming. Hand in all your work to your instructor. Your instructor may ask you to perform this activity in groups of three or four.

    - Brainstorm the names of parts and subparts.
    - Choose the most significant parts.
    - Arrange the parts into a logical pattern.
    - Name and define each part in the first sentence.
    - Describe each part in a paragraph.
    - Create a visual aid of your mechanism, complete with appropriate callouts (use keywords from the text as callouts).

2. In class, develop a mechanism description through a visual aid. Hand in all your work to your instructor. Your instructor may ask you to perform this activity in groups of three or four.

    - Draw a visual aid of your mechanism. Use callouts to point out key parts.
    - Name each part that the audience needs to understand.
    - Select a logical pattern for discussing the parts.
    - Name and define each part in a sentence.
    - Describe each part in a paragraph; be sure to discuss each part named in the callouts.

3. In class, develop a process description through brainstorming. Assume that you need either to demonstrate that a problem exists or to provide cause–effect theoretical background. Hand in all your work to your instructor. Your instructor may ask you to perform this activity in groups of three or four.

    - Brainstorm the names of as many steps and substeps as you can.
    - Arrange the steps into chronological or cause-effect order.
    - Define the end goal of each step in one sentence.

4. In class, develop a brief process description through visual aids. Assume that your audience needs a basic understanding of the process in order to discuss it at a meeting. Hand in all your work to your instructor. Your instructor may ask you to perform this activity in groups of three or four.

- Draw a flow chart of the process, or else a diagram of the parts interacting, as in the following diagram:
- Write a brief paragraph for each step.

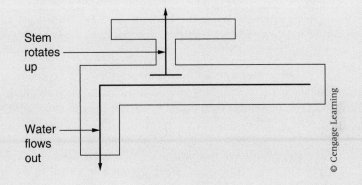

▶ You Revise

5. Either singly or in groups of two to four, analyze these paragraphs for consistency in presentation. Rewrite the paragraphs to eliminate passive voice. What identity does the author of this text appear to express? Do you like that? Rewrite the paragraphs to achieve a different identity.

Date: March 28, 2012
To: Dan Riordan
From: Tadd Hohlfelder
Subject: To give an overview of the Norclad case-out process

As the new supervisor it is important for you to be familiar with the Norclad case-out process. I compiled this information while reviewing the department last week.

The purpose of the process is to construct complete window units from separate clad awnings, casement, and picture assemblies. After construction, they are tagged and moved to shipping.

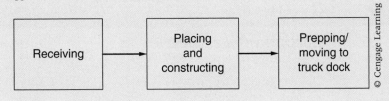

When subassemblies are received in case-out they must be checked in to be sure all are accounted for. Each of the four tables has a stack of orders which need to be built. There is not one correct method for checking in assemblies. For example, some workers check their subassemblies separately as they build them. Others choose to check them all in at the beginning of the shift.

Once they are checked in, the subassemblies for a particular order are placed on a table. All components of the unit are also placed on the table. The components are listed on the order sheet. One order might call for a 6¾-inch extension with screens and storm panes. Another may just have the standard 4⁹⁄₁₆-inch extension without other options. After correct placement, the unit can be built. We use several types and sizes of air-powered nail guns to construct our units.

After construction, the unit must be prepped and sent to shipping. We are currently using plastic strips to protect the units while shipping. The strips are placed on the facing edges of the units. Cardboard pieces are then stapled on the corners of the unit. On the work order for each unit there are three tags attached: green, white, and red. The green tag must be attached to the finished unit to identify it for shipping. The remaining tags are left on the order, which is placed in a bin on your desk upon completion. The units are loaded on racks and moved to the truck dock using either pallet jacks or fork trucks.

6. Analyze the strategy of the "help" example in these two paragraphs. Do you like the way it is introduced in each section? Do you like using the same example throughout the description? Decide what kind of identity this author achieves by presenting the example as she does. Does she make you feel confident? Rewrite the paragraph so that you create a different identity for the author.

OBTAINING BACKGROUND INFORMATION

Obtaining background information means to gather information from the customers in order to get a feeling of their overall knowledge of computer software. This information is usually gathered through interviews or questionnaires. The information we get is used to design a user interface that will be easy for the customer to learn.

We can use the simple example of getting help when trying to understand what a certain function does. For example, if the customer is familiar with making the help text appear on the screen by typing "help," we design ours to work in a similar fashion.

DEFINING THE SEMANTICS

Defining the semantics of the user interface means to come up with a clear view of all the tasks the customer must do to make the system perform.

When performing these tasks, two customers will, more than likely, take two different paths through the interface to obtain the same outcome. We must discuss each of these paths for each task, then create and describe them within our system.

If you refer back to the help text example, you may understand this better. If two users are looking for help on a particular topic, one may go to a site map to find it, and the other straight to the topic by typing in the subject of their question in the search box. Both customers will get the same result, but each did it in his or her own way.

7. Redo this paragraph so that your supervisor can take it to a committee that needs to know what happens in this step.

HANDLE

The handle is made from kiln dried walnut rough cut lumber 1"×8"×11". Lumber is cropped to 30" lengths. Cut to 30" lengths. Jointed to get a true edge. Ripped to 1¼" widths. 1¾" sections are cropped off starting ¼" from one end leaving $^1/_{16}$" between marks. 15 sections are turned down to ¾" diameter one end and $^5/_8$" diameter other end. Diameter is 100% inspected and lengths are marked off. Sections are cut to 1¾" lengths. $^5/_{16}$" diameter hole is drilled 1¼" deep. Drill depth is inspected randomly. All surfaces are sanded and 100% inspected.

▶ You Analyze

8. Compare these two versions of the same paragraph. Which version gives you more confidence in the writer? In groups of three or four, discuss the stylistic features that cause confidence.

   A. To affix the bacteria means to "glue" them to the slide so that they are permanently mounted. Basically there are three steps. First, the lab assistant lights the Bunsen burner and grasps the slide on each end. Second, the assistant dips the loop of the sterile poker into the culture and smears the liquid onto the center of the slide. Third, the assistant passes the slide (wet side up) several times through the flame.

   B. Affix the bacteria to the slide. First, the lab assistant places a sterile poker in the culture of bacteria. Then he or she places the loop on the center of the clean glass slide and smears it around in a tiny circle. The lab assistant then heat fixes the slide with the bacteria on it by passing it through the flames of the Bunsen burner two or three times.

9. In class (or in small groups if your instructor prefers), compare a paragraph from the document on the product tester selection policy (p. 224) with a paragraph from the Tamrac ZipShot description (pp. 217–218) or the Business Process description (p. 218). How are the paragraphs organized? What

is the function of the first sentence of each? Which paragraph seems to more effectively convey its message to the audience? Be prepared to briefly present your findings to the class.

Writing Assignments

1. Annotate a screen so that a reader understands the significance of the items and menus available. You should be able to complete this assignment using a word processing program (e.g. Word or Pages) and textboxes. Consider this a "generalized mechanism description." Your goal is to give the readers access to the screen so that they feel comfortable using the site. Probably one of the social media sites, such as a blog creation site, will work best for this assignment.

1. Assume that you must describe a problem with a process at your workplace. Describe the process in detail, and then explain the problem and offer a solution. Use a report format with heads and a visual aid. As you work, you can use the worksheet in this chapter. Use Exercise 4 or 5 to start your work.

2. Write a brief description of the steps you took to solve a problem. Assume that your audience is someone who must be assured that your solution is based on credible actions, but who does not know the terms and concepts you must use. For instance, you could explain the process you used to test an object or the process you used to select a vendor for a product your company must purchase.

3. Write an article for a company newsletter, describing a common process on the job. Use a visual aid. Sample topics might include the route a check follows through a bank, the billing procedure for accounts receivable, the company grievance procedure, the route a job takes through a printing plant, or the method for laminating sheets of materials together to form a package. Fill out the worksheet in this chapter. Use Exercise 4 or 5 to start your work. Your article should answer the question "Have you ever wondered how we . . . ?"

4. Write several paragraphs to convince an audience to purchase a mechanism or to implement a process. The mechanism might be a machine, and the procedure might be a system, such as hiring new personnel. Describe the advantages that this mechanism or process offers over the mechanism or process currently in use. Fill out the worksheet in this chapter. Use a visual aid. Use Exercise 2 or 3 to start your work. Choose a mechanism or process you know well, or else choose from this list (for X, substitute an actual name).

    the lens system of brand X camera

    the action of brand X bike gear shift

the X theory of product design

the X theory of handling employee grievances

the X retort process

how brand X air conditioner cools air

how brand X solar furnace heats a room

5. Expand into a several-page paper one of the brief descriptions you wrote in Exercises 1–4. Your audience is a manager who needs general background. (Alternate: your audience is a sixth-grade class.) Bring a draft of the paper to class. In groups of two to three, evaluate the draft in terms of these concerns:

   a. Does the introduction present the purpose of the mechanism process and provide a basis for your credibility?

   b. Does the introduction present a preview of the paper?

   c. Does each new section start with an effective keyword?

   d. Are the details sufficient to explain the part or step to the audience?

   e. Is the visual aid correctly sized, clear, and clearly referred to?

   f. Do callouts in the visual aid duplicate key terms in the text?

   g. Is the style at a high enough quality level?

6. Write a learning report for the writing assignment you just completed. See Chapter 5, Writing Assignment 8, page 137, for details of the assignment.

Web Exercises

1. Describe a Web browser homepage (Chrome, Firefox, Internet Explorer, one of the many search engines) as if it were a mechanism. Name and explain the function of each part of the page.

2. Search the Web using the keywords "process description" (or, to change your results slightly, add a company name—"process description" Ford). Analyze the results you find both for the way they are organized and the degree to which your search results are pertinent to your search. Then, following your instructor's directions, use the example you found to create a process description. (Alternate: write an analytical report explaining the organizations and roles you found.)

3. Describe the process of finding some type of information (for instance, air fares or technical data relevant to your major or job focus). Name and explain the sequence of steps that a person must follow in order to find results efficiently. (Note continuation of this exercise in Chapter 9, Web Exercise 2.)

Works Cited

How Things Work in Your Home. New York: Holt, 1987. Print.

Brown, Lois Brown. "Tuning Protocols." National Staff Development Council. Web. 20 May 2012. <www.pds-hrd.wikispaces.net/file/view/Tuning+Protocols. pdf>. Orig. Pub. *Journal of Staff Development* 20.3 (Summer 1999).

Jordan, Michael P. *Fundamentals of Technical Description.* Malabar, FL: Robert E. Krieger, 1984. Print.

Perlman, Mike. "Accessory Review: Tamrac Zipshot Tripod." 21 Sept. 2012. DPReview (Digital Photography Review). Web. 9 Nov. 2012 < http://www. dpreview.com/articles/6145161331/accessory-review-tamrac-zipshot-tripod>.

Cousins, Jay and Tony Stewart. "What Is Business Process Design and Why Should I Care?" RivCom Ltd. 2002. Web. 18 May 2012. <http://www. rivcominc.com/resources/papers-and-presentations/>.

Lousada, Patricia. *Chocolate.* New York: DK Publishing, 1999.

"Product Tester Selection Policy." *Illuminated Ink.* 10 Nov. 2003. Web. 10 Nov. 2003. <www.IlluminatedInk.com>.

Technical Communication Applications

Sets of Instructions

CHAPTER CONTENTS

CHAPTER 9 IN A NUTSHELL

The goal of a set of instructions is to enable readers to take charge of the situation and accomplish whatever it is that they need to do.

Introduction

▶ Tell the end goal of the instructions (or do that in the title).

▶ Define any terms the readers might not know; if necessary, explain the level of knowledge you expect.

▶ List tools they must have or conditions to be aware of.

Body Steps

▶ Explain one action at a time.

▶ Tell the readers what they need to know to do the step, including warnings, special conditions, and any "good enough" criteria that allow them to judge whether they have done the step correctly.

Format

▶ Use clear heads.

▶ Number each step.

▶ Provide visuals that are big enough, clear enough, and near enough (usually directly under or next to) the appropriate text.

▶ Use lots of white space that clearly indicates the main and the subordinate sections.

▶ Write the goal at the "top" of the section—so the readers can skip the rest if they already know how to do that.

Tone

▶ Be definite. Make each order explicit. If the monitor "must be placed on top of the CPU," don't say "should."

▶ Discover what readers feel is arbitrary by asking them in a field test.

Sets of instructions appear everywhere. Magazines and books explain how to canoe, how to prepare income taxes, and how to take effective photographs; consumer manuals explain how to assemble furniture, how to program DVRs, and how to make purchased items work. On the job you will write instructions for performing many processes and running machines. This chapter explains how to plan and write a useful set of instructions.

Planning the Set of Instructions

To plan your instructions, determine your goal, consider your audience, analyze the sequence, choose visual aids, and follow the usual form. In the following discussion, the subject is setting up a high-speed Internet gateway with a wifi access point, for wireless devices in your home to connect to broadband Internet. Broadband access is provided either by your telephone service using DSL (digital subscriber line), or through your cable service using a cable modem. For the purposes of these instructions we will be using DSL service for our examples.

Determine Your Goal

Instructions enable readers to complete a project or to learn a process. *To complete a project* means to arrive at a definite end result: The reader can complete a form or assemble a toy or make a garage door open and close on command. *To learn a process* means to become proficient enough to perform the process without the set of instructions. The reader can paddle a canoe, log on to the computer, or adjust the camera. In effect, every set of instructions should become obsolete as the reader either finishes the project or learns to perform the process without the set of instructions.

Consider the Audience

When you analyze your audience, estimate their knowledge level and any physical or emotional constraints they might have.

Estimate the Audience's Knowledge Level

The audience will be either absolute beginners who know nothing about the process, or intermediates who understand the process but need a memory jog before they can function effectively.

The reader's knowledge level determines how much information you need to include. Think, for instance, about telling beginners to turn on a computer. They will not be able to do this because they will not know where to look for the power switch. For an intermediate, however, "turn it on" is sufficient.

Globalization and **Instructions**

Knowing your international audience means knowing their cultural norms. You need to know how to present ideas to them. Martin Schell warns writers that in some countries the phrase "Turn Power On" is an essential first step for any set of instructions, whereas in other cultures this step is assumed and unnecessary, and, in some cases, it is insulting to include such an obvious step.

A European professional communicator commented on this point in this way: "American instructions very often seem 'over—everything' compared to our mentality. Everything is so outspoken. On the other hand, the information is very often necessary. Maybe the trick is to chunk and organize the text so much that it is easy to jump and find the relevant things."

The same rules of clear and concise writing that you apply to other documents are especially important in writing instructions. Keep your prose simple and succinct. As much as possible, use the present tense, stay away from contractions, pare down your writing to one idea per sentence and avoid using jargon and clichés (Dehaas).

Some problems to be especially aware of include the following:

1. Noun phrases can be extremely difficult to decipher. Using them often causes ambiguity. Begin with a simple phrase initially and then introduce a more difficult compound phrase after the reader has acquired the relevant background to understand the phrase. Early in your text avoid common (to Americans) constructions like *disabled vehicle garage* or *potato chip bag depot*. Spell it out: garage where you park the vehicles for the disabled. (Bhatia).

2. Another language feature that causes problems is the difference between simple and progressive tenses, which American English speakers understand intuitively, but speakers of English as a second language might not. For instance the sentence "Sales are surging; consumers are buying" means something different than "Sales surge; people buy." The first is a description of current events; the second is a general truth. Sometimes the difference can be translated into something that makes little sense: "the writer creates the manuscript; the editor fixes it" could be translated as "the writer is creating the manuscript; the editor is fixing it." The way to prevent the mistranslation is to add a time indicator: "After the writer creates the manuscript, the editor fixes it" or "While the writer is creating the manuscript, the editor is fixing it."

3. In instructions for lay people, the natural voice is the active voice. If an operator is instructed as follows, "Put the lever in the 'on' position," there is never any doubt about what the operator is expected to do. On the other hand, the passive voice is useful in instructions, too. The same sentence in the passive voice would be useful if the operator has to check a sequence of parameters on a machine, among which one would be "The lever should be positioned in the 'on' position." If that is so: fine. If not, the operator knows

that he or she has to do it for the operation to succeed. Dehaas states that by changing the instructions to the active voice, you will make the instructions more precise. Another aspect of the active/passive voice is focus. Consider the following two sentences:

1. *The gas turbine drives the engine.*
2. *The engine is driven by the gas turbine.*

The focus in the first sentence is on how the gas turbine works; the second sentence belongs to a text on engines. Please also note that the examples are not instructions, but information.

4. Even with strict conventions, instructions may vary from one culture to another. Some other cultures think American texts number everything, even to an extent where it may seem silly. In one collaborative exercise between two university classes studying technical writing/technical translation, an American student created the text "How to parallel park your car in 10 steps!" European students responded: "You cannot parallel park a car in 10 steps. It is an ongoing movement!" (Mousten et al. 319).

Identify Constraints

Emotional and physical constraints may interfere with the audience's attempts to follow instructions. Many people have a good deal of anxiety about doing something for the first time. They worry that they will make mistakes and that those mistakes will cost them their labor. If they tighten the wrench too hard, will the bolt snap off? If they hit the wrong key, will they lose the entire contents of their document? To offset this anxiety, include tips about what should take place at each step and about what to do if something else happens. Step 7 in the example "Installing High-Speed Internet," page 258, explains what it means when the red lights flash—nothing is wrong; the DSL device is making the proper connections.

The physical constraints are usually the materials needed to perform the process, but they might also be special environmental considerations. A Phillips screw cannot be tightened with a regular screwdriver; a three-pound hammer cannot be swung in a restricted space; in a darkroom, only a red light can shine. Physical constraints also include safety concerns. If touching a certain electrical connection can injure the reader, make that very clear. Step 4 in the example on page 258 tells readers that it doesn't make any difference which Ethernet port on the DSL gateway device the cable gets plugged into. Whichever one chosen achieves the same result.

Examples for Different Audiences

To see how the audience affects the set of instructions, compare the brief version below with "Preparation" (p. 255). That section explains the steps in detail, assuming that the beginner audience needs detailed "hand-holding" assistance. The brief example below, designed for an intermediate audience, simply lists the sequence of steps to jog the reader's memory.

Instructions for an Intermediate

1. Install the port filters.
2. Connect the data cable to the gateway.
3. Attach the computer to the gateway using an Ethernet cable.
4. Plug the power cable into the gateway.

Analyze the Sequence

The sequence is the chronological order of the steps involved. To analyze the sequence, determine the end goal, analyze the tasks, name and explain the tasks, and analyze any special conditions. (See the sample flow chart of an analysis in Figure 9.1, p. 251).

Determine the End Goal

The end goal is whatever you want the reader to achieve, the "place" at which the user will arrive. This goal affects the number of steps in your sequence because different end goals will require you to provide different sets of instructions, with different sections. In the preceding example, the end goal is establish a wireless home network connected to high-speed Internet, and the document ends at that point. Other end goals, however, are possible. For instance, if the goal were "The user will connect optional devices," the sequence would obviously include more steps and sections.

Analyze the Tasks

You have two goals here: determine the sequence and name the steps. To determine the sequence, you either go backward from the end goal or perform the sequence yourself. If the end goal is to install high-speed Internet access, the question to ask is "What step must the user perform before the service is activated?" The answer is "Register your device with your service provider." If you continue to go backward, the next question is "What does the user do before registering the device with the service provider?" As you answer that question, another will be suggested, and then another—until you are back at the beginning, unboxing the gateway device and identifying the pieces. You can also do the process yourself; as you do it, record every act you take. Then perform the task a second time, following your written

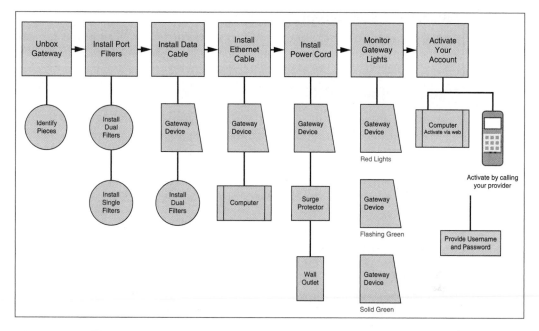

Figure 9.1 Flow Chart of an Analysis

notes exactly. You will quickly find whether or not you have included all the steps.

Name and Explain the Tasks

Having decided on the sequence, you name each task and explain any subtask or special information that accompanies it. The Installing High-Speed Internet example has three subtasks labeled *a* to *c* under the task "Register your device with your service provider," and many of the steps include explanations for the audience. For instance, step 8b (p. 258) tells the user to contact the provider by web or phone to activate service.

Analyze Conditions

You must also analyze any special conditions that the user must know about. For instance, step 2 of "Preparation" (p. 257) explains where the dual filter goes in contrast to where the single filters go. Safety considerations are very important, and safety warnings are an essential part of many instructions. *If it will hurt them or the machine, tell the audience.* Warn the user to use a surge protector instead of plugging into a bare wall outlet.

Example of Process Analysis

An easy way to conduct your analysis is to make a flow chart of the process. Put the steps in boxes and any notes in circles (see Figure 9.1, p. 251).

Choose Visual Aids

Visual aids either clarify or replace the prose explanation. In Figure 9.5 in point 1 (p. 257) of the Preparation section, the photograph replaces text; to describe the difference between an RJ45 and an RJ11 cord would take far more than the words used to convey the instruction.

Figure 9.2 below *clarifies* the text—and reassures the readers that their actions are correct—by showing what the screen will look like as the actions occur on a computer.

Here are a few guidelines for choosing visual aids:

- Use a visual aid to orient the reader. For instance, present a drawing of a keyboard with the return key highlighted.
- Use a visual aid to show the effect of an action. For instance, show what a screen looks like *after* the user enters a command.

At the TO: prompt type their NAME, not their real name but their e-mail user name, and then press enter. The SUBJ: prompt will appear. (See Figure 1 —note that you do not have to type in all capital letters, though you may.) On our system, the user name is generally the last name and the first one or two letters of the first name.

```
mail>SEND
TO: SMITHJ
SUBJ: Learning e-mail
Enter your message below
```

Figure 1.
On-Campus E-mail Address

© Cengage Learning

Figure 9.2   Visual Aid that Clarifies

- Decide whether you need one or two visual aids for the entire process or one visual aid per step. Use one per step if each step is complicated. Choose a clear drawing or photograph. (To determine which one to use, see Chapter 7.)
- Place the visual aid as close as possible to the relevant discussion, usually either below the text or to the left.
- Make each visual aid large enough. Do not skimp on size.
- Clearly identify each visual aid. Beneath each one, put a caption (e.g., *Figure 1. E-Mail Address* or *Fig. 1. E-Mail Address*).
- Refer to each visual aid at the appropriate place in the text.
- Use *callouts*—letters or words to indicate key parts. Draw a line or an arrow from each callout to the part.

Follow the Usual Form for Instructions

The usual form for a set of instructions is an introduction followed by a step-by-step body. The introduction states the purpose of the set of instructions, and the steps present all the actions in chronological order. The models at the end of this chapter illustrate these guidelines. Make a style sheet of all your decisions.

For steps and visual aids, use these guidelines:

- Place a highlighted (underlined or boldfaced) head at the beginning of each section.
- Number each step.
- Start the second and following lines of each step under the first letter of the first word in the first line.
- Use margins to indicate "relative weight"; show substeps by indenting to the right in outline style.
- Decide where you will place the visual aids. Usually place them to the left or below the text.
- Use white space above and below each step. Do not cramp the text.

For columns, the decisions are more complex. Basically, you can choose one or two columns, but their arrangement can vary, and each will have different effects on the reader. Figure 9.3 (p. 254) presents several basic layouts you can choose. You can place visual aids below or to the right or left of the text. To the left and below are very common places. Generally, you place to the left (text or visuals) whatever you want to emphasize.

Writing the Set of Instructions

A clear set of instructions has an introduction and a body. After you have drafted them, you will be more confident that your instructions are clear if you field-test them.

Figure 9.3 Different Column Arrangements for Instructions

Write an Effective Introduction

Although short introductions are the norm, you may want to include many different bits of information, depending on your analysis of the audience's knowledge level and of the demands of the process. You should always

▶ State the objective of the instructions for the reader.

Depending on the audience, you may also

▶ Define the process.
▶ Define important terms.
▶ List any necessary tools, materials, or conditions.
▶ Explain who needs to use the process.
▶ Explain where and/or when to perform the process.
▶ List assumptions you make about the audience's knowledge.

A Sample Introduction to a Set of Instructions

In the following introduction, note that the writer states the objective ("These instructions enable you to set up a high-speed Internet gateway with a wifi access point"), defines the topic, lists knowledge assumptions, and lists materials.

Write an Effective Body

The body consists of numbered steps arranged in chronological order. Construct the steps carefully, place the information in the correct order, use imperative verbs, and do not omit articles (*a, an,* and *the*) or prepositions.

INSTALLING HIGH-SPEED INTERNET

INTRODUCTION

End goal
Background

These instructions enable you to set up a high-speed Internet gateway with a wifi access point. This will allow you to share files over a wireless network without the need to install Ethernet cables throughout your home. It will also allow you to connect wifi-only devices to the Internet including smartphones, tablets, laptops, and Internet-ready televisions.

Knowledge assumption

Materials list

These instructions assume that you have subscribed to DSL service and have purchased the hardware gateway device kit provided through your local telephone company. Make sure you have wireless cards in all the computers that will connect to the wireless network.

© Cengage Learning 2014

Figure 9.4   Introduction to a Set of Instructions

TIP ||||

Two Style Tips for Instructions

1. Use imperative verbs.

   An imperative verb gives an order. Imperative verbs make clear that the step must be done. Notice below that "should" introduces a note of uncertainty about whether the act must be performed.

   Say

   Make sure you have wireless cards.

   Rather than

   You should make sure you have wireless cards.

2. Retain the short words.

   Use *a, an, the* in all the usual places. Eliminating these "short words" often makes the instructions harder to grasp because it blurs the distinction between verbs, nouns, and adjectives.

   No short words:

   Unbox gateway device and identify pieces against instruction set.

   Short words added:

   Unbox the gateway device and identify all the pieces against the instruction set.

Construct Steps Carefully

To make each step clear, follow these guidelines:

- Number each step.
- State only one action per number (although the effect of the action is often included in the step).
- Explain unusual effects.
- Give important rationales.
- Refer to visual aids.
- Make suggestions for avoiding or correcting mistakes.
- Place safety cautions before the instructions.

Review the Examples on pages 263–267 to see how the writers incorporated these guidelines. An example of how to write the body follows.

Sample Body

Here is the body of the set of instructions that follows the introduction on page 255.

PREPARATION

1. Unbox the gateway device and identify all the pieces against the instruction set. All photographs are not shown to save space.

   a. Gateway device (DSL router and wireless access point)
   b. Power cord (black)
   c. Ethernet cable (blue cord with RJ45 Ethernet connectors on the ends)
   d. Data cable (gray phone cord with RJ11 telephone connectors on the ends)

RJ 45 connector (right). RJ 11 (left)

   e. Single port filters for the household telephones or fax machines
   f. Dual port filter for the DSL gateway and a phone line
   g. Instructions, setup guide, and passcode

2. Install port filters into the telephone jacks. These filters are used to prevent the phone line from interfering with your Internet connection and prevent the DSL line from damaging your telephone or fax machine.

   a. Install the dual filter into the phone jack closest to the primary computer where you'll be setting up your gateway.
   b. Install single filters on every other phone jack in use.

INSTALLATION

1. Plug the data cable (gray phone cord) into the DSL port on the dual filter.
2. Plug the other end into the port on the gateway device. The gray color of this cable matches the DSL port on the filter and the DSL port on the gateway device.
3. Plug the blue Ethernet cable into the Ethernet port on the back of your computer.

(Continued)

Instruction
Explanatory comment
Special conditions

Action
Action
Explanatory comment
Action

Action

Note

4. Plug the other end into any of the Ethernet ports on the back of the gateway device. There are probably four or more Ethernet ports on the back of the device. It doesn't matter which one you connect to.

Action

5. Plug the black power cord into the back of the gateway device.

Action

Special note

6. Plug the other end of the power cord into a wall outlet. To protect your computer equipment, you may want to use a surge protector instead of a bare wall outlet.

Action

Note

7. Monitor the green indicator lights on the DSL device, watching for a solid green line. The lights will flash red, then intermittent green, while the device is making its connections. Do not unplug the power cord or the data cable during this time, or your device may not activate.

Action

8. Register your device with your service provider in order to activate it.
   a. Locate the specific activation instructions from your provider.
   b. Visit the provider's website to activate your account, or call the provider's activation line.
   c. Provide the individualized username and password.

Special note

The DSL gateway provides a solid broadband Internet connection to your local network. It also manages your home network, allowing devices to be connected via wired Ethernet cables or through wireless signals. The DSL gateway also provides a security firewall between your local network and the Internet.

CONNECTING OPTIONAL DEVICES

Introduction to subsection

Special note

After completing the previous steps, your network is active. Now you can determine what additional devices to connect to your home network. Some devices can connect to your network through wired Ethernet cables, while others connect using wifi signals. Some can connect either way.

Head for special condition

Wired Connection

Printers, desktop computers, and some other devices may require an Ethernet connection.
1. Find an Ethernet cable that reaches between the device and the DSL gateway.
2. Plug one end of the Ethernet cable into the back of the DSL gateway.
3. Plug the other end of the Ethernet cable into the back of the device.
4. Restart the computer or the connected device to get a network address from the gateway.

Head for
special
condition

Wireless Connections

Tablets, smartphones, and e-book readers typically use wireless connections.

1. Find the wireless network name (SSID) and numeric security key on the bottom or side of the gateway device.
2. On the wireless device, go into **settings**.
3. Choose the **wireless settings** option.
4. Select the network you want to connect to.

Note

   The default network name will be the same as the SSID printed on the gateway device.

5. Choose **connect**.
6. Enter the security key at the prompt.

Note

On the device, the broadcast icon in the system tray or top bar will flash until a connection is established.

Head for
special
condition

Wired or Wireless Devices

Laptops and network-ready television devices have both a wired Ethernet port and built-in wifi. Some devices like desktop computers or network printers may be modified to include wifi capability. As a general rule, the wired connection will be faster than the wifi connection. In order to connect these devices, refer to the previous instructions.

© Cengage Learning 2014

Figure 9.5   Body of a Set of Instructions

Field-Testing Instructions

A field test is a method of direct observation by which you can check the accuracy of your instructions. To perform a field test, ask someone who is unfamiliar with the process to follow your instructions while you watch. If you have written the instructions correctly, the reader should be able to perform the entire activity without asking any questions. When you field-test instructions, keep a record of all the places where the reader hesitates or asks you a question.

Worksheet for Preparing Instructions

☐ **Assess the audience for these instructions.**

- Estimate the amount of knowledge the audience has about the process. Are they beginners or intermediates?
- What will you tell the readers in the introduction? What will you assume about them? What do they need to know? What can they get from your instructions? How do they decide if they want to read your instructions? What will make them feel you are helpful and not just filling in lines for an assignment? How will you orient them to the situation?

☐ **What is the end goal for your readers?**

☐ **Analyze the process.**

- Construct a flow chart that moves backward from the end goal.
- Use as many boxes as you need.

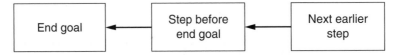

☐ **List all the conditions that must be true for the end goal to occur.** (For instance, what must be true for a document to open in a word processing program? The machine is turned on, the correct program is selected, and a new document is opened.)

☐ **List all the words and terms that the audience might not know.**

☐ **List all the materials that a person must have in order to carry out the process.**

☐ **Where do the readers need a visual aid to "give them permission," or to orient them to the situation, or to show them something quickly that is easy to see but hard to describe in words?**

☐ **Draw the visual aids that will help readers grasp this process.** Use visuals that illustrate the action or show the effect of the action.

☐ **How will you arrange this material on the page so that it is easy for readers to read quickly, but also to keep their place or find it again as they read?**

□ **Construct a style sheet.** Choose your head system, margins, columns, method of treating individual steps, and style for writing captions.

□ **Convert the topic of each box in the flow chart into an imperative instruction.** Add cautions, suggestions, and substeps. Decide whether a sequence of steps should be one step with several substeps or should be treated as individual steps.

□ **How will you tell them each step? How—and where—will you tell them results of a step? How—and where—will you tell them background or variations in a step?**

□ **Why should you write them a set of instructions in the first place?** Why not write them a short report or an article? A report tells the results of a project, an article informally explains the concepts related to a project, and a set of instructions tells how to do the project.

Worksheet for Evaluating Instructions

□ **Evaluate your work. Answer these questions:**

- Does the introduction tell what the instructions will enable the reader to do?
- Does the introduction contain all the necessary information on special conditions, materials, and tools?
- Is each step a single, clear action?
- Does any step need more information—result of the action, safety warning, definitions, action hints?
- Do the steps follow in a clear sequence?
- Are appropriate visual aids present? Does any step either need or not need a visual aid?
- Are the visual aids presented effectively (size, caption, position on page)?
- Does the page layout help the reader?
- Are all terms used consistently?

TIP ||||

Information Order in a Step

If your step contains more than just the action, arrange the items as action-effect. In the following example, the first sentence is the action, the second sentence is the effect.

Press **Enter** on the integrator.
After you press **Enter**, the exposure time in units will appear in the top left display window.

If your step contains a caution or warning, place it first, before you tell the audience the action to perform.

1. CAUTION: DO NOT LIGHT THE MATCH DIRECTLY OVER THE BUNSEN BURNER!

Light the match and slowly bring it toward the top of the Bunsen burner.

Examples

The four examples that follow exemplify sets of instructions.

Example 9.1

Instructions
for a Beginner

INSTRUCTIONS: HOW TO USE THE MODEL 6050 PH METER

Introduction

This set of instructions provides a step-by-step process to accurately test the pH of any given solution using the pH Meter Model 6050. The pH meter is designed primarily to measure pH or mV (millivolts) in grounded or ungrounded solutions. This set of instructions assumes that the pH meter is plugged in and that the electrode is immersed in a two-molar solution of potassium chloride.

Materials Needed

- Beaker containing 100 ml of 7.00 pH buffer solution
- Beaker containing 100 ml of 4.00 pH buffer solution
- Thermometer
- Squeeze bottle containing distilled water
- Four squares of lint-free tissue paper

How to Program the pH Meter

1. Press the button marked pH (A in Figure 1) to set the meter to pH mode.
2. Set pH sensitivity by pushing the pH sensitivity button down to .01 (B in Figure 1).

3. Gently remove the pH electrode (C in Figure 1) from the plastic bottle in which it is stored, and rinse it gently with distilled water from your squeeze bottle.

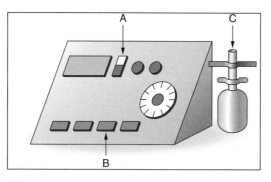

Figure 1
Sargent-Welch pH meter Model 6050

4. Carefully lower the electrode into the beaker containing the pH 7.00 buffer solution.
5. Set temperature control.
   a. Using the thermometer, take the temperature of pH 7.0 buffer solution.
   b. Turn the temperature dial (D in Figure 2) to the temperature reading on the thermometer in degrees Celsius.
6. Set electrode asymmetry (intercept) by rotating the dial marked "intercept" (E in Figure 2) until the digital display (F in Figure 2) reads 7.00.
7. Raise the electrode from the 7.00 pH buffer solution, rinse gently with distilled water from your squeeze bottle, and dry tip of the electrode using lint-free tissue paper.

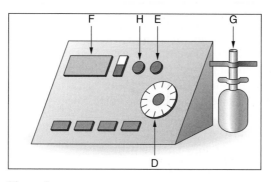

Figure 2
Sargent-Welch pH meter Model 6050

(Continued)

8. Lower the electrode (G in Figure 2) into the buffer solution of pH 4.00 to set the lower pH limit.
9. Set the response adjustment (slope) by rotating the dial marked "slope" (H in Figure 2) until the digital display reads 4.00.
10. Raise the electrode from the 4.00 pH buffer solution.
11. Rinse the electrode gently with distilled water from your squeeze bottle.
12. Dry the tip of the electrode using lint-free tissue paper.

Example 9.2

Two Sets of Instructions for Intermediate Readers

HOW TO CREATE A PDF FROM AN MS WORD FILE (IN MS WORD 2007)

1. Open the file.
2. Select *Save As*.
3. Select *Adobe PDF*.
    OR
4. Open the file and select *Print*.
5. In the *Print* menu find *Print to Adobe PDF* (the command varies but all of them say PDF).
6. Select *OK*. The file is saved as a PDF on your computer.

HOW TO ADD AUDIO COMMENTS TO A PDF

Equipment: A head set with a microphone makes your comments clearer. You can also just turn on the computer microphone and speak to the computer.

1. Open Acrobat Professional.
2. Select *File/Open*.
3. Open the file to which you plan to add comments.
4. Select *Review and Comment,* then *Comment and Markup Tools*.
5. Select *Record Audio Comment*. The audio comment icon appears.
6. Position the icon at the spot where you wish to add a comment.
7. Right Click. The **Sound Recorder Menu** appears.
8. Click the red **Record** dot to begin recording. Speak your comment. Click the **Record** dot again to end recording.
9. Save the document.
10. Repeat the process as often as you wish.

Example 9.3

Professional
Instructions
for a Large
Beginner
Group

HOW TO RETURN ALL THE TESTING
MATERIALS SO WE DON'T MESS UP

1. You MUST return everything that you received. If for some reason something is missing, you need to contact me, Fred, or Joan and we will have to fill out a "Materials Not Returned List." It is a very big deal to have missing materials, so if you are missing something please make every effort to find it.

2. Pack together all *Non-scorable items* (preprinted scripts, kits for special sections) and all *Unused Test Materials* (test booklets that were not used because they were overage, the student is no longer here, the student bled or worse on it, the test was given in error).

   a. Check the materials that you are returning against the material identification numbers that you received.

   b. For the unused/unscorable test booklets, write a huge **DO NOT SCORE** in permanent marker across the front cover.

3. *Scorable Materials* will go together in groups according to grade, cluster, and level.

   a. Each pile of test booklets needs a *School Identification Sheet* in front of it. The School Identification Sheets that you have are blue and you can make as many copies as you need of it.

   b. Here is the tricky part: no more than 25 booklets can be bound into a pile. If you have more than 25 of any one kind of test, you need to break it up into smaller groups. Example: we have 74 Grade 6–7 Level M booklets. You will have two piles of 25 and one of 24, each with their own header completely filled out. Rubber band each pile of books with their header.

   c. The Scorable Materials DOES include tests that are incomplete but that we still want scored. So, students who moved to another district can be in this pile. Students that took part of the test but were absent for another part are also part of this pile. Make sure that those absences are indicated on the test booklet.

Return the materials in the boxes you were sent.

Example 9.4

Professional
Instructions

Understanding the HEAD and BODY

Most webpages are divided into two sections: the HEAD and the BODY. The HEAD section provides information about the URL of your webpage as well as its relationship with the other pages at your site. The only element in the HEAD section that is visible to the user is the title of theç webpage (*see page 39*).

(Continued)

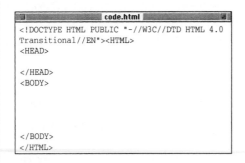

```
                  code.html
<!DOCTYPE HTML PUBLIC "-//W3C//DTD HTML 4.0
Transitional//EN"><HTML>
<HEAD>

</HEAD>
<BODY>




</BODY>
</HTML>
```

*Every HTML document should be divided
into a HEAD and a BODY*

To Create the HEAD Sections

1. Directly after the initial !DOCTYPE and HTML tags (*see page 37*), type **<HEAD>**.
2. Create the HEAD section, including the TITLE (*see page 39*). Add META information (*see pages 290–293*) and the BASE (*see page 113*), if desired.
3. Type **</HEAD>**.

The BODY of your HTML document contains the bulk of your webpage, including all the text, graphics, and formatting.

```
         file:///.../basic.dumps/basic.html/headbody.html



```

*With no title and no contents, a browser
has to scrape together a little substance
(in the form of a title) from the file name
of the HTML document.*

To Create the Body

1. After the final</HEAD> tag and before anything else, type **<BODY>**.
2. Create the contents of your webpage.
3. Type **</BODY>**.

Tip

For pages with frames, the BODY section is replaced by the FRAMESET.

Source: Elizabeth Castro, *HTML4 for the World Wide Web: Visual Quickstart Guide*, p. 38, © 1998. Republished by permission of Pearson Education, Inc. Publishing as Peachpit Press.

Exercises

▶ You Create

1. Construct a visual aid that illustrates an action. For instance, show a jack properly positioned for changing a tire. Then write the instructions that would accompany that visual.

2. Write a set of instructions for a common activity, such as wrapping a package, tying a shoe, or programming a digital video recorder. Choose one of the columnar formats shown in Figure 9.3 on page 254. Have a classmate try to perform the process by following your instructions. Discuss with the class the decisions you had to make to write the instructions. Consider word choice, layout, visual aids, sequence of steps, etc.

3. Make a flow chart or decision chart of a process. Choose an easy topic, such as a hobby, a campus activity, or some everyday task. In class, write the instructions that a person would need to perform the process. Depending on your instructor's preferences, you may either use your own chart or exchange charts with another student and write instructions for that student's chart.

▶ You Revise

4. Rewrite the following steps from the instructions for changing a car's oil:

   1. Get drainage pan and place it under the oil pan of the car.
   2. Grab a crescent wrench and locate the oil plug, on one side of the oil pan.
   3. Use the crescent wrench to turn the plug counterclockwise (ccw) until it comes out and oil drains out.
   4. While this is draining, grab a filter wrench and locate oil filter.
   5. Turn the oil filter counterclockwise with the filter wrench until it comes off and the oil drains into the drainage pan.

5. Read the following two sets of instructions about setting up social media sites. Follow them to check them for accuracy. Using screen captures insert visual aids (if you are performing this exercise without a computer, draw rough sketches).

   ### Twitter Account

   1. Create a Twitter account at www.twitter.com. Choose the name by which you'll be known on Twitter, enter your e-mail address, and create a password. Click **Sign up for Twitter**.
   2. Choose your Twitter name, and your username. Select **Create my account**.
   3. On the Welcome screen, choose to watch the tutorial or click Next.

4. Confirm your e-mail address.

5. Create your profile. Use the **Edit your profile** button and fill in the information form.

6. To send a Tweet click on the **Home** button (upper left corner) to access the tweet composition screen. Create your Tweet in the **Compose new Tweet** area.

Facebook for Company Account

Setting up an Account for a Business

1. Create your business account at www.facebook.com. Select the link for setting up a page for a celebrity, band, or business. It is beneath the **Sign Up** button. The **Create a Page** page opens.

2. Choose what kind of organization your company would be listed as. Most likely you will choose one of the options in the top row. The easiest to use is **Company, Organization** or **Institution**. The **Create a Facebook Account** page appears.

3. To create an account for your company, fill in the information as required on the page.

4. Follow the confirmation instruction by clicking on the link sent to your e-mail by Facebook. The **Set Up** page appears.

5. If you wish to place a photo on your page, upload it from your computer. Click **Next**.

6. Supply general information about your company. Click **Save Info**.

7. Open the **Admin Panel**. Select **Build Audience**. Invite e-mail contacts or create an ad.

8. Post on your company's Facebook page wall. Create **Status** posts, upload **Photos** or **Video**, or announce **Events** or **Milestones**.

6. Convert the following paragraphs into a set of instructions:

First I went to the website www.uwstout.edu/place/studentwebregistration*. html to read the general instructions given by the Co-op and Placement Office to reach the new website and log in to my profile. I was directed to www. uwstout.edu/place/ and was instructed to click on the *"Students create or update your eRecruiting profile"* link to enter the log-in screen. The username is the last five digits of my ID number with the word *"stout"* added on the end, and the password is the last four digits.

I next chose to enter my personal information by clicking on the *"Edit your profile"* link. There are then options to update *"Personal Info, Academics, Future Plans,"* and *"Administration"* links to choose. In each of these, I either typed in information about myself, my education, my qualifications, and plans or chose information from drop-down lists containing possible options. As each section was completed, more information about me was saved in my profile.

> After finishing entering personal information, I chose to post my résumé online. I clicked on the *"Documents"* link and read the instructions there to upload files. After clicking on *"Upload Documents"* and choosing *"Résumé,"* I was instructed to type in the name of the file containing my résumé or search for it among the folders on my computer and push *"Upload."*

7. Rewrite all of the items in Exercise 4 from the point of view of a "chatty help" columnist in a newspaper. Use paragraphs, not numbered steps.

8. Rewrite Example 9.3 (p. 265) as a report structured in paragraphs.

▶ Group

9. In groups of two to four, depending on your instructor's choice, post a YouTube video set of instructions. You can deal with any topic, from appropriately planting trees to adding audio comments to a PDF file. In other words you can film a person performing an action or you can capture a sequence of actions a person needs to follow on screen. Carefully review the instructions for posting a video before you begin. If your instructor so requires, hand in regular progress reports (see Chapter 10). You could also set up a wiki to share questions, problems, solutions, and accomplishments with one another.

10. Compare Example 9.2 (p. 264) with Example 9.3 (p. 265). In groups of three or four, discuss the differences in tone and in the presentation of the action. Report to the class which document you prefer to read and why.

11. For Writing Assignment 1 or 2, construct a flow chart of the process. Explain it to a small peer group who question you closely, causing you to explain the steps in detail. Revise the chart based on this discussion.

12. For Writing Assignment 1 or 2, create a template for your instructions, including methods for handling heads, introduction, steps, visual aids, captions, and columns. Review this template with your peer group, explaining why you have made the choices you have. Your peer group will edit the template for consistency and effectiveness.

13. For Writing Assignment 1 or 2, bring the final draft of your instructions to your peer group. Choose a person to field-test. With your instructor's permission, field-test each other's instructions. Note every place where your classmate hesitates or asks a question, and revise your instructions accordingly.

14. Bring an article instruction to class. Computer and household magazines offer the best sources for these articles. In groups of three or four, analyze the models and decide why the authors decided to use the article method. Depending on your instructor, either report your analysis to the class, or as a group rewrite the instructions, either with a different tone (say, as a coach) or as a numbered set.

Writing Assignments

1. Write a set of instructions for a process you know well. Fill out the worksheet and then write the instructions. Use visual aids and design your pages effectively, using one of the columnar formats shown on page 254. Pick a process that a beginning student in your major will have to perform or choose something that you do as a hobby or at a job, such as waxing skis, developing film, ringing up a sale, or taking inventory.

2. Divide into groups of three or four. Pick a topic that everyone knows, such as checking books out of the library, applying for financial aid, reserving a meeting room, operating an LCD projector, replacing a lost ID or driver's license, or appealing a grade. Then each team should write a set of instructions for that process. Complete the worksheet on pages 260–261. When you are finished, decide which team's set is best in terms of design, clarity of steps, and introduction.

3. Write a learning report for the writing assignment you just completed. See Chapter 5, Writing Assignment 8, page 137, for details of the assignment.

Web Exercises

1. Give instructions to a beginner on how to create a webpage using a wizard, template, or Web-authoring tool with which you are familiar.
2. Convert the process paper that you wrote for Chapter 8, Web Exercise 3, into a set of instructions for a beginner.

Works Cited

Bhatia, Vijay K. *Analyzing Genre: Language Use in Professional Settings*, Longman, 1993, Print.

Castro, Elizabeth. *HTML for the World Wide Web*. Berkeley, CA: Peachpit, 1998, Print.

Dehaas, David. "Say What You Mean." *OHS Canada*. Web. 28 May 2012. <http://www.ohscanada.com/training/saywhatyoumean.aspx>.

Lawler, Sean. "9 Tips for Writing Instructions Your Speakers Will Understand." Big Ideas Blog. 2010. Web. 28 May 2012. <http://blog.omnipress.com/2010/08/9-tips-for-writing-instructions-your-speakers-will-understand/>.

Mousten, B. *Communication in English for Science and Technology.* Aarhus School of Business, 2007. Print.

Mousten, B., Humbley, J., Maylath, B., and Vandepitte, S. "Communicating Pragmatics About Content and Culture in Virtually Mediated Educational Environments." Chapter 12 in St. Amant, K. and Kelsey, S. IGI Global, 2012. Print.

Schell, Martin A. "Frequently Asked Questions About Globalization and Localization." *American Services in Asia.* Web. 28 May 2012. <www.globalenglish.info/faq.htm#two>.

Starke-Meyerring and Wilson, M., eds. *Designing Globally Networked Learning Environments—Visionary Partnerships, Policies, and Pedagogies.* Sense Publishers, 2008. Print.

St. Amant, K., and Kelsey, S. *Computer-Mediated Communication across Cultures—International Interactions in Online Environments.* IGI Global, 2012. Print.

Informal Reports and E-mail

CHAPTER CONTENTS

CHAPTER 10 **IN A NUTSHELL**

Informal reports. Informal reports are usually short (one to ten pages). Their goal is to convey the message in an understandable context, from a credible person, in clear, easy-to-read text. They do not have "formal elements" like title pages, table of contents and other elements covered in Chapter 12.

Informal report structure. Basically anything can be called an informal report. If it is short and contains no formal elements, it is an informal report. Probably the most important structure to learn is the IMRD (Introduction, Method, Results, and Discussion). This structure is as much a way of thinking as it is a report format.

The *Introduction* explains your goal and why this situation has developed.

The *Method* outlines what you did to find out about the situation. It establishes your credibility.

The *Results* establish what you found out, the information the reader can use.

The *Discussion* describes the implications of the information. It gives the reader a new context.

Informal report strategies. Key strategies include

▶ Explain your purpose—what your reader will get from the report.

▶ Use a top-down strategy.

▶ Develop a clear visual logic.

▶ Provide the contents in an easy-to-grasp sequence and help the reader out by defining, using analogies, and explaining the significance to the person or organization.

The day-to-day operation of a company depends on e-mails and informal reports that circulate within and among its departments. These documents report on various problems and present information about products, methods, and equipment. The basic informal format, easy to use in nearly any situation, has been adapted to many purposes throughout industry.

This chapter explains the elements of informal reports, and the types of informal reports, including IMRD reports, analytical reports, progress reports, outline reports, and summaries.

Basic Strategies for Informal Reports

Informal reports are those that will not have wide distribution, will not be published, and are (usually) shorter than ten pages (General Motors). The basic strategies are

create an effective introduction,
develop a consistent visual presentation
Follow the expected "thought path," if there is one.

Introduction

Introductions orient readers to the contents of the document. You can choose from several options, basing your decision on the audience's knowledge level and community attitudes. To create an introduction, you can do one of three things: provide the objective, provide context, or provide an expanded context.

Provide the Objective

The basic informal introduction is a one-sentence statement of the purpose or main point of the project or report, sometimes of both. This type of introduction is appropriate for almost all situations and readers.

| Objective of the project | To evaluate whether the Mertes Hardware should install an Iconglow retail point of sale system |
| Objective of the report | To report on investigation of the feasibility of installing an Iconglow retail point of sale system in the three Mertes hardware stores. |

If this statement is enough for your readers, go right into the discussion. If not, add context sections as explained later.

Provide Context

To provide *context* for a report means you explain the situation that caused you to write the report. This type of introduction is an excellent way to begin informal reports. It is especially helpful for readers who are unfamiliar with the project. Include four pieces of information: cause, credibility, purpose, and preview. Follow these guidelines:

▶ Tell what caused you to write. Perhaps you are reporting on an assignment, or you may have discovered something the recipient needs to know.
▶ Explain why you are credible in the situation. You are credible because of either your actions or your position.
▶ State the report's purpose. Use one clear sentence: "This report recommends that Mertes Tile should install an Iconglow retail point of sale system."
▶ Preview the contents. List the main heads that will follow.

Here is a sample basic introduction.

Cause for writing Source of credibility Purpose Preview	I am responding to your recent request that I determine whether Mertes Hardware should install an Iconglow retail point of sale system. In gathering this information, I interviewed John Broderick, the Iconglow Regional Sales Representative. He reviewed records of current business computer system and personnel who work it in various capacities. This report recommends that Mertes Tile install the Iconglow system. I base the recommendation on cost, space, training, and customer relations.

Special Case: Alert the Reader to a Problem.   Sometimes the easiest way to provide context is to set up a problem statement. Use one of the following methods:

▶ Contrast a general truth (positive) with the problem (negative).
▶ Contrast the problem (negative) with a proposed solution (positive).

In either case, point out the significance of the problem or the solution. If you cast the problem as a negative, show how it violates some expected norm. If you are proposing a solution, point out its positive effect. Here is a sample problem–solution introduction.

Negative problem and its significance Proposed solution Positive significance Purpose of report	Our current point of sale system has an average of two system break downs a month. It does not provide adequate inventory control data. The interface of current system causes customer wait times of up to ten minutes in line. The break downs and the interface problem cause annoyed customers, some of whom simply leave items and go elsewhere. Because of the inadequate inventory control, clerks in one site are unable to determine if an item is available in another site. An Iconglow POS system would eliminate breakdowns, dramatically reduce long line waits and create easy-to-access inventory data. This report recommends that we purchase the Iconglow system.

Develop a Consistent Visual Presentation

Visual presentation was explained in Chapter 7. Use those concepts as you write an informal report. Remember, the most important strategy is "be consistent." Pay attention to three items in particular.

Headings
Page numbers
Identifying visual aids

Headings

Informal reports almost always contain heads. Usually you need only one level; the most commonly used format is the "side left." Follow these guidelines:

▶ Use a word or phrase that indicates the contents immediately following.
▶ At times, use a question for an effective head.
▶ Place heads at the left margin, double-spaced above and single-spaced below. Use boldface or all caps. Or choose a "heading style" from your word processor's *Styles* menu. Usually bold face is easiest to add and to read.
▶ Capitalize either every word or else use "down style" (capitalize only the first word and any proper nouns).

Note: If you use one of your word processor's heading styles, don't go overboard. Many options include various blocks of colors in varying intensities. Just use the black on white varieties.

▶ Do not punctuate after heads (unless you ask a question).

Side left,
boldface

Double-space

Will the New System Save Money?

The new Iconglow system will pay for itself within ten months. Currently, employees spend 87 hours a month updating files. The new system will reduce that figure to 27, a savings of 60 hours. These 60 hours represent a payroll savings of $900.00 a month. We will need to purchase capabilities for six stations at a cost of $1500 each, or a total of $9,000. The savings alone will pay for the system in ten months. This amount of time is under the one-year period allowed for recovery.

Is There Enough Space for the System?

Double space

The new interface screens will easily fit in their allocated spaces at the retail counters. The current office computer system will accommodate the software upgrade and will not have to be replaced. Workflow both in the office and at the check counter will not be hampered by adding this system.

(Continued)

Double-space

> ## Will the Point of Sale System Affect Customer Relations?
>
> The Iconglow system will allow employees to process customer issues such as coupons more quickly, reducing customer wait time dramatically. Employees will also be able to inform customers whether out-of-stock items are available at one of our other locations. Both changes will reduce customer complaints and increase loyalty as we compete against the big box stores in the area.

Figure 10.1 Heading Example

Page numbers

Page numbers are very helpful if you or others have to refer to the document. If the document is one page, then you don't need page numbers, but if it is two pages or more, add them. The easiest way to add number is simply use the "Insert" menu on your word processor. Follow these guidelines:

- ▶ If you use just page numbers, place them in the upper right corner or in the bottom center.
- ▶ If you use "Insert page numbers" the processor will put the number in the "Header," the area at the top of the page. In some situations it is helpful if you add "header information" such as report title and date. Generally, the page number goes to the far right and other information (report title and/ or date) appears to the left.

Iconglow Recommendation   12/24/14   2

Consistently Identify Visual Aids

Handle all these aspects in the same fashion

Choose your caption word—for instance, Figure, Chart, Graph. While these words can indicate differences that are important in formal reports, they are not as important in informal reports.

Place your visuals in the same relative location, at the left-hand margin or in the center.

Place the caption either above or below.

Develop a "caption style" and use it consistently. Notice the capitalization, use of numerals and period in this example: Figure 1. Quarterly Sales

Types of Informal Reports

Writers use informal reports in many situations. This section introduces you to several variations.

IMRD Reports

An IMRD (*Introduction, Methodology, Results, Discussion*) report is a standard way to present information that is the result of some kind of research. This approach can present laboratory research, questionnaire results, or the results of any action whose goal is to find out about a topic and discuss the significance of what was discovered. You could write an IMRD about deciding what food to purchase at a fast food store. Actually this type of report is really a way of thinking. It causes you to tell a story about your project in a way that most readers will find satisfying because tell the reader what you wanted to find out, how you went about it, what you found out, and what those findings mean.

▶ For the *introduction*, present the question you investigated (the goal of the project) and the point of the paper. It is helpful to give a general answer to the question. Consider these questions:
  ▶ What is the goal of this project?
  ▶ What is the goal of this report?
▶ For the *methodology section*, write a process description of your actions and why you performed those actions. This section establishes your credibility. Explain such things as to whom you talked, actions you took and why. This description should allow a reader to replicate your actions. Consider these questions:
  ▶ What steps or actions did you take to achieve the goal or answer the questions? (Explain all your actions. Arrange them in sequence, if necessary.)
  ▶ Why did you perform those actions?
▶ For the *results section*, tell what you discovered, usually by presenting a table or graph of the data. If a visual aid is all you need in this section, combine it with the discussion section. If you add text, tell the readers what to focus on in the results. Honesty requires that you point out material that might contradict what you expected to discover. Consider these questions:
  ▶ What are the results of each action or sequence?
  ▶ Can I present the results in one visual aid?
▶ In the *discussion section*, explain the significance of what you found out. Either interpret it by relating it to some other important concept or suggest its causes or effects. Relate the results to the problem or concerns you mentioned in the introduction. If the method affects the results, tell how and suggest changes. Often you can suggest or recommend further actions at the end of this section. Consider these questions:
  ▶ Did you achieve your goal? (If you didn't, say so, and explain why.)
  ▶ What are the implications of your results? For you and your goals? For other people and their goals?
  ▶ What new questions do your results raise?

Introduction

Current digital cameras include many options that affect the look of the captured image. These options are so numerous that many first-time owners of digital cameras never explore them. At a recent gathering of a Learning in Retirement class, attendees admitted that they not only never looked at the options, they did not know which ones were contained in their camera nor the effect of using them. Because many of the photographers in this age group wish to take portrait photographs, this report explores which menu option might best help them achieve quality photographs.

Method

The camera used for this experiment was a Canon 7D. While this is a top-of-the-line model, its options are similar to other less expensive models. The ISO was set to 1000. A Canon 18-135 mm lens was used, set at 100 mm, a common length for taking portraits. Natural lighting illuminated the indoor scene. A female model sat approximately 8 feet from the camera. Within approximately two minutes, images were taken with the camera set at Standard, Portait and Faithful. These three settings would give a range of quality. Other settings such as Black and White and Macro were not used in this test. After the images were taken, two professional photographers evaluated them for quality, based on color accuracy and image clarity.

Results

The three images are displayed in the following text.

Standard   *Portrait*   *Faithful*

Discussion

Both photographers agreed that the image created with the Faithful setting best represented the subject in terms of color accuracy and image clarity. Based on this conclusion facilitators of photography workshops should introduce photographers to the Faithful menu option and encourage its use. As a result of this experiment, further tests should be conducted in order to determine the helpfulness and ease of use of the many other options found in camera menus.

Figure 10.2 Digital Camera IMRD Report

Brief Analytical Reports

Brief analytical reports are very common in industry. Writers review an issue with the goal of revealing important factors in the issue and of presenting relevant conclusions. The two reports given in the following text illustrate varied uses of this form.

Objective

The purpose of this report is to inform you of a malfunction in the ventilation system and to propose a solution to this problem.

Summary

After receiving numerous complaints of illness from many of the employees in the Painting Department, I have inspected the ventilation system and found it inadequate. The fresh air volume entering the room is well under the requirement for safe working conditions. I have found that the problem does not lie with the fan itself, but in the drive system design. I have decided that the most economical solution is to redesign the drive system of the main ventilation fan. In this report I will address both the problem and the solution in detail.

Problem

The ventilation system that is currently in use is not providing proper fresh air volume to the painting area. The result of this problem has been numerous complaints of headaches and nausea among the employees who work in this department. After conducting extensive research, I have found that the volume of fresh air entering the area is well under the OSHA standard for safe working conditions. This lack of ventilation is not caused directly by the fan but the system that drives it.

Upon my inspection of the drive system of the fan, I found two problems: severe misalignment between the motor and fan sheaves and an improper drive speed ratio. The severe misalignment is causing extensive wear and decreased performance. The misalignment causes the belt to fit improperly into the sheave groove. The improper drive speed ratio is caused by a miscalculation in the design of the drive system. The ratio is a numeric value that compares the speed of the faster shaft to the speed of the slower shaft. This ratio is controlled through the use of sheaves with different diameters for the motor and the fan.

Solution

I propose the redesign of the drive system for the main ventilation fan. This process would include realigning the motor with respect to the fan and correcting the drive speed ratio. The realignment of the motor can be achieved by adding an adjustable base to the foot of the motor. The adjustment is made through the use of elongated holes in the base. Positioning the sheaves parallel to each other allows proper contact between the belt and the sheave thus there is better transmission of power to the fan and belt life is increased. Correction of the speed ratio is easily achieved through the substitution of the current sheaves with those of the proper diameters. Many combinations of diameters are available that will fulfill the requirements for increasing the speed of the fan. This is a relatively economical solution to a very serious problem.

© Cengage Learning 2014

Figure 10.3   Lack of Proper Ventilation in the Painting Department

INTRODUCTION

Dietetics professionals must be aware of the resources available on the World Wide Web. These resources include nutrition education materials, current legislation information, job opportunities, government programs, disease/disorder information.

FINDING CREDIBLE RESOURCES

Keyword searches (e.g. "dietetics", "dietitian," "nutrition education") produce literally millions of resources. As you search use the CARS (Credibility, Accuracy, Reasonableness, Support) method to determine whether the information provided is appropriate for use in one's practice. For a complete explanation of the CARS method see "The CARS Checklist"). The CARS method was utilized reviewing websites related to dietetics. All websites described in this report passed the CARS examination. A summary of the findings are given in Table 1.

WEB RESOURCES AVAILABLE
Government Sites

The most detailed and reliable information found on the Internet came from government resources. Nutrition.gov (www.nutrition.gov) is the official nutrition site for the United States government. This site links to all nutrition-related information, ranging from food safety and security to diabetes and disease management. The best part of this site is that it links to all federal nutrition programs. These sites give important information as well as provide the ability to download and print forms. Nutrition education, tools and resources are also available on this site. Another government site, U.S. Senate Committee on Agriculture, Forestry, and Nutrition (agriculture.senate.gov), provides archived federal bills as well as current legislation regarding nutrition, forestry, and agriculture.

Professional Sites

Professional websites that provide many links and extremely reliable information. "Food and Nutrition You Can Trust" created by the Academy of Nutrition and Dietetics (www.eatright.org). This site contains a wide variety of resources beneficial to dietitians, ranging from current nutrition issues to patient education materials.

Dietetics.co.uk (dietetics.co.uk), based in the United Kingdom, is a message board forum for dietitians across the world. Dietetics professionals can post and reply to message boards dealing with all aspects of the dietetics profession, including enteral/parenteral feedings, professional issues, nutrition assessment and screening, freelance and private practice dietetics, and more.

Dietetics.com houses links to state and national dietetic associations, antiquackery information, and other basic information to help the dietetic professional.

Dietetic Career Searches

The Academy of Nutrition and Dietetics offers Search Jobs, a career link page that is a national database of current openings in the field of dietetics. A search can be narrowed by choosing an area of discipline or choosing by location. However, the career link is not extensive at this point and does not offer many positions.

Jobs in Dietetics (jobsindietetics.com) offers a nationwide career search. There is a membership charge for utilizing this Web resource. Its member-only approach makes it impossible to summarize its usability or quality.

Results of Query

Table 1 summarizes the previous paragraphs. The websites were placed into three categories: professional, government, and career search. It was noted whether the website provided outside links. The availability of links on pages was taken into consideration in the overall rating of the quality and usefulness of the websites. Based on the CARS analysis, each site was scored with a one to five rating with five being the highest quality and most beneficial to dietetic practitioners.

CONCLUSION

There are many beneficial resources available to dietetic professionals on the World Wide Web. When viewing websites, it is important to keep the CARS (Credibility, Accuracy, and Reasonableness, Support) method in mind. Professional, government, and dietetic career search capabilities can be found and utilized easily using the Internet. Becoming familiar with this process will enhance the dietetic professional.

Resources

"Food and Nutrition You Can Trust." Academy of Nutrition and Dietetics. 2012.22 May 2012<www.eatright.org>.

"Job Search." Academy of Nutrition and Dietetics.2012.22 May 2012. <http://www.healthecareers.com/eatright/search-jobs/>.

"CARS Checklist." Student Success. 2012. Web. 22 May 2012. http://novella.mhhe.com/sites/0079876543/student_view0/research_center-999/research_papers30/conducting_web-based_research.html.

TABLE 1

Summary of Dietetics Resources on the Web

Name of Site	Web Address	Category	Links Available?	Rating (5 = best)
Food and Nutrition You Can Trust	www.eatright. org	Professional	Yes! Many	5
U.S. Senate Committee on Agriculture, Forestry, and Nutrition	agriculture. senate.gov	Government	Yes! Some not nutrition related	4
Nutrition. Gov	www.nutrition. gov	Government	Yes! Many	5
Job Search, Academy of Nutrition and Dietetics	www. adacareerlink. org	Dietetic Career Search	Yes! Other medical career searches	3
Jobs in Dietetics	jobsindietetics. com	Dietetic Career Search	No	2
Dietetics.Com	dietetics.com	Professional	Yes	3
Nutrition and Dietetics Forum	dietetics.co.uk	Professional	No	3

Dietetics.Com. 2010. Web. 22 May 2012. <http://dietetics.com>.
Jobs in Dietetics. Web. 22 May 2012. <http://jobsindietetics.com>.
Nutrition and Dietetics Forum. 2012. Web. 22 May 2012<http://dietetics.co.uk>.
Nutrition.Gov. Web. 22 May 2012<http://nutrition.gov>.
U.S. Senate Committee on Agriculture, Forestry, and Nutrition. 2012. Web.
    22 May 2012. <http://agriculture.senate.gov>.

© Cengage Learning

Figure 10.4   Credible Resources Available for Use by Dietitians

Progress Reports

Progress reports inform management about the status of a project. Submitted regularly throughout the life of the project, they let the readers know whether work is progressing satisfactorily—that is, within the project's budget and time limitations. To write an effective progress report, follow the usual process.

Evaluate your audience's knowledge and needs. Determine how much they know, what they expect to find in your report, and how they will use the information. Select the topics you will cover. The standard sections are the following:

▶ Introduction
▶ Work Completed
▶ Work Scheduled
▶ Problems

In the Introduction, name the project, define the time period covered by the report, and state the purpose: to inform readers about the current status of the project. In the Work Completed section, specify the time period, divide the project into major tasks, and report the appropriate details. In the Work Scheduled section, explain the work that will occur on each major task in the next time period. In the Problems section, discuss any special topics that require the reader's attention. An e-mail "cover letter" accompanied the report below, which was sent as an attachment. The text was "Dr. Riordan, attached is my March progress report on my construction manual project. If you have any questions, please contact me. Julia"

Summary

I am working on developing a user manual for the Universal Test Machine, TestWorks QT, for the UW-Stout Construction Lab. Tests have been modified for the students' use, which will require a new set of instructions. The instructions will be designed to help guide the student through machine setup, starting up the TestWorks QT software program, running the test, and proper shutdown.

Work Completed

The client and I have decided that the manual will be hard copy, 5.50∞ 80, bound with a plastic spiral. Each step of the process will include an illustration. The client decided that the manual would include instructions for three different types of tests: Bending, Compression, and Tensile. I have written instructions for the Compression test. Additional information will be added after the client reviews the instructions. Thursday, March 28, I met with the client in the construction lab where approximately 15 photos were taken for the Tensile, Compression, and Bending tests. All photos will use JPG format.

Work Scheduled

Digital photos of the Bending, Compression, and Tensile tests will be viewed and enhanced using Photoshop. I plan to have the photos prepared and sent to the client by Wednesday to view and approve.

The next client meeting occurs on April 6. After that meeting I will develop the written instructions. The client will receive the rough draft of the instructions by April 15.

(Continued)

Problems

The Universal Test machine is scheduled to be used for classes during the only hours I have free to conduct usability testing in the instructions. At present no students have agreed to serve as usability testers. I will work with the lab instructor to resolve these issues.

© Cengage Learning

Figure 10.5   Progress Report Example

Summaries and Abstracts

Summaries tell readers the main points of an article or a report. Readers may use summaries to decide whether to read the entire article or report, to get the gist of the article or report without reading it, or to preview the material before reading it. Usually the term *abstract* refers to a one-paragraph document; *summary* refers to a multi paragraph document. A special type of summary, the executive summary, is discussed in Chapter 12.

> ❱ To create a summary or abstract
> ❱ Write the main point of the article in one sentence
> ❱ List the topics or sections of the article
> ❱ Add the key detail or point discussed in each section
> ❱ Be brief (usually one paragraph to one page)
> ❱ Define terms and acronyms if needed for your audience

The abstract given below (called an informative abstract) tells the reader exactly what points the article contains, not just the topics but the key idea for each topic.

This study examines batch respirometry as a screening tool to identify problematic papermaking additives that could disrupt the biological treatment of mill effluent. The method rapidly evaluates the toxicity of paper additives by tracking oxygen consumption of the respiring microorganisms in the activated sludge. Batch screening tests of 20 paper additives indicated that three paper dyes, a cleaner/solvent, and a microbiocide were the most toxic to the respiring biomass, while polymeric additives had no significant impact. A four-month pilot study with an orange dye confirmed the validity of the rapid respirometric method coupled with microscopic examination. The results also show that biological treatment systems can recover from the impact of harmful additives.

Figure 10.6   Professional Abstract

Source: From Keech, Gregory W., Phillip Whiting, and D. Grant Allen. Effect of paper machine additives on the health of activated sludge. TAPPI Journal 83.3 (March 2000): 86–90.

Theories of ethics typically emphasize either good character (asking "Who will I be?") or right behavior (asking "What will I do?"). Studies of ethics in technical communication have typically focused on the analysis of behavior, offering heuristics for deciding ethical dilemmas. Interviews with 48 technical communicators, however, reveal little exercise of such analytical processes. In making moral choices on the job, the majority look to feelings, intuition, and conscience. Ethics might be more effectively taught through a narrative perspective, especially by identifying models of moral courage and integrity.

Figure 10.7 From Dragga, Sam.

Source: From Dragga, Sam. "A Question of Ethics: Lessons From Technical Communicators on the Job." Technical Communication 6:2 (1996): 161–178. Reprinted by Permission of Taylor and Francis (http://www.tandfonine.com)

Background or Conceptual Reports

A background or conceptual report gives readers information that they need in order to orient themselves to a situation or topic. Basically such a report is an explanation. These reports can explain any topic, from cloud formations, to plumbing systems, to volcano eruptions. These reports give readers access to the topic, with the intent that they are reasonably satisfied that they understand the basic or important issues involved. To write such a report consider the audience's knowledge level. In almost all cases the audience knows little about the topic. Often tone is especially important. Consider the all-too-frequent case of being diagnosed with any form of cancer. Almost always the patient knows little or nothing about the characteristics and causes of his or her cancer, and often the patient is very concerned about the potential fatal consequences of the disease. But to give a less emotional example to people learning new technologies, for instance a social media site such as Twitter or Facebook, need aspects of the site explained. For instance, the privacy settings on photographs are a concern to many who post photos. A clear explanation of each choice will allow users to understand the system and make choices that they are comfortable with.

The following example clearly employs a reader-friendly tone, clear nontechnical language, and convincing details to give the reader the background necessary to understand light pollution. Notice for instance, the use of common words like "brightening of the sky" and "worst of which," words that indicate an expert is speaking like "The key factor," details like defining all the terms, using a common example like suddenly encountering a car with its bright headlights on, and providing a visual aid that illustrates all the terms in the article.

What is light pollution?

Light pollution is an unwanted consequence of outdoor lighting and includes such effects as sky glow, light trespass, and glare. An illustration of both useful light and the components of light pollution are illustrated in Figure 2. Sky glow is a brightening of the sky caused by both natural and human-made factors. The key factor of sky glow that contributes to light pollution is outdoor lighting.

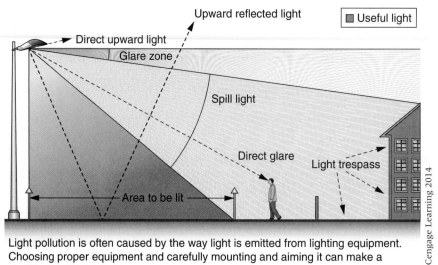

Upward reflected light

Direct upward light

☐ Useful light

Glare zone

Spill light

Direct glare

Light trespass

Area to be lit

© Cengage Learning 2014

Light pollution is often caused by the way light is emitted from lighting equipment. Choosing proper equipment and carefully mounting and aiming it can make a significant difference.

Figure 2. *Example of useful light and light pollution from a typical pole-mounted outdoor luminaire*

Light trespass is light being cast where it is not wanted or needed, such as light from a streetlight or a floodlight that illuminates a neighbor's bedroom at night making it difficult to sleep.

Glare can be thought of as objectionable brightness. It can be disabling or discomforting. There are several kinds of glare, the worst of which is disability glare, because it causes a loss of visibility from stray light being scattered within the eye. Discomfort glare is the sensation of annoyance or even pain induced by overly bright sources. Think of driving along a dark road when an oncoming car with bright headlights suddenly appears. The sudden bright light can be uncomfortable and make it difficult to see. Discomfort and even disability glare can also be caused by streetlights, parking lot lights, floodlights, signs, sports field lighting, and decorative and landscape lights.

Figure 10.8 Background Report

Source: Adapted from "What is Light Pollution?" Lighting Research Center. Rensselaer. Web. October 25, 2012 http://www.lrc.rpi.edu/programs/nlpip/lightinganswers/lightpollution/lightpollution.asp

TIP ||||

Create an Operational Definition

An *operational definition* gives the meaning of an abstract word for one particular time and place. Scientists and managers use operational definitions to give measurable meanings to abstractions. The operational definition "creates a test for discriminating in one particular circumstance" (Fahnestock and Secor 84). For instance, to determine whether a marketing program is a success, managers need to define success. If their operational definition of success is "to increase sales by 10 percent" and if the increase occurs, the program is successful. In this sense, the operational definition is an agreed-upon criterion. If everyone agrees, the definition facilitates the discussion and evaluation of a topic.

Outline Reports

An expanded outline is a common type of report, set up like a résumé, with distinct headings. This form often accompanies an oral presentation. The speaker follows the outline, explaining details at the appropriate places. Procedural specifications and retail management reports often use this form. The brevity of the form allows the writer to condense material, but of course the reader must be able to comprehend the condensed information. To write this kind of report, follow these guidelines:

▶ Use heads to indicate sections *and* to function as introductions.
▶ Present information in phrases or sentences, not paragraphs.
▶ Indent information (as in an outline) underneath the appropriate head.

REPLACE GYMNASIUM FLOORING?

December 10, 2014

Researcher: Aaron Santana

Purpose: Evaluate whether Athletic Department should install new flooring in the Memorial Gymnasium.

Method: Interview Athletic Director, Athletic Trainer, vendors. Used these criteria:
- Cost—not to exceed 250K
- Time—less than three months
- Benefits—personal health and overall usage must be impacted

Conclusions: Gym floors meet all criteria
- Cost is less than 250K allocated
- Time to install is less than three months
- Benefits—fewer injuries; likelihood of increased usage with greater durability.

Recommendation: Install the new floor.

© Cengage Learning

Figure 10.9   Sample Outline Report

E-Mail

E-mail is a major method of communicating. People use e-mail to send everything from birthday greetings to intergovernmental communications. A major problem with e-mail, however, is that there is so much of it. People regularly report getting 25–1000 e-mails a day. While it is easy for you to send an e-mail, it is not always easy to get your audience to open it and respond to it (Brogan). How do you fashion your e-mail so it doesn't sit unread in your recipient's in-box or, worse, is dumped altogether? To create effective e-mails, consider your audience and use the elements of e-mail effectively.

Consider Your Audience

Your audience (singular or plural) probably gets lots of e-mail daily. In addition the audience probably does not have a lot of free time. They may be willing to answer you, but they would like the answer to be one they can construct easily ("15"). To engage that "willing to help" characteristic, follow these guidelines:

- Arrange your e-mail so that the audience can quickly grasp what you want from them. Put the most important points at "the top" of the message.
- Personalize the e-mail with a quick personal comment ("Nice to see you last weekend") at the end. Add your name to the end, even if your "signature name" also appears (Type "Gwen" even if your signature name appears as "Gwendolyn P. Goldman") (Burstein; "15")
- Use the appropriate level of formality. Don't write as if the recipient were your best friend, if she is not. Don't use "text style" ("How r u?").

Use the Elements of E-mail Effectively

Pay close attention to how you handle e-mail elements such as the subject line, address, attachments, and paragraphs. This section explains best practices in the use of e-mail elements (based on "15"; Brogan: Burstein" E-mail: Shannon; Joshi: "ITS").

Write a Clear Subject Line

Experts who have studied e-mail find that the subject line is the most important item when trying to connect with the intended reader. Messages are often displayed in a directory that lists the sender's name, the date, and the subject. Many readers choose to read or delete messages solely on the basis of the subject line, because they can't possibly take the time to respond to so much mail. Your message will more likely be opened if the subject line connects with the reader's needs. If the subject line does not engage the reader, he or she will often simply delete the message unread. Here are some tips:

- Start with an information-bearing word. Say "Budget meeting scheduled Monday, 10 a.m. Rm103 " rather than "Budget meeting" or "Meeting." Or "Hi—meet me after your class?" rather than "Hi."

▶ Keep the subject line relatively short. This tip could conflict with the previous one, so be judicious in your phrasing of the information-bearing word or phrase.

▶ People often open messages with RE in the subject line (so don't change the subject when you reply). In a subject line, state content—"Response to your 7-25 budget request."

▶ Make the subject line a short summary of your message. (Nielsen; Rhodes; "ITS")

Use the To and CC Lines Effectively

The To line should contain only the names of persons who you are asking to do something. In the CC line, list people who should know about the message, or who are getting the e-mail simply for information purposes ("ITS").

Check Addresses

Many e-mail addresses are remarkably similar. It is quite easy to make a typing mistake, so that the e-mail intended for jonessu goes instead to joness or jonesu. Although this is often a minor annoyance, it can be a major embarrassment if the content is sensitive or classified ("ITS").

Consider Whether to Send an Attachment

Attachments take more time to download and often easily become separated from the original e-mail. In addition, many attachments can't be opened at all by the receiver, especially if they were created in another platform or by an application not owned by the receiver.

If you do send an attachment, be sure that the document contains such information as a title and the name of the person who sent it. Sometimes this information appears only in the e-mail; if the e-mail is deleted, the attachment becomes difficult to make meaningful. If the attachment is long, consider posting it on a website, or company wiki (if that option is easily available to you) and sending your recipients an e-mail with the URL to that space ("ITS"). In order to avoid "losing" an attachment, or to ensure that there are no problems opening the document, paste the contents directly into the e-mail. Note, however, that this strategy makes the e-mail long, so in the introduction establish the context for the content. Be sure to give the attachment a meaningful filename. If the attachment is opened directly from the e-mail, the context for it is clear. But if the e-mail is gone and the attachment resides in a directory with many other files, the filename must be meaningful. Say "jonesresume" rather than "resume," or "ABCapplicationform" rather than "ABCaf."

Keep Messages Short and to the Point

Research has established that readers categorize e-mails. "To-do" messages require some action from the recipient. Often, these messages stay in in-boxes as a reminder to the recipient of work to do. "To read" messages usually are

E-mailing Reports

Often the goal of an e-mail is to send a report to an audience. Suppose the report is three pages long. It is too long to be effective in the body of the e-mail, and the e-mail program might strip out all the formatting you have inserted to help readers, like bold face.

If recipients decide to print the report, it is much easier to print the attachment, which is a word processing document and contains the report and any visual aids, but not all the To/From/Subject and other routing material contained in the e-mail.

How should you handle this?

Turn the e-mail into a cover letter. In several sentences name the report (and include its filename), its contents, and why the reader(s) are receiving it.

Hi all, attached is the First Quarter 2013 sales report (2013FQSalesNW) for the Northwest region. It condenses all the sales data by retail item. We will discuss this report at our meeting, Tuesday, July 22, 2014.

long documents that take time and effort to read. Although the content could be important, the length causes recipients to delay reading them. "Indeterminate" messages are those whose significance is not clear to the reader. Like long messages, these messages are usually not read, but left in in-boxes so that when there is time enough, the reader will make the effort to read the message and determine the significance (Rhodes).

Establish the Context

In the body of the e-mail, repeat questions or key phrases. Briefly explain why you are writing, then go on with your message. If a person has sent out 20 messages the day before, he or she might not easily remember exactly what was sent to you. Offer help. Remember that you are not in a dialogue in which the other person can respond instantaneously to your statements, so avoid the temptation to use one-line speeches. For instance, don't just write one word— "No"—but explain the topic you are saying "No" to. One respondent to an e-mail survey said, "X is unbelievable in that he never puts in the context of what he is replying to. He always comes up with these one-line responses, and I have no idea what it is that he's talking about" (Rhodes).

Remember to Use Paragraphs

E-mail's format has a kind of hypnotic quality that encourages people to write as if they were speaking. And, of course, in speech there are no obvious paragraphs. However, remember that e-mail is text that a person reads, so chunk into manageable paragraphs. Use keywords at the beginning of units in order to establish the context of the sentence or paragraph that follows.

Signal the End

Because e-mail exists in scrolling screen form, there is no obvious cue to its end, unlike a hard copy where you always know when you are on the last page. Therefore, signal the end by typing your name, with or without a closing. You may also use the words *the end* or a line of asterisks.

Avoid Mind Dumps

The point of e-mail is to satisfy the reader's needs as concisely as possible. Do not ramble. Plan for a moment before you start to write. If you have "on-line fear," the same strange emotional response that often makes people give awkward, rambling messages on an answering machine, type your message first on a familiar word processing program, when you have time to gather your thoughts and get them down coherently. Edit in the word processor, then upload and send (know the capabilities of your system).

Don't Type in All Caps

The lack of variation in letter size makes the message much harder to grasp and gives the impression that you're shouting.

Get Permission to Publish

E-mail is the intellectual property of its creator. Do not publish an e-mail message unless the creator gives you permission.

Be Prudent

Technically (and legally), the institution that provides you the e-mail service (such as your university or employer or governmental agency) owns the e-mail you are sending and receiving. As a result, any number of people can access individual e-mails if they have some reason to. Be careful about sending sensitive or personal information. In addition, remember that any e-mail is easily forwarded—and so is any attachment—without your knowledge. Although you might think that the sensitive meeting notes that you send to the committee chair will remain only on her computer, or that the personal comments you make about another person will stay buried in an in-box, it is all too easy for these messages to be forwarded, deliberately or accidentally, to others.

Use E-mail as a Cover Letter for Attachments

When you send a document as an attachment to an e-mail, briefly reference the document in the e-mail. The sample e-mail given in the following text illustrates how to handle this situation. The writer explains briefly aspects of the attachment to notice, then includes comments about a previous discussion on feasibility reports, and other professional and personal items for the future.

Dan,

See attached for the resume I sent to Dr. Franklin about a month ago. I generalized some of the items, removing specifics. You can be consistent in the way you modify the other resume contents—or discard if this isn't helpful!

Thanks for the extra details about the feasibility report. We've talked about writing it up on our four hour drive home, so you may have it even sooner.

Please do keep me posted about a potential visit.

Heading back to W soon—appointments at the bank are excellent excuses to get back early.

Laurel

Figure 10.10 Email "Cover Letter"

Ethics and E-Mail

Being aware of the ethical guidelines for conduct when sending communications over the Internet is important in remaining professional and courteous. The most important thing to remember is "If it is unethical in real life, it's unethical in e-mail" (Brenner). Brenner suggests a number of actions that are unethical. They include intentionally omitted someone from a To list (for instance in an e-mail to a group), causing delay by intentionally sending a message late so that time-critical material does not reach the recipient in time, intentionally writing a vague e-mail in order to slow things down.

Since e-mail can be edited a number of commentators (Brenner, Lynmar) say it is unethical to delete wording in an e-mail that you have received so that when you send it on the message is changed. The ethical way to indicate wording deletions is to use ellipsis (…) or <snip>. It is also unethical to send on e-mails to people who not intended recipients unless you obtain permission from the original sender.

While it is not an ethical concern, you should be aware that e-mails don't go away. They are stored on a computer somewhere, especially if you use a company e-mail. Many companies have policies that state that the e-mail is the property of the company. In other words, confidential information cannot be guaranteed to remain confidential.

Another concern is sending inappropriately emotional e-mails. E-mails should retain a professional, courteous tone. The best advice is to think before you hit "send." Remember e-mail can be forwarded easily and your nasty comments could easily end up embarrassing you.

Worksheet for Planning a Project

☐ **Write the question you want answered.**

☐ **Create a research plan.**

    a. List topics and keywords that might help you find information on your question

    b. List a method for finding out about those topics. Tell which specific acts you will undertake. E.g., "Explore Compendex and Ebscohost using X and Y as keywords." or "Talk to all employees affected by the change, using questions X and Y."

Carry out your plan.

Worksheet for IMRD Reports

☐ **Write an introduction in which you briefly describe the goal of your project and your goal in this report. Give enough information to orient a reader to your situation.**

☐ **Write the methods statement.**

- Name the actions that you took in enough detail so that a reader could replicate the acts if necessary.
- Use terms and details at a level appropriate to the reader, but necessary for the subject.
- Explain *why* you chose this strategy or actions.

☐ **Name the actual results of the actions. This section might be very short.**

☐ **Tell the significance of your actions.**

- What do the results indicate?
- How do the results relate to the audience concerns?
- What will you do next, as a result of this project?
- Did you accomplish your goals?
- How is this important to your classmates, in this class?

☐ **Develop a style sheet for your report.**

- How will you handle heads?
- How will you handle chunks?
- How will you handle page numbers?
- How will you handle visual aids?
- How will you handle the title/report heads?

☐ **Develop an idea of how you will present yourself.**

- Will you write in the first person?
- Will you call the reader "you"?
- Will you write short or long sentences?
- Are you an expert? How do experts sound? What will you do to make yourself sound like one?

Key question: Why should I believe you? Why are you credible?

Worksheet for **Informal Reports**

☐ **Identify the audience.**

- Who will receive this report?
- How familiar are they with the topic?
- How will the audience use this report?
- What type of report does your audience expect in this situation (Lengthier prose? Outline? Lots of design? Just gray?)

☐ **Determine your schedule for completing the report.**

☐ **Determine how you will prove your credibility.**

☐ **Outline the discussion section. Will this section contain background? Divide the section into appropriate subsections.**

☐ **For IMRDs, outline each of the three body sections.**

- Clearly distinguish methods and results.
- In the discussion, relate results to the audience's concerns.

☐ **Prepare the visual aids you need.**

- What function will the visuals serve for the reader?
- What type of visual aid will best convey your message?

☐ **Prepare a style sheet for heads (one or two levels), margins, page numbers, and visual aid captions.**

☐ **Select and write the type of introduction you need**

- To give the objective of the report.
- To provide brief context.
- To provide expanded context.

Select the combination of introductory elements you will use to give the gist of the report to the reader. Write each section.

Worksheet for Evaluating IMRDs

Read your report or a peer's. Then answer these questions:

☐ **Introduction**

- Is the goal of this project clear?
- What is the basic question that the writer answers?

☐ **Methods**

- Does the writer tell all steps or actions that he or she followed to achieve the goal or answer the question? Often there are several sequences of them.
- Is it clear why the writer took those steps or actions?

☐ **Results**

- Does this section present all the things that the writer found out?
- If there is a visual aid, does it help you grasp the results quickly?

☐ **Discussion**

- Does the discussion answer the question or explain success or failure in achieving the project's goal?
- What are the implications of the results? Implications mean (1) effects on various groups of people or their goals, or (2) perceptions about the system (e.g., Web search engines, Web authoring programs) discussed in the report.

What New Questions Do The Results Cause?

Examples

Examples 10.1–10.3 show three informal reports. These reports illustrate the wide range of topics that the informal report can present. Note the varied handling of the introduction and of the format of the pages. The goal in all the reports is to make the readers confident that they have the information necessary to make a decision.

Example 10.1

IMRD Report

Determination of Caffeine Variability in Coffee

Erin Molzner
Colby Vorland
University of Wisconsin-Stout
May 10, 2012

Introduction

Caffeine is the most widely consumed stimulant in the world, with approximately 75% coming from coffee (1). Although early research suggested an increased risk for various chronic diseases due to caffeine ingestion, prospective studies on coffee consumption have not confirmed this (2). However, if caffeine doses in coffees vary significantly from expected, subjective arousal and performance effects may not be achieved, or undesirable effects may result (e.g. anxiety, nausea) (3). A recent study showed that the amount of caffeine can indeed vary dramatically between different and within the same brands of specialty coffee (purchased from coffee shops) of the same volume (4). To verify these findings in a different setting, samples of specialty coffee from vendors on the University of Wisconsin-Stout campus, as well as homebrewed samples prepared at different volumes were tested to see if caffeine from the same coffee grounds is consistent over different batches.

Materials and Methods

Samples

In this experiment, samples were collected each day from Monday, April 16, 2012 to Friday, April 20, 2012 (Table 1). Due to closure, the last sample from the Library was not obtained.

TABLE 1
Specialty Coffee Samples Collected

Location	Coffee Name
Brew Devils	Starbucks Caffè Verona
HC2 Jarvis Express	Trescerro 100% Colombian
Swanson Library Express Cart	Trescerro Mexican Altura Regular

© Cengage Learning 2014

In addition, 6 samples of coffee were prepared using Starbucks Medium Breakfast Blend grounds using a drip-brew coffee maker (Mr. Coffee, Model DRX23, Sunbeam Products, Inc.) and paper filters (Brew Rite, 3.25" (8.25 cm) base) according to package directions (Table 2). Refrigerated water (Shur Fine Purified Drinking Water) was used to make all samples. Tablespoons of grounds were leveled off.

TABLE 2
Homebrewed Coffee Samples Prepared

Samples	Preparation	Equivalent To
3	6 fl. Oz. water, 2 T. grounds	˜1 cup of coffee
3	36 fl. Oz. water, 12 T. grounds	˜6 cups of coffee

© Cengage Learning 2014

Standards

Standards were made using caffeine dissolved in water at varying concentrations of 1, 0.75, 0.5, 0.25, and 0.1 mg/ml.

The 1 mg/mL standard was prepared by adding 0.025 g caffeine to a 0.025 L volumetric flask and filling to the mark with water. The calculated molarity was 0.0051 M (194.19 mol/g $\times$ 1/0.025 L $\times$ 0.025 g). From this concentration, the other standards were prepared using dilution with the $C_1V_1 = C_2V_2$ equation. Calculations are shown below.

> 0.75 mg/mL: $(0.0051 \text{ M})(x) = (0.0039 \text{ M})(0.025 \text{ L})$; $x = 0.01875$ L (from the 1 mg/mL standard added to a 0.025 L volumetric flask and filled to the mark with water).
> 0.50 mg/mL: $(0.0039 \text{ M})(x) = (0.0026 \text{ M})(0.025 \text{ L})$; $x = 0.01667$ L
> 0.25 mg/mL: $(0.0026 \text{ M})(x) = (0.0013 \text{ M})(0.025 \text{ L})$; $x = 0.01250$ L
> 0.10 mg/mL: $(0.0013 \text{ M})(x) = (0.0005 \text{ M})(0.025 \text{ L})$; $x = 0.01000$ L

All samples and standards were placed into HPLC sample vials after being filtered using a 0.45 μm syringe filter and kept refrigerated until ready for experimental analysis. These samples were then evaluated for their caffeine content using a Shimadzu HPLC fitted with a C18 column as the stationary phase and methanol/water (70:30) with 1% acetonitrile as the mobile phase (flow rate, 1 ml/min). Injections were made in volumes of 5 μl. The total run time for each sample was 4.5 minutes. Each sample and standard was tested 3 times. The linear equations when graphing the standards by height and concentration (y = mx + b) were used to solve for the unknown sample caffeine concentrations.

Student's t tests were used to compare group means, with F-tests for variance determining equal or unequal variance. One-way ANOVAs compared samples from the same location throughout the week, and the different homebrewed samples to test for variation. All calculations were performed in Microsoft Excel 2010. Significance was set at P = 0.05.

(Continued)

Results

Two standard curves were created because solvent volume was insufficient to run all samples (Figure 1, Figure 2). The R^2 of each was 0.9914 and 0.9893 for the curve with specialty coffee samples and homebrewed samples, respectively.

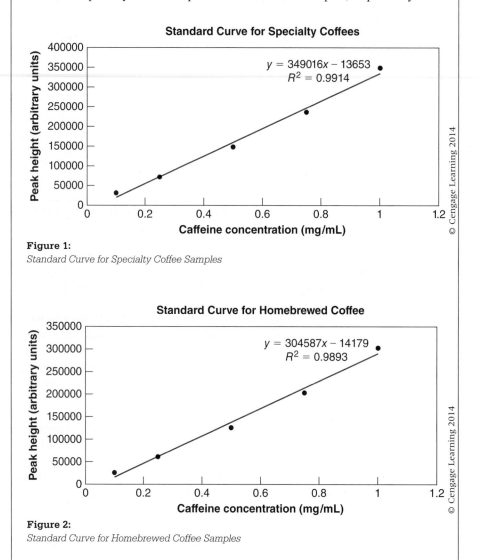

Figure 1:
Standard Curve for Specialty Coffee Samples

Figure 2:
Standard Curve for Homebrewed Coffee Samples

There was significant variation in caffeine concentration in all three locations when samples were measured on different days (P<0.001 for all; Figure 3).

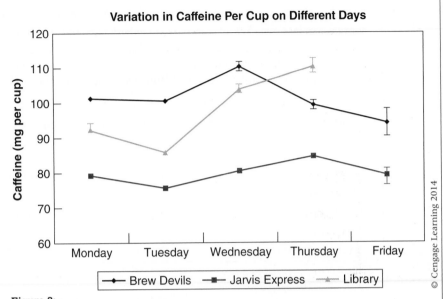

Figure 3:
Variation of Caffeine in Specialty Coffees Per Cup Over One Week

There was also significant variation in caffeine concentration between the home-brewed coffees for both one cup and six cup preparations ($P<0.001$ for each; Figure 4).

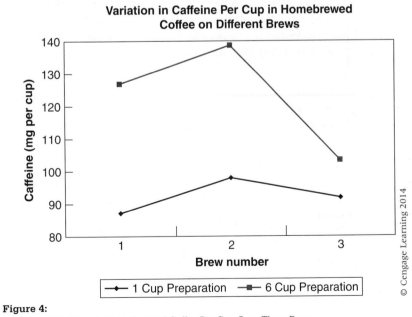

Figure 4:
Variation of Caffeine in Homebrewed Coffee Per Cup Over Three Brews

(Continued)

The average caffeine content in each location is shown below. Results are reported for one standard cup (5 fl. oz.) and for a standard "medium" size (16 fl. oz.).

Samples from both Brew Devils and the Library both contained more caffeine than samples from Jarvis (P<0.0001 and P=0.047, respectively; Table 3). There was no difference between caffeine at Brew Devils and the Library (P=0.64).

TABLE 3

Average Caffeine Content in Specialty Coffee Samples

Location	Average Caffeine Per 5 fl. oz. cup	Average Caffeine Per 16 fl. oz. cup
Brew Devils	101 ± 5.8 mg*	324 ± 19 mg
Swanson Library Express Cart	98 ± 11.1 mg**	315 ± 35 mg
HC2 Jarvis Express	80 ± 3.3 mg	255 ± 11 mg

*p<0.0001 compared to Jarvis

**p=0.047 compared to Jarvis

© Cengage Learning 2014

Homebrewing 6 cups of coffee resulted in an increase in the amount of caffeine per 5 fl. oz. cup compared to brewing only 1 cup (P = 0.0475; Table 4).

TABLE 4

Average Caffeine Content in Homebrewed Samples

Average Caffeine per Homebrewed Cup When 1 Cup Was Brewed	Average Caffeine per Homebrewed Cup When 6 Cups were Brewed
92 ± 5.5 mg	123 ± 18 mg*

*p=0.0475 for the difference

© Cengage Learning 2014

The range (difference between the highest and lowest measure from different days) within each group is shown below (Table 5).

TABLE 5

Range of Caffeine Content in Specialty Coffee Samples

Location	Range Per 5 fl. oz. cup	Range Per 16 fl. oz. cup
Swanson Library Express Cart	25 mg	79 mg
Brew Devils	16 mg	51 mg
HC2 Jarvis Express	9 mg	29 mg

© Cengage Learning 2014

The range within each preparation group is shown below (Table 6).

TABLE 6

Range of Caffeine Content in Homebrewed Coffee Samples

Preparation	Range Per Cup	Estimated Range Per 16 fl. oz. cup
1 Brewed Cup	11 mg	35 mg
6 Brewed Cups	35 mg	113 mg

© Cengage Learning 2014

Discussion

The results of the current experiment demonstrated a significant variation in the amount of caffeine between different brews over one week of samples collected in three locations on the UW-Stout campus and between six homebrewed samples with two different preparation methods. In the specialty coffees per 16 fl. oz., the highest caffeine difference between days was 79 mg, and 113 mg was the highest observed in the homebrewed coffees using Starbucks Breakfast Blend. A previous study found up to a 305 mg difference per 16 fl. oz. cup of Starbucks Breakfast Blend on different days (4). It is unknown why such substantial variations exist, and even more so when brewing a larger volume of coffee within the monitored and standardized conditions of the homebrewing in this experiment. It is possible that for specialty coffees, doses of coffee grounds and water are not consistent in every brew, and this information should be gathered in future research. In the homebrewed experiments, the samples were brewed back-to-back, thus the machine may not have had adequate time to cool to a similar initial temperature to heat the water each time. It would be of interest to verify these results with more samples and different grounds while allowing adequate time between brewing. Although it is possible that an inconsistent dose of caffeine results in an under- or overstimulation of the central nervous system, it is unknown whether this is a perceived problem in coffee drinkers. Many other variables, such as consumption speed and time after and size of the last meal, may influence perceived alertness.

Another notable result is that the average caffeine concentration per homebrewed cup was greater when six cups were brewed at once compared to just one. It is possible that with higher brewing volumes and longer times, water temperature is warmer and extracts more caffeine (5). Additionally, more water may be absorbed into the grounds or evaporated and condensed within the machine, lowering the volume of water that filters into the pot. These hypotheses can be easily tested in future experiments by measuring water temperature, final coffee volume, and volume of water absorbed into the grounds.

In this experiment, homebrewed coffee samples contained on average 108 ± 21 mg per 5 fl. oz. cup, and specialty coffee samples 93 ± 12 mg. The homebrewed

(Continued)

samples had more variation in caffeine between them and fewer samples, and this difference only approached significance (P=0.056). These values are reasonably close to the proposed standard of 85 mg per 5 fl. oz. (6).

Limitations in this experiment include that, due to cost, it was not possible to use a different filter for each coffee sample when transferring to HPLC vials. While measures were taken to use the same filter for similar samples, it is possible that contamination from filtering influenced results. Additionally, we were unable to achieve complete peak separation on the HPLC output, so the results assume that the height of the caffeine peak consistently and accurately represents caffeine content.

Conclusions

Variation in caffeine concentration in coffee was observed in all three vendor locations on the UW-Stout campus as well as in both homebrewing preparations. The fact that the highest variation was detected in homebrewed samples, which were measured carefully, suggests that factors beyond small variations in the amount of grounds and water may influence caffeine extraction from grounds. It was also found that brewing a higher volume of coffee produced more caffeine per cup than a lower volume. As water temperature may have an underlying influence on both of these results, future studies should monitor this.

Acknowledgements

Special thanks to Dr. John Kirk and Shane Medin for suggestions and technical assistance in the laboratory.

References

1. Chou, T. (1992). Wake Up and Smell the Coffee.Caffeine, coffee, and medical consequences.*West J Med., 157*(5), 544–553.
2. Lopez-Garcia, E. (2012). Coffee consumption and risk of chronic diseases: changing our views. *The American Journal of Clinical Nutrition, 95*(4), 787–8. doi:10.3945/ajcn.111.033761
3. Kaplan, G. B., Greenblatt, D. J., Ehrenberg, B. L., Goddard, J. E., Cotreau, M. M., Harmatz, J. S., &Shader, R. I. (1997). Dose-dependent pharmacokinetics and psychomotor effects of caffeine in humans.*Journal of Clinical Pharmacology, 37*(8), 693–703. doi:9378841
4. McCusker, R. R., Goldberger, B. A., & Cone, E. J. (2003). Caffeine content of specialty coffees.*Journal of Analytical Toxicology, 27*(7), 520–2.
5. Spiro, M., &Selwood, R. M. (1984). The kinetics and mechanism of caffeine infusion from coffee: The effect of particle size. *Journal of the Science of Food and Agriculture, 35*(8), 915–924. doi:10.1002/jsfa.2740350817
6. Barone, J. J., & Roberts, H. R. (1996). Caffeine consumption. *Food and Chemical Toxicology, 34*(1), 119–129. doi:10.1016/0278–6915(95)00093-3

Example 10.2

Short
Recommen-
dation

RECOMMENDATION OF TOOLING SCHEDULE

This letter is a follow-up of our discussion pertaining to the Storage Cover tooling schedule, which we discussed briefly during your visit at MPD on Thursday, June 27.

There are three main options available to expedite the pilot run date from October 2 up to the week of October 14.

1. Postpone the tool chroming until after the pilot run.
2. Postpone any major tool modifications until after the pilot run.
3. Expedite the tool building.

I believe that option 1 is the best choice. There will not be any additional costs for this option, and quality parts will still be produced for the pilot run. Option 2 may not be a reliable option because we cannot judge until after sampling what modifications may be necessary. Option 3 will carry additional cost due to overtime labor.

I am also optimistic, yet concerned, about the September 9 sample date (two samples at your facility, hand drilled and bonded, no paint). My concern is with delays through customs for shipping parts from the tool shop in Canada to the United States for assembly and back to Canada.

We are taking every step possible to stay on schedule with the Draft #3 tooling schedule, which you have a copy of. The enclosed tooling schedule (Draft #4) shows the projected pilot run date if option 1 above is employed. I welcome your comments or suggestions on any of the above issues. Thank you.

enc: tooling schedule draft #4

Example 10.3

Problem-
Solution Report

To: Dan Riordan   November 6, 2014

From: Michael Mundy

Subject: Procedures for displays failing in the field.

Objective

Determine why failed vendor displays are not being repaired under warranty and propose sending them directly to the vendor under a blanket purchase order.

Summary

Displays received by the Failure Verification lab of department BB7 are not being returned to the vendor for warranty repair because there is no blanket purchase order. When a display fails, it should be returned directly to the vendor from the field.

(Continued)

Current Procedure

When a display manufactured by a vendor fails in the field, it is returned to our lab by a Customer Engineer for failure verification. As shown in Table 1, only two of the displays we verified as bad were returned to the vendor for warranty repair. We have not returned more displays because there is no blanket purchase order.

As of 11/4/87, our department is storing 72 failed displays. Our technicians do not have the time or the equipment to repair the displays. We are losing money and wasting storage space by not having the vendor fix the displays.

TABLE 1

Vendor Displays Failing in the Field

Number of Displays

Part #	Received by Lab	Verified as Bad	Returned to Vendor
101A150	14	13	0
101A154	17	17	0
101A231	16	15	1
101A416	10	9	0
101A537	18	8	1

© Cengage Learning 2014

Proposed Procedure

When a display manufactured by a vendor fails in the field, the Customer Engineer should send the display directly to the vendor using a blanket purchase order. This new procedure will benefit us in three ways. It will reduce the flow of parts through our lab, giving our technicians more time to work on other problems. It will free up storage space for the department. It will insure the vendor pays for repairing failed displays.

An alternative would be for our lab to send the displays to the vendor. Table 1 shows we have verified the Customer Engineer's diagnosis of a display's failure in almost all cases. So there is no need for our lab to perform failure verification.

Exercises

❱ You Create

1. Create an objective/summary introduction for the Iconglow report on pages 275–276.

2. Create a different introduction for the analytical report on purchasing a Procedure for Displays report, pp. 303–304.

3. Write a methodology statement that explains how you recently went about solving some problem or discovered some information. When you have finished, construct a visual aid that shows the results of your actions. Compare these statements and visuals in groups of two to three.

4. Write the introduction for the material you wrote in Exercise 3.

5. Write the discussion section for the material you wrote in Exercise 3.

❱ You Analyze

6. Because introductions imply a lot about the relationship of the writer to the reader, analyze the introductions of the reports in Examples 10.1–10.3 to determine what you can about the audience-writer relationship. How is that relationship affected when you change the introduction as you did in Exercise 2?

7. Read the introduction and body of another student's paper from one of the Writing Assignments later in the chapter. Does the discussion really present all the material needed to support the introduction? Are the visual aids effective? Is the format effective?

❱ Group

8. In groups of three or four, create an IMRD report on collaborative software. This report could focus on one type, such as a wiki or an application like Google Docs. Create a task list and assign each member a task and deadline. Create visual aids that will help your readers answer the question: What collaborative software should I and my group use in this class or in any situation?

9. In groups of three, read the introduction of each person's paper from one of the Writing Assignments later in this chapter. Decide whether to maintain the current arrangement; if not, propose another.

10. a. In groups of two or three, decide on a question that you will find the answer to. A good example is how to use some aspect of e-mail, the library, or the Web. Before the next class, find the answer. In class, write an IMRD that presents your answer.

I   question you wanted to answer and goal of this paper
M relevant actions you took to find the answer
R   the actual answer
D the implications of the answer for yourself or other people with your
    level of knowledge and interest

b. In groups of two or three, read each other's IMRD reports. Answer these
    questions:

Do you know the question that had to be answered?
Could you perform the actions or steps given to arrive at the answer?
Is the answer clear?
Is the discussion helpful or irrelevant?

11. In groups of two or three, compare the differences in Figure 10.2 (p. 278)
    and Example 10.1 (pp. 296–302). Focus on differences in tone and in
    presentation of the actions. Report to the class which document you prefer
    to read and why. What principles would affect your choice to write an IMRD
    or a set of instructions?

▶ You Revise

1. Rewrite this text for a more professional tone. Evaluate the table for effec-
    tiveness; if necessary create a table or graph.

HAS TO BE UNDER $75

Introduction

One of the biggest things for our team is going to be the fact of money and how
much these sweat suits will be costing. First off none of us have much money to
begin with and we also have limited funds for our sport as is. I feel that if you can
set a limit that the whole team can agree is reasonable while at the same time
getting a quality product. I feel that this is actually one of the most important
criterion.

Research

First of all to get the price set at $75 was very easy, at practice one day when
we were discussing the sweat suits, all I had to say was that price and everyone
agreed right on the spot. The next part of the research was the tough part. The
three places I ended up going to look for this price was eastbay.com, askjeeves.
com, and to Fleet Feet. The research that I did turned out pretty helpful for this
project although a lot more work then I had intended.

Results

When I first started looking at sweat suits on eastbay, it was a big disappoint-
ment, the selection was really good and they had quite a bit of stuff but I knew

that there was no way that anyone on our team could afford those. All the suits were at least $100 or more and I won't even mention some of the prices that were listed. When I went to askjeeves, I ran into basically the same problem. All that it was really giving me were links to really expensive name brand products. Although name brands are very dependable there is always another no name brand that can be just as durable for a cheaper price, so pushed on in my journey. When I went into Fleet Feet I just had a feeling that they would at least help me find something if they didn't have it there with them. Sure enough they gave me a catalog along with a website that I could visit. When I went to the website which is hollowayusa.com, I found what I was looking for right away. When I saw it, it was just too good to be true, the pants were listed at $35.90 and the top is listed at $37.90. Which is a grand total of $73.80. You can see the different prices I ended up finding on the different sites located in Table 1.

Conclusion

When I saw the sweats and then saw the price of them I knew that they had a good shot of getting accepted. This price is a perfect price and a price that everyone on the team had agreed on before I even started looking for the suits. When I let the team know the price they were all very excited.

Writing Assignments

1. In groups of 2–4 write a report explaining to your class the benefits of using an application that allows you to save web pages to an account so that you can read them later. At this writing a number of these exist with varying degrees of usefulness. Consider such characteristics as ability to alphabetize entries, to print them, to send them to all devices (to a smart phone or a tablet), to highlight passages. As you investigate these applications you will discover other important characteristics to call attention to.

2. Granted a "generation gap" in the use of technology, explain a "very current" new social media technology to a generation (probably older) other than yours. You might use Ping or Pinterest but because new technologies appear with dazzling swiftness, use whatever is current as you implement this assignment. Alternative: explain a function of an older technology such as Facebook, for instance, how to post pictures of grandchildren so that the photos are only available to selected family members and friends.

3. In groups of 2–4 investigate the potential harm of social media posts to job seekers. Explain to your audience the types of postings in social media that companies look upon unfavorably.

4. Create a concept report in which you explain one of the collaboration aspects of a social media site, for instance, groups in a site such as Facebook or LinkedIn. Your instructor may ask you to prepare a presentation to give to the class rather than write a paper.

5. Write an informal report explaining the implications of the e-mail and social media policies of (a) your college or university, (b) your corporation, or (c) an e-mail provider such as Gmail or Roadrunner, or even Facebook and LinkedIn. Do the policies contain statements of unethical behavior? Do they discuss confidentiality and ownership of the messages you send?

6. Write an informal report in which you use a table or graph to explain a problem and its solution to your manager. Select a problem from your area of professional interest—for example, a problem you solved (or saw someone else solve) on a job. Consider topics such as pilferage of towels in a hotel, difficulties in manufacturing a machine part, a sales decline in a store at a mall, difficulties with a measuring device in a lab, or problems in the shipping department of a furniture company. Use at least one visual aid.

7. Write an IMRD report in which you explain a topic you have investigated. The report could be a lab report or a report of any investigation. For instance, you could compare the fastest way to reproduce a paper, by scanning or retyping, or give the results of a session in which you learned something about navigating on the Internet, or present the results of an interview you conducted about any worthwhile concern at your school or business. Your instructor may combine this assignment with Writing Assignment 3.

8. Bring a draft of the IMRD you are writing to class. In groups of two or three, evaluate these concerns:

   a. Is the basic research question clear?
   b. Does the method make you feel like a professional is reporting?
   c. Could you replicate the actions? Could other people?
   d. Does only method—and not results—appear in the method section?
   e. Is the method statement written like instructions or a process description? Which is best for this situation?
   f. Are the results clear? Are they a clear answer to the original question?
   g. Does every topic mentioned in the discussion section have a clear basis of fact in the methods or results section?
   h. Is the significance the writer points out useful?
   i. Does the visual aid help you with the methods or result section? Would it help other people?
   j. Is the tone all right? Or is it too dry? Too chatty? Too technical?
   k. Does the formatting of the report make it easy to read?

9. Rewrite the IMRD from Writing Assignment 2 from a completely different framework—for instance, a coach explaining the subject to a high school team. After you complete the new IMRD, in groups of three or four, discuss the difference "author identity" makes and create questions to tell writers how to choose an identity.

10. Convert your IMRD report from Writing Assignment 2 into an article for a newsletter.

11. Convert your IMRD report from Writing Assignment 2 into a set of instructions. After you complete the instructions, in groups of three or four, construct a list of the differences between the two, especially the method statement. Alternate: In groups of three or four, construct a set of guidelines for when to use instructions and when to use IMRD. Hand this list in to your instructor.

12. Write an outline report in which you summarize a long report. Depending on your instructor's requirements, use a report you have already written or one you are writing in this class.

13. Write a learning report for the writing assignment you just completed. See Chapter 5, Writing Assignment 8, page 137, for details of the assignment.

Web Exercise

Write an IMRD that explains a research project on the effectiveness of a search strategy on the Web. Choose any set of three words (e.g., plastic + biodegradable + packaging). Choose any major search engine (Yahoo!, Google, Bing). Using the "advanced" or "custom" search mode, type in your keywords in three sequences—plastic + biodegradable + packaging, packaging + plastic + biodegradable, biodegradable + packaging + plastic. Investigate the first three sites for each search. In the IMRD, explain your method and results and discuss the effectiveness of the strategy and of the search engine for this kind of topic.

Large group alternative: Divide the class into groups of four. All members of the class agree to use the same keywords, but each group will use a different search engine. After the individual searches are completed, have each group compile a report in which they present their results to the class orally, via e-mail, or on the Web.

Works Cited

Brenner, Rick. "Email Ethics." Point Lookout. 2005. Web. 29 May 2012. <http://www.chacocanyon.com/pointlookout/050406.shtml>.

Brogan, Chris. "Writing More Effective Email." Chris Brogan. 2008. Web. May 18, 2012. < http://www.chrisbrogan.com/writing-more-effective-email/>.

Burstein, Daniel. "Email Marketing: How your peers create an effective email message." MarketingExperiments Blog. 2011. Web. May 18, 2012. < http://www.marketingexperiments.com/blog/research-topics/email-marketing/effective-email-messages.html>.

"Email Etiquette: 101 Email Etiquette Tips." WordPress Concierge Consulting. Web. 18 May 2012. <http://www.101emailetiquettetips.com/>.

Fahnestock, Jeanne, and Marie Secor. *A Rhetoric of Argument.* New York: Random House, 1982. Print.

General Motors. *Writing Style Guide.* Rev. ed. GM1620. By Lawrence H. Freeman and Terry R. Bacon. Warren, MI: Author, 1991. Print.

Joshi, Yateendra. "Writing effective e-mail messages 1: The subject-line." 2012. Editage Blog. Web. 18 May 2012. <http://blog.editage.com/?q=writing-effective-email-messages-1-the-subjectline>.

"ITS E-mail Tips." E-mail and Network Services. Yale University. 17 Aug. 2002. Web. 13 Oct. 2003 <www.yale.edu/email/emailtips.html>.

Lynmar Solutions. "E-mail Ethics and Good Practice: A Sample Policy." Web. 28 May 2012, <www.lynmarsolutions.co.uk/files/emailethics.pdf>.

Northern Illinois University Electronic Mail (E-Mail) Policy. Northern Illinois University. 2010. Web. 30 May 2012. <http://www.its.niu.edu/its/policies/email_pol.shtml>.

"15 Tips for Writing Effective Email." Think Simple. 18 May 2012. <http://thinksimplenow.com/productivity/15-tips-for-writing-effective-email/>.

Shannon, Eric. "Don't suck at e-mail." JustJobs Academy. 2011. Web. 18 May 2012. <http://academy.justjobs.com/dont-suck-at-mail>.

Developing Websites/ Using Social Media

CHAPTER CONTENTS

CHAPTER 11 **IN A NUTSHELL**

Websites and Web documents are important methods of conveying information. Creating effective websites requires careful planning, drafting, and testing.

To plan effectively, you need to consider your audience. Determine who they are. Of course, on the Web they could be anyone in the world, but that's too broad. A helpful way to create a sense of your audience is to define a role for them, as if they were actors in your "Web play." Are they customers? Students? Curious surfers?

In addition to considering your audience, you need to plan a flowchart and a template. The flowchart is a device that indicates how you will link your material together. For instance, if you have four files and if you want your reader to link from any one to any other one, your flowchart would look like Figure 11.1. Each line is a link and each box is a webpage.

Your template is a design of your site's look. It shows how you will place various kinds of information (title, text, links, visuals) so that your reader can easily grasp the sense of your site.

Websites create special concerns for writing. Good Web text is scannable (easy to find key ideas), correct (no spelling, grammar mistakes), and consistent (all items treated in a similar fashion).

Visuals must be legible, but not so large that they take up most of the screen or take a long time to load.

Websites must be tested to make sure links work, visuals appear, and the site displays consistently in various browsers.

Technical Communicators must know how to write for corporate blogs, microblogs, and wikis. These Social Media are often interconnected, with posts on one of the media leading readers to posts on the other media, including corporate websites.

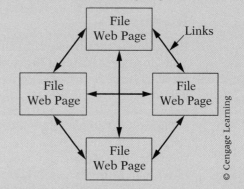

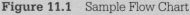

Figure 11.1   Sample Flow Chart

© Cengage Learning

Since the late 1990s, the Web has quickly become as ubiquitous as it is central to many—if not most—aspects of our personal and professional lives, mediating how we maintain relationships with friends and family, how we work and entertain ourselves, and, perhaps most important of all, how we create and share information. While such connectedness has many advantages, the relentless pace at which web technologies change creates increasingly diverse and complicated writing situations for those looking to publish their ideas online. On the job or freelancing, professional and technical communicators encounter a wide range of web-related writing—designing, coding, and composing entire websites, writing content for a site where the design is managed with a Content Management System (CMS) (and therefore requiring little to no coding), or simply managing a company's presence on one or more social and/or professional networks. The technological skills required across these situations may vary, but perhaps the greatest aspect dictating how successful a piece of online writing will be remains constant—that is, the writer's understanding and planning around rhetorical elements of purpose, audience, and form.

Now more than ever, technical communicators must know how to translate traditionally print-based planning of writing to digital screens and online spaces. While the same general process of planning, drafting, and editing the final product applies to the web, the tools required differ from print-based thinking so as to account for variables like how people read online and the non linear manner in which websites can be navigated.

This chapter covers the basics of writing in online spaces, as well as the process of planning, drafting, and testing web-based documents.

Basic Web Concepts

Three basic concepts that will help you create effective websites and documents for readers are hierarchy, Web structure, and reader freedom.

Hierarchy

Hierarchy is the structure of the contents of a document. All websites and Web documents have a hierarchy, that is, levels of information. The highest level is the *homepage,* a term that can apply either to an entire site or to a document. Lower levels are called *nodes;* the paths among the nodes are the *links.*

Figure 11.2 shows three levels in the hierarchy, each giving more detail. *Writing* is the most general category. *Technical* and *fiction* are the two subcategories of writing. *Reports* and *novels* are subdivisions, respectively, of their types of writing. More levels could be added. For instance, reports could be broken down into feasibility reports and proposals.

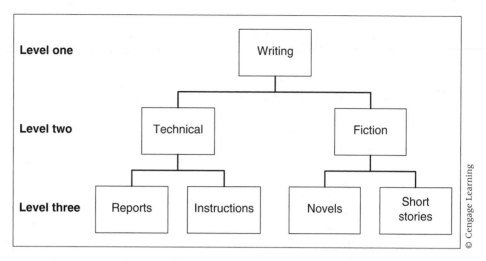

Figure 11.2   Three-level Hierarchy

Web Structure

"Web structure" means that the document contains hyperlinks (or "links") that allow readers to structure their own reading sequence. When the reader clicks the cursor on a link, the browser opens the screen indicated by the link. This feature allows readers to move to new topics quickly and in any order. This arrangement is a radical departure in organizing strategy. The author gives the readers maximum freedom to choose the order in which they will view the site or read the document.

To see the difference between traditional and Web structures, consider these two examples. If a document has seven sections and a traditional (or "linear") structure, then a reader will progress through the sections as shown in Figure 11.3. But if the same seven sections have a Web structure, then they

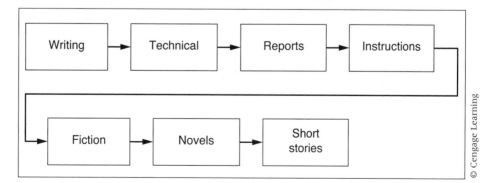

Figure 11.3   Linear Structure

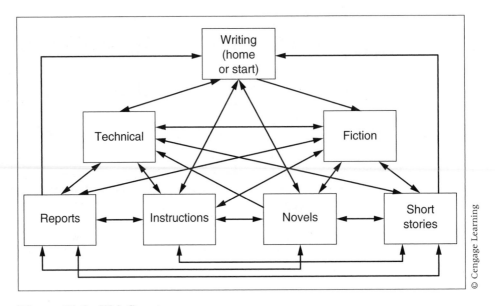

Figure 11.4 Web Structure

would look like Figure 11.4 with each line a link. Once readers arrive at the start, or home, screen, they may read the document in any order they please.

The two Web homepages in Figures 11.5 and 11.6 show two ways that authors used Web structure. Both have a lengthy report, which they want the readers to be able to read without having to scroll through various screens. Maertens (Figure 11.5, p. 315) has divided the report into five linkable sections. The homepage provides an index and brief abstract of each section. The reader can choose any section and link to it. Currier (Figure 11.6, pp. 316–317) has provided "anchor links" that move the screen to that part of the document without scrolling.

To "think Web" is a radical departure in organizing strategy. You can give readers maximum freedom to choose in what order they read your document.

Reader Freedom

Reader freedom is the degree to which the reader of the website can easily select the order in which he or she will read sections of a document. Whereas hierarchy imposes control on the reader's freedom, Web structure provides freedom. The Web author must find a way to combine the two. Figures 11.7 to 11.9 demonstrate how authors can use the two concepts to affect the way readers view the document or site.

Tips for Making a Good Website

The following report contains some ideas that I feel are important to consider if one is going to produce a good website. These are just some of the ideas I have learned through my Technical Writing class. Remember, you are free to do whatever you want, but the following ideas may be helpful in achieving your goal.

Go to

- <u>**Planning Is Key**</u>—Planning is essential in order for you to produce a good . . .
- <u>**Establishing a Purpose for Your Website**</u>—Before you embark on this voyage . . .
- <u>**Implementing Links in Your Website**</u>—In order to make your site easy to use . . .
- <u>**Handling Visuals and Text**</u>—These seem to spice up your website, but . . .
- <u>**Interactive Websites**</u>—People do not just want to see a webpage, they want to . . .

Return to

- <u>**Matt Maertens Homepage**</u>
- <u>**Technical Writing Index Page**</u>
- <u>**English Department Student Projects Page**</u>

Let me know what you think! e-mail me **maertensm@uwstout.edu**

Clicking on a link causes that page to open on the screen.

© Cengage Learning

Figure 11.5   Linked Section Report

In Figure 11.7, the reader starts at the home page and can progress only to one of the two level 2 nodes, and then to one of the level 3 nodes. To arrive at *reports,* the reader must click to *technical* and then to *reports.* To get to *instructions* from *reports,* the reader must first go back to *technical.* To get to *novels* from *reports* the reader must follow the path back to *writing* and then click forward to *novels.*

Figures 11.8 and 11.9 show progressively less control by the author and more control for the reader. In the hierarchy shown in Figure 11.8, the reader can move directly from *reports* to *instructions* without clicking back to *technical.* In the hierarchy shown in Figure 11.9, the reader can move to any document from any other document. A reader could click from *reports* to *fiction* and then to *instructions.*

The Maertens model in Figure 11.5 appears to have a tightly controlled hierarchy, as shown in Figure 11.10. But if he would supply links between each section, then the document would have a hierarchy of little control and great reader freedom, as in Figure 11.11 (p. 319).

How to Create a Great Webpage

HTML stands for Hypertext Markup Language. This language allows the computer to read your document and makes it appear in Web format. Once you get the hang of this computer language, the possibilities are endless. However, good quality writing and construction of your website could determine how many people will stop and stay long enough to check out your site. The following areas are important to consider in order to create a great website:

Before You Begin|Here Goes|Introduction|Chunks and Heads|Graphics

BEFORE YOU BEGIN

Before you start typing anything into the computer, sit down and plan out what you are going to do. Designing a website is 90% planning and 10% actual work. Figure out the purpose of your website. Things can get out of control fast, so knowing your boundaries is essential.

It is incredibly helpful to draw a map of your website. When creating links, things can get confusing. Good links enable you to get in, through, and out of the document easily. Nobody likes to get trapped in a webpage with no way out. Having a map of links in front of you can make linking your pages easier. You will save yourself a lot of time and hassle, if you work out what you are going to do before you do it. **Top**

HERE GOES!

Now that you know what you are going to do, it is time to acquire the things necessary to get it done. You will need a computer, a simple word processing program, a list of HTML commands and what they do, and a browser. Type your information in a simple word processor, such as Simple Text, or save it in your present word processing program as a text file. You will need knowledge of HTML commands to convert your page into a language the computer can read. A browser allows you to open up your document on the Web to see what it looks like. **Top**

Introduction

An introduction is an important part of the website. The introduction contains a lot of necessary information, such as the reason for the site and why it would be of interest to the viewer. In the introduction, the purpose of the website should be clearly stated. If viewers don't know what the site is about or how it will benefit them, they are not going to stick around. **Top**

Anchor

Chunks and Heads

People like information presented in small chunks. It is easier for them to digest. It also gives you a better chance of holding their attention. People don't have the desire or patience to read through a lot of unbroken text. It is also a good idea to use heads. They are also useful in breaking up the page. Heads inform the reader about each section's content, giving them the option of whether they want to read it.   **Top**

Anchor

GRAPHICS

When incorporating graphics into your webpage, there are several things to consider. Determine how long it takes the graphic to load. If it takes too long, try to make the graphic smaller. If that doesn't help, you should cut it out. People don't like to wait very long for a graphic to load. Another thing to consider about graphics is if they will appear when the document is loading or if the viewer will have to click on a graphics icon to bring up the picture. Either one is acceptable. It is up to the designer which way to go. The Web can provide additional options for incorporating a graphic into you site.   **Top**

My Set of Instructions|How to Reply in Eudora

Nikki's Index|Technical Writing Home Page

Figure 11.6   Anchor Link Strategy

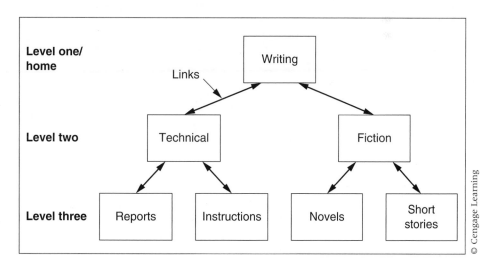

Figure 11.7   Little Reader Freedom

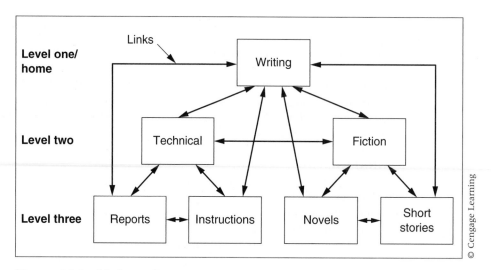

Figure 11.8 Moderate Freedom

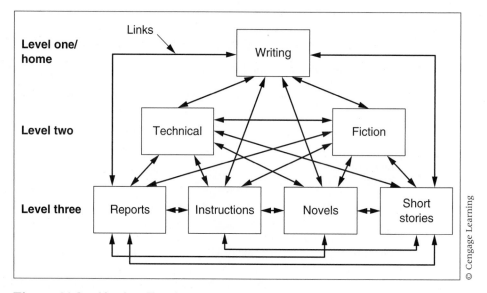

Figure 11.9 Absolute Freedom

Thus, the hierarchy of a site—and, as will be discussed later, the page layout—can shape a user's experience by attempting to dictate how a user encounters information. Yet, despite all the planning that may go into a site's hierarchy, navigation, and layout, we are left with the major caveat that readers take a more active role in online reading, which ultimately makes

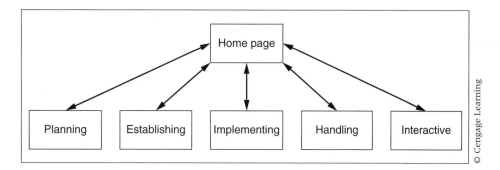

Figure 11.10   Controlled Hierarchy

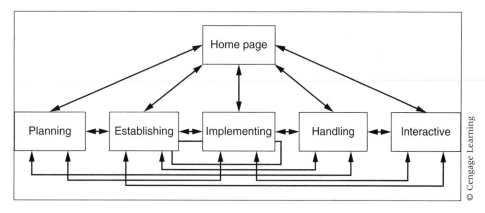

Figure 11.11   Web Structure Allowing Reader Freedom

their interaction with the text less predictable than in the context of a print publication. Unlike other media that are more passive, "lean-back" types that lead and/or guide the reader or viewer, web users are very much engaged and active in their experience, oftentimes seeking out particular information and unwilling to follow a set information path or sequence, making the web more of a "lean-forward," engaging medium (Nielsen, "Writing"). In this sense, the web invites users to create their own experience and follow their own path, something that web writers ought to keep in mind as they design hierarchies, pages, and content (Nielsen, "Writing").

Guidelines for Working with Web Structure

The two possibilities of rigid hierarchical organization and loose Web organization mean that the writer must choose to insert enough links to be flexible, but not so many that the reader is overwhelmed with choices. William Horton suggests that an effective strategy is to "layer" documents, "designing them so that they can serve different users for different purposes, each user getting the information needed for the task at hand" (178).

Horton suggests that each level of the hierarchy is a layer and that each layer provides more detailed information. Use this principle in these ways:

- ▶ In higher layers, put information that everyone needs. To learn about writing, everyone needs the definitions on the homepage, but only a few readers need the concepts explained on the instructions page.
- ▶ Control reader's paths. Figure 11.7, for instance, indicates that readers have complete control in either of the two major categories, technical writing and fiction, but that accessing the material in the other category will require the reader to "start over." The author has assumed that readers in fiction will want to know more about that category, so it should be easy to get around in it. However, it is less likely that they would want to compare items in fiction with the items in technical writing, so the path to it is restricted.

Planning a Website or Web Document

The range of Web-based projects a writer might take on is quite broad, but all of them start with careful planning. A content strategy (Halvorson 83), or a collection of guidelines about a project's purpose/goals, audience, genre, delivery and maintenance, is a useful place to start.

Decide Your Goal

Your site or document should have a "mission statement"—for instance, "To explain the purposes and services of the campus antique auto club." Make this statement as narrow and specific as possible. It will help you with the many other decisions that you will have to make.

Analyze Your Audience

Ask the standard questions: Who is the audience? How much do they know? What is their level of expertise? (See Chapter 2 for more information on audience analysis.)

Do not answer: Anyone who comes onto the Web. Such a broad answer will not allow you to make decisions. Narrowly focus these answers—they are people who are interested in antique cars, the university in general, or clubs in particular. They know a little or a lot; they will have experience or not. A website aimed at an audience of people who have restored antique cars is quite different from a website aimed at students who have some interest in old cars and wish to join a university group.

One helpful way to think about audience is to develop specific **personas** that are representative of the types of people that might visit the site. A persona is a biographical snapshot of a person, usually including a name, age, profession, interests, possible reasons for visiting your site, questions they may be seeking to answer, technological ability, and a scenario in which the user might visit the site. Adding a photo makes personas even more realistic, and, when combined with names and the other information, they allow planning teams to discuss in highly concrete terms how specific user groups might react to a site. Understanding how various users might interact with the site will help you make better decisions about how to present the site's information.

Evaluate the Questions the Audience Will Ask

Speculate on general questions: What is the purpose of the club? When does the club meet? What are the club activities? What antique cars does the club have? Can I learn about where to purchase an antique car? Can I learn how to restore an antique car? What are the bylaws of the club?

You can decide which of these questions you want to, or should, answer. Your audience decisions will help you. For instance, if you feel that only car buffs will look at the site, then there is no need to provide rudimentary information about cars.

Determine Genre Guidelines

In addition to determining a site's purpose and audience, a content strategy should also include some statement of the **form or genre** the site will take as well as **guidelines** that the content creation will follow. From personal online portfolios to corporate portals to various types of blogs, writers have numerous genres and subgenres to choose from to suit the project's needs. The number of choices can be intimidating, to say the least.

For as daunting a task it can be to determine a specific category and create a site or page from scratch, one advantage of Web-based writing is that writers have an ocean of examples to study and learn from in planning their own content. Performing a brief **genre analysis** of relevant and/or seemingly best-in-class sites can generate a set of guidelines for a project's Content Strategy document. Analyzing, comparing, and contrasting a handful of sites across a few areas—rhetorical purpose and audience, content, structure, and linguistic features—can lead to helpful insights that jump-start a project's planning.

Evaluate and Select a Delivery Technology

Aside from rhetorical considerations, equally important are the technological decisions figuring into writing online, perhaps the most common of which is the selection of a means of publishing. Since 2000, Content Management

Systems (CMS) of various types and complexity have proliferated on the Web, thereby giving those unacquainted with HTML and other Web languages instant, push-button publishing on a global level and with little to now technological expertise. Coding one's own site is obviously an option, though choosing a prefabricated CMS is increasingly attractive, especially in scenarios when writers with little coding knowledge need to lead short- or long-term maintenance of the site. In short, a Content Strategy needs to have a well-reasoned statement of the key web technologies that will be used to publish and maintain the site, be they desktop programs like DreamWeaver or Web-based CMS applications like WordPress or Drupal. Investigate these options carefully and determine which one best suits the project's parameters.

Plan for Maintenance

Often overlooked in the creation of websites and Web content—and especially in freelance situations—is the short- and long-term maintenance of the information. The planning stage is an ideal time to determine what elements of the site are static and which ones will need to be updated, as well as who will actually be responsible for the work. Will the original content and/or site creators be around, or will someone else be stepping in? And will the people in charge in the future have the technological know-how to update and maintain the site? Accounting for a site's maintenance in advance can play a significant role in a site's long-term success.

Drafting for Screens

Creating a website or document takes several drafts. Creating clear content, effective structures for reader freedom, and accessible pages seldom happens in one draft. As you create content for your site, keep in mind the reading habits of those visiting your site, orient your reader, and write in a scannable style. As you draft and refine the visual design of the site and/or pages, make use of wireframes and mock-ups, consider available organizational schemes, establish credibility, use visuals effectively, and document key design decisions.

Online Reading Habits

Before one begins writing for the Web, it can be helpful to understand the *reading* habits of people in online spaces. A 2006 study of how users read websites and, specifically, *where* they look on the screen while they're reading, revealed a habit of page scanning best described as an F-shaped pattern (Nielsen, "F-shaped . . ."). Specifically, users begin scanning the top-most

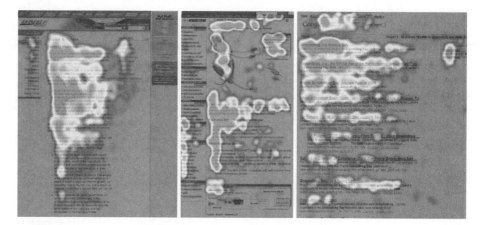

Figure 11.12   The F Pattern
Source: Nielsen

horizontal section, followed by a short horizontal section below that, after which they scan the left-hand vertical space; an approximation of this F-pattern can be seen from their eye-tracking heat map in (Figure 11.12), which shows were the users looked most often.

The implications of these reading habits for writing tasks are three-fold: the study's lead investigator suggests: (1) users scan online texts as opposed to reading them word-for-word; (2) the main point of any page must be communicated early on and where the user is more likely to see it; and (3) writing useful headings, chunking information into paragraphs and lists, and being as efficient as possible (within the genre's guidelines) will lead to the most easily readable webpages (Nielsen, "F-shaped..."). Allowing these basics to guide one's work will promote Web writing that better anticipates how user interact with the screen and the larger site.

Page Layout, High- and Low-Fidelity Wireframes, and Mock-ups

Designing pages for online reading is simplified by first drafting high- and low-fidelity wireframes and then scaling up to full mockups before actually coding the site. Gradually adding more and more detail to a site's layout and design allows for incremental feedback from key project team members and other stakeholders. Additionally, it helps ensure only minor design changes will need to be made once the page is actually set in code.

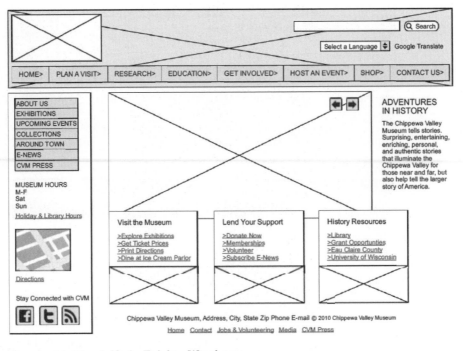

Figure 11.13   A High–Fidelity Wireframe

Wireframes are visual outlines for a page's layout and are useful not only for Web design and writing, but also for print and video applications. Using basic lines and blocks, one can identify regions of the screen and indentify their purpose. **Low-fidelity** wireframes lack detail and are the most basic visual outlines, while **high-fidelity** wireframes add a bit more information, though stop short of describing the site's overall visual style. A sample high-fidelity wireframe can be seen in Figure 11.13.

Once both low- and high-fidelity wireframes have been generated and discussed by key decision-makers, and after any necessary revisions have been made, one can produce a full **mock-up,** which adds color, graphics, content, and any other high-fidelity elements that make the page look as close as possible to the intended final design. Importantly, a mock-up can and probably should be created outside of the Web-publishing tools intended for the final product; rather, the mock-up could be made by hand using a collage style, fashioned in a presentation program like PowerPoint or Keynote, or created in a graphics editing suite like Photoshop—anything that provides flexible visual tools while not requiring coding or dealing with a CMS. Focusing on adding the more polished visual elements while steering clear of Web-based tools keeps the drafting incremental and allows reviewers to discuss the design

Figure 11.14 Website Mockup

instead of getting caught up in technological issues. As with the wireframes, the mock-up should be critiqued and tweaked until it is ready to go into production using the author's application of choice. A sample mockup can be seen in Figure 11.14.

Organizational Schemes and Navigation

Organization of a site and its navigation are two concerns that go hand-in-hand in the planning and drafting stages—how one structures the site will usually dictate what links are presented. In structuring the site's pages (or even the pages' information), and in designing the navigational options, it can be useful to ask the following question: Which organizational scheme will best suit the site's purpose **and** suit the needs and habits of users? A number of schemes exist, and each is often uniquely suited to particular types of information and/or audiences. Some examples include organizing by chronology or time, location or place, people, task or action, and subject or topic.

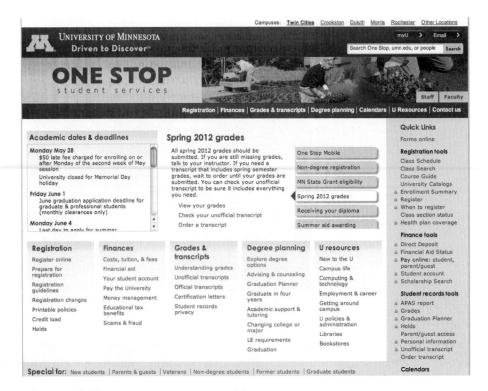

Figure 11.15   Various Organizational Schemes
Source: University of Minnesota.

Providing more than one scheme on a page can provide flexibility for different types of users. For example, the University of Minnesota's student, faculty, and staff portal, OneStop (see Figure 11.15), provides a number of organizational schemes—place, people, topic, and task—that allow users access to the site's function from a number of avenues. While not necessarily the case in the example, adding more links can obviously crowd the page, too, meaning the writer has some important balancing decisions to make about user needs, site requirements, and the layout's final look.

Because organization and navigation are often of critical importance for a site, it would be sensible to revisit the project's personas and/or larger content strategy document to ensure the chosen approach is a good fit.

Navigation

With an organizational approach in mind, one can then begin thinking about how to represent the users navigational choices on the screen. Menu bars on the top-most horizontal section and along the left-hand vertical space are

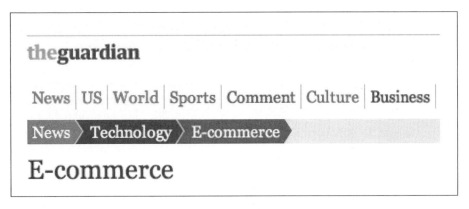

Figure 11.16 A Sample Bread Crumb Trail
Source: The Guardian

the most common locations for creating navigation, though there are always exceptions to those standards; regardless, they are good places to start with your design.

In addition to considering placement of the navigation tools on the page, you also need to help the user maintain a sense of where they are on the site. A simple breadcrumb trail approach can help the user have a sense of their location (see Figure 11.16). Similarly, once the user is past the homepage, how much of the rest of the site tree will be represented in the navigation? Will they see a list of everything on the same level in the structure, or perhaps everything below that page? What do they need to see? These are more important questions to ask as the navigation is refined.

Document Design Decisions

As you progress through the drafting stage, documenting both small and large design decisions can help ensure that your final product is supported by sound reasoning. Moreover, compiling key decisions into a single document can streamline discussion with other team members, management, and clients, allowing all participants to examine the justification behind components like color schemes, logo selection, page layout, and a site's written style, among other considerations.

Maintaining a Design Justification Table, much like the one seen in Figure 11.17, can simplify the process of cataloging key decisions and their underlying reasoning. Keeping it organized according to issues like layout, navigation, organization, and visual elements can help you keep your project decisions organized.

Page Element	Design Decision and/ or Description	Justification
Layout: *Search Box*	Search box is located in upper right-hand corner of every page	Consistent placement helps make every page more uniform and predictable. Also, eye tracking heat map data suggest the upper right-hand corner is where most users look for search info.
Layout: *Logo*	...	...

Figure 11.17   Design Justification Table

Testing

Your Web document must be usable. Readers must be able to navigate the site easily and access the information they need. In order to ensure easy navigation and access, you must test your document. Either perform the test yourself or have another person do it. Testing consists of checking for basic editing, audience effectiveness, consistency, navigation, the electronic environment, and clarity.

Basic Editing

Web documents include large amounts of text that must be presented with the same exactness as text in a paper document. If the editing details are handled effectively, the credibility of the site is increased.

Check your site for stylistic elements.

- Spelling
- Fragments, run-ons, comma splices
- Overuse of there are, this is
- Weak pronoun reference
- Scannable presentation, including use of the inverted pyramid style

Audience Effectiveness

Your site must put the audience into a role. Checking for audience role is highly subjective.

- Is it apparent what role the audience should assume, either from direct comments or from clear implications in the way the audience is addressed?
- Do all parts of the site help the audience assume that role?

Consistency

Readers find sites easier to navigate and access if all items are handled consistently (Pearrow). The document has many text features that should be repeated consistently in order to establish visual logic. These features include font, font size, color, placement on the screen, and treatment (bold, italics, all caps). In addition, a document has many visual features that must remain consistent in order to establish the visual logic. These features include size and placement on the screen.

Check these textual items for font, font size, color, placement on the screen, and treatment:

▶ Titles
▶ Headings
▶ Captions for figures
▶ Body text
▶ Lists
▶ Links

Check these visual items for size and placement on the screen:

▶ Clip art
▶ Photographs
▶ Tables and graphs
▶ Screen captures

Navigation

To check for navigation is to investigate whether all the links work and whether the path through the material makes the material accessible. To check whether all the links work is easy. Simply try each link. Make note of any that are "broken," that is, do not lead to any document. To determine if the path through the material makes the material accessible is more subjective. An effective way to investigate accessibility is to ask questions of yourself or your tester (Pearrow):

▶ Starting at the homepage, can you find information X quickly?
▶ From any point in the site or document, can you easily return to the homepage or top?
▶ Does the homepage give the reader an overview of the purpose and contents of the site or document?
▶ Does the title of any page repeat the wording of the link that led to it?
▶ At any time is the user annoyed? For instance, does a visual take a long time to open?

Ethics and Websites

For websites, as for all sources of information, technical communicators must act ethically. As with all documents, it is unethical to manipulate data, use deliberately misleading or ambiguous language, exaggerate claims, or conceal information that your users need in order to make good decisions (Summers and Summers 127).

However, communicators must also act ethically in regard to all the elements of websites: their design, code, graphics, and text. A major issue is plagiarism, using someone else's work without permission or acknowledgment. To state it succinctly, all aspects of websites are protected by copyright law. You are not permitted to copy anything from a website and use it on your website unless you either have permission or clearly acknowledge the source. As one expert says, "A work is copyrighted by the author at the moment of creation." A copyright notice or symbol, or "official" registration of a copyright, is not necessary—the rights automatically exist (Pfaff-Harris).

Another expert says this: "Simply put, by the time you see a webpage posted on the Web or a Usenet message posted in a newsgroup, it . . . is protected by copyright. You are not permitted to copy it, even if there is no copyright notice on the page or message" (Bunday). The best way for a communicator to ensure that she or he is using resources ethically is simply to ask the administrator of a website or the author of an e-mail message for permission to use the material and to give credit where credit is due. Often an e-mail request is all that is needed.

From another point of view, creators of websites can encourage ethical action by placing copyright notices on their sites, even though such notice is not required in order to be protected by the copyright law. Such a notice includes a copyright symbol, the years of creation and last modification, the name of the copyright holder, and the phrase "All rights reserved" (Murtaugh).

Another, somewhat different ethical strategy is to help viewers situate themselves. Summers and Summers suggest that you should include information on the creator and purpose of the site, and on the date it was last updated. This type of information will "help users evaluate the quality of the content you have provided" (Summers and Summers 130).

For a helpful overview of websites and ethical issues, see the article by Pfaff-Harris.

The Electronic Environment

The electronic environment of your site or document is the way in which it interacts with the reader's viewing equipment—modem, computer, and browser software. The basic guideline is that all your material should appear on screen quickly and be designed as you intended.

Check these electronic aspects:

- How long does the site or document take to load? Ten seconds appears to be the limit readers will wait before they get annoyed and click away to another site. Answer this question by using different access methods. Load the site over a 56K modem and over a T1 line. The differences are often large. Some programs (such as Front Page) provide a menu item that gives this information.
- Does the browser used affect whether the features appear? The two major browsers are Internet Explorer and Netscape Communicator. In most cases, but not all, a site will appear exactly the same regardless of which browser the viewer uses. For various program coding reasons, more sophisticated elements such as tables, frames, and videos sometimes work in one browser but not in the other. Checking a site with both browsers will ensure that readers see the items that you intended in the manner in which you intended.

If you must include features that only one browser supports, alert viewers to that effect ("Use Internet Explorer in order to view all the features of this site."). If you are using an application that viewers can download a version of for free (e.g., Shockwave), include a link to that site.

Clarity

The site or document must appear clearly on the screen. A viewer must be able to read all the elements.

To check for clarity, answer the following questions:

- Do all the visual aids appear? If not, edit the Web document to make sure that they do.
- Can all the text be read? Sometimes, inexperienced Web authors use color combinations that make text hard to read (black text on a blue background, yellow text on a white background). Revise the color (see "Focus on Color," pp. 171–178).
- Are the visual aids clear? If visual aids are fuzzy or illegible, edit them in a software program (such as Photoshop) that allows you to resample the image.

Worksheet for **Planning** a **Website** or **Document**

☐ **Identify the audience and the role they will play.**

☐ **Identify questions the audience will have about the content.**

☐ **Identify probable nodes for the site or document.**

☐ **Create a flow chart that indicates hierarchy and paths.**

☐ **Plan paths that give readers the freedom they need.**

☐ **Plan features of site (both screen and text items) that will facilitate the way readers find the information that they need.**

☐ **Create a screen template that groups similar information into distinct locations, including placement of visual aids.**

☐ **Choose font and color for heads and text.**

☐ **Determine which visual aids you need to convey your information.**

☐ **Choose a neutral background (light blues are good).**

☐ **Write text in manageable chunks.**

Worksheet for **Evaluating** a **Website**

Homepage
- Does the title make the content clear?
- Does the introduction tell you the purpose of the site?
- Can the homepage fit on one screen only?
- Did the site load quickly?

Navigation/Links
- Does every link work?
- Does every link have the same wording as the title of the page it links to?
- Are links coded or designed so that similar links look the same?
- Are all links of the same type always in the same location on each page?
- Do readers have sufficient freedom to access sections?
- Is the level of freedom too restrictive or annoying?
- Did you get lost navigating the site? If so, where?

Style
- Does the word choice indicate an exact awareness of the audience's knowledge and expectations?
- Are all the words spelled correctly?
- Is the grammar correct?

Text/Screen
- Is the background the same color in each document?
- Is the title of the page in the largest type on that page?
- Are all similar objects placed in the same place and do they have the same size?
- Are items on the page "clumped" so that the most important are together, set in bigger type and placed higher up?
- Can an audience easily read each document or is the font too small or too big or too busy?
- Does every document have the same font, including size?
- Is the text always aligned left?
- Is the format of each page consistent with every other page in the site?

Information
- Can readers easily figure out your plan of organization?
- Is the plan easy to find in the visual design you present?
- Does each section use appropriately convincing examples to inform the reader?
- Does each section start with a clear introduction that lists both the purpose and the parts of that section?
- Does each section contain a list of references with title of page, URL, and date visited?

Visual Aid Design
- Does each image load?
- Does the picture or table add a dimension of detail or interest not available in words?
- Is each visual aid in the same relative place on the page?
- Is the visual aid one that really helps a reader (and is not a waste of space and reader time)?
- Is there a clear cross-reference to each visual aid from the text?
- Does each visual aid have a caption?
- Is each visual aid roughly the same size?

Examples

Here are the report sections from the homepage presented in Figure 11.5, another informational Web report, and two sets of instructions. Note that Example 11.3 contains links to many other sections that are lower in the report's hierarchy. Those sections are not reproduced here.

Example 11.1

Report Sections That Can Be Linked to Homepage

PLANNING IS THE KEY TO SUCCESS

Much like any other project that you may tackle, creating a good website involves a little bit of planning. Many things have to be taken into account when you begin writing your own webpage.

The following should be considered:

1. Appearance
2. Actual size
3. Links
4. Visuals
5. Audience
6. Design

Go to other sections of this report:

List of links includes all sections, except planning.

Establishing a Purpose—Before you embark on this voyage . . .
Implementing Links—In order to make your site easy to use . . .
Handling Visuals and Text—These seem to spice up your website, but . . .
Interactive Websites—People do not just want to see a webpage, they want to . . .

ESTABLISHING A PURPOSE FOR YOUR WEBSITE

Before you decide what you are going to put up on your webpage, it is important that you define what purpose your website intends to serve. A lot of people have a webpage just for the sake of having one. Their homepage is just kind of there for others to look at. There is no way to interact with the webpage. Webpages should have the purpose clearly stated on them, so the reader knows what they are getting into.

Go to other sections of this report:

Planning Is the Key—Planning is essential in order for you to . . .
Implementing Links—In order to make your site easy to use . . .
Handling Visuals and Text—These seem to spice up your website, but . . .
Interactive Websites—People do not just want to see a webpage, they want to . . .

IMPLEMENTING LINKS IN YOUR WEBPAGE

People need to be able to navigate their way around your website once they are in it. They should be able to move back, forward, and to other sites if they want to. The more links a webpage has, the more freedom the reader has to pick and choose whatever it is he or she wants to read.

One of the biggest complaints I've heard from Web users is the fact that they often feel "trapped" inside of webpages, with no way out except to use the "Back" key. However, one must decide carefully the number of links to include, and which ones to exclude as well.

Go to other sections of this report:

Planning Is the Key—Planning is essential in order for you to . . .
Establishing a Purpose—Before you embark on this voyage . . .
Handling Visuals and Text—These seem to spice up your website, but . . .
Interactive Websites—People do not just want to see a webpage, they want to . . .

Wording of links repeats keywords in titles of other sections.

HANDLING VISUALS AND TEXT

Much like the traditional medium of written communication, webpages, too, benefit from visual aids. Visuals tend to catch the eye of the reader and take the place of text as well. However, when placing visuals on the Web, they must be planned just as carefully as they are when they are placed on paper. The same technical writing rules apply to visuals on the Web.

Text needs to be thought about, too. Especially on the Web, short "chunks" are necessary to keep the reader's interest. Because a computer screen seems to be smaller than what we would actually see on a regular sheet of paper, readers seemed to be turned off by large blocks of text on the Web. They are forced to keep scrolling down in order to get everything. Also, the bolding of heads and increasing their font sizes makes these items dominant over all other text, as they should be.

Go to other sections of this report:

Planning Is the Key—Planning is essential in order for you to . . .
Establishing a Purpose—Before you embark on this voyage . . .
Implementing Links—In order to make your site easy to use . . .
Interactive Websites—People do not just want to see a webpage, they want to . . .

(Continued)

INTERACTIVE WEBSITES

It is nice to have a webpage that allows readers to interact with the webpage somehow. Some examples of this may include a webpage that allows readers to e-mail the authors with questions or comments. Including your e-mail address on your webpage allows others to give you some input and constructive criticism about your website.

Another way a webpage could be interactive is through links that allow you to order something or inquire about something. Many companies today have their catalogs on the Web, and people can order things directly from the Internet. It is amazing what kind of feedback you can receive, or how your sales can increase, if you make your website interactive.

Go to other sections of this report:

Planning Is the Key—Planning is essential in order for you to . . .
Establishing a Purpose—Before you embark on this voyage . . .
Implementing Links—In order to make your site easy to use . . .
Handling Visuals and Text—These seem to spice up your website, but . . .

Example 11.2

Informational Web Report, Using Anchor Link Strategy

IMRD: RESEARCH REPORT

<u>**Introduction**</u> / <u>**Method**</u> / <u>**Results**</u> / <u>**Discussion**</u>

Introduction

With the advanced use of the electronic job-search, it is becoming difficult to ignore the increasingly important role of the online résumé. Since computers are becoming increasingly user-friendly, even those with relatively little computer experience are becoming familiar with this technique. One eminent problem, however, is the question of how users can get the information they need to the end they desire. This page will focus on how to translate a current résumé into one that is readily accessible for use in this type of a search. We aim to overcome the problem that there are limitless ways for an employer to request such résumés: as an attachment, as text format, as HTML, and still others limiting the characters per line to 80. A solution is a résumé done in HTML, since anyone with Web access can get it open—in most cases, if the reader is getting your e-mail, he or she also has a browser installed. The only problem that remains is that of the fonts you choose—the user may have different default fonts set. Add a line "Best read in xxx font"—is there another way?   **Top**

"Top" links return reader to links at beginning of document.

Method

I went on to create an electronic résumé, to be either included or posted on the Web. To do this, I saved my résumé, which was previously in Word format, as HTML. I inserted horizontal lines, as well as targets and anchors. Then I re-opened the copy of my résumé that had been saved in Word and saved it this time in .txt format. I then changed the page layout from a contemporary design to a more standardized paragraph form.   **Top**

Results

Attempt	Result
Word to HTML	When I changed my résumé from Word to HTML format, the result was that I lost much of the formatting I had done, because HTML does not support it. To force my résumé to look essentially the same, I inserted tables. Moreover, the targets and anchors were inserted to help eliminate the problem of not being able to view all of the information on the screen, as you would if the résumé were in front of you. The horizontal lines further helped to separate each section of information for the reader's understanding.
Word to text	When I changed my résumé from Word to .txt format, the result was that I lost much of the formatting I had done. In this case, I had even fewer options and ended up deciding on a different layout for my résumé altogether, simplifying it greatly. This, however, includes the issue of having 80 characters per line, as it was easy to change the page layout so that, despite any line specification, it would match. **Top**

Discussion

Because specifications vary from employer to employer, there are limitless ways of sending a résumé electronically. Hence, there is no perfect or universal résumé—a relatively frustrating result of this process. However, it is useful at this time to have a copy of your résumé in multiple formats so that they are readily available despite the request—particularly in .txt format, as this can be attached, sent as an e-mail, and easily translated to HTML. A solution to this problem is, just as a paper résumé has standards, to have standards for the electronic résumé. This will take time and the acceptance of the résumé in this medium. The best way to be safe right now is to save your résumé as text.   **Top**

Example 13.3

Instructions on
a Webpage

Links lead to
more detailed
discussions.

DEVELOPING ELECTRONIC RÉSUMÉS

Developing an electronic résumé is a useful technique, particularly when doing electronic job searches. It gets your credentials across to the reader, as well as allowing the freedom to move around efficiently. These instructions will lead you step by step through the process of creating an electronic résumé.

RECOMMENDATIONS

- *It is recommended that you create your résumé before beginning this process.* Typically, résumés require a great deal of information, and the beginner will find it simpler to understand the contents of these instructions without being concerned with this information. <u>Traditional résumé sections</u>

ASSUMPTIONS

These instructions assume that the user

- Has a basic knowledge of how to operate a personal computer.
- Has a basic knowledge of how to operate basic features of a word processor program.
- Has not previously created an HTML-formatted résumé.
- Has a basic concept of a résumé.

 Operational assumptions:

- The user is currently operating in Microsoft Word.
- The user has previously saved a copy of the résumé in Word format.

DIRECTIONS

1. Open your résumé as you would normally.
   - Check to ensure that you have it saved in <u>Word format.</u>
2. Save your résumé as an <u>HTML document.</u>
3. As a result of the previous step, *your document will lose all formatting* that was in place in the Word document. To get the results of formatting in your new HTML document,
   - <u>Insert a table.</u>
   - <u>Insert horizontal lines to visually separate section bodies of the résumé.</u>
4. <u>Insert a horizontal menu</u> between your name and the first section of your résumé. This menu should be horizontal, with the main sections of your résumé and index items. <u>See example.</u>
5. <u>Create internal hyperlinks</u> from each index item in the menu to the respective section body within the résumé.
   - This will let the user view the sections of your résumé that they wish to see.

6. Other formatting to enhance your résumé:
   a. <u>Horizontal menus</u> at the end of each résumé section (<u>example</u>)
      • Add internal hyperlinks for usability.
7. Add a Letter of Application.
   a. Save as a separate HTML document.
   b. <u>Insert a hyperlink</u> at the bottom of each page
      • So that the reader can move back and forth between your Résumé and Letter of Application.
8. Open a copy of a Web browser. Open your document and view it to ensure that it is functioning properly.
9. *Your resume is now ready!* You can now
   • Send it as an attachment to anyone with a browser that can view it.
   • Contact your local Web administrator to post it on your website.
   • Do an electronic job search and post your résumé on employers' sites.

Globalization/Localization and Websites

As Internet use grows worldwide, so does the number of non-English-speakers who use it. In order to reach a wider audience and/or client base, websites written in English must be able to be translated for use in non-English-speaking countries. Preparing a website for localization in other countries, languages, and cultures presents certain challenges.

One of the most common problems in website translation is the expansion and contraction of text. European languages often take up as much as 30 percent more space than English on a webpage. For example, Italian, Czech, German, and Greek will all expand to over 100 percent of English text, whereas Arabic, Hebrew, and Hindi will contract to less than 100 percent ("Text"). The languages that rely on characters (e.g., Japanese, Chinese, Korean—referred to as "double-byte" languages) can take up more or less space. It is difficult to predict exactly how a double-byte translation will affect your screen design until the actual translation is in place. This in turn interferes with spacing in the text and effective positioning of visual aids. Keep this contraction and expansion in mind when designing your screens, composing content, setting table properties, and creating graphics. Give extra space around items such as control buttons, menu trees, and dialog boxes (Macromedia).

When creating tables on your websites, base the properties on the longest translation that will be used on your site. If you use percentages rather than pixels, you can set your tables to 100 percent width. This will allow the table and the text to expand according to the user's browser—the text and the graphics

(Continued)

will appear the same to users viewing it in French as it will to those using the English-language version (Macromedia).

The symbols you choose are also critical. Avoid culturally specific symbols. Graphics with multiple meanings, pictures of body parts, religious symbols (e.g., a cross), or even symbols such as a stop sign may cause confusion, misunderstanding, and may even be construed as offensive (Pereira). Both the American A-Okay sign (thumb and forefinger forming a circle) and the thumbs-up sign are considered obscene gestures in parts of Europe, Middle Eastern countries, Brazil, and Australia. A website that utilized these symbols would signal to users in those areas that you hadn't done any research on their culture and customs.

Be sure to budget time for adjustments, even if you think you've optimized the site for localization.

For further reference:

The DreamWeaver Support Center at <www.adobe.com/support/dreamweaver/ manage/localization_design/localization_design03.html> has a great example, based on a French chocolate manufacturer's website, of text expansion and contraction and other issues with website design.

Social Media and Technical Writing

An earlier chapter introduced the idea of technical communication as interactive communication. In the past, technical writers entered into a give-and-take with readers, even though texts were static. The writers intended for readers to be able to do something as a result of the procedures or instructions they had written. Although writers were producing texts like instructions or handbooks, the point was to enable the reader to act, whether putting together a toy for a child, or carrying out maintenance on an important piece of equipment. This paradigm resulted in a "dialog of distance," in which the reader responded by acting upon the static text produced by the writer.

The Internet has changed all of that.

The Internet and Social Media

Since 2000 the technologies known as Social Media have grown exponentially. Social Media is "a group of Internet-based applications that build on the ideological and technological foundations of Web 2.0, and allow the creation and exchange of user-generated content" (Kaplan & Heinlein, 61). In other words, social media allow creators to interact with one another by posting, commenting and changing each other's work. Now (in 2012) the "two-way" communication possibilities have evolved into graphically pleasing, easy-to-use technologies including social networking sites (Facebook and LinkedIn), microblogs (Twitter), blogs, and information sharing tools like wikis, forums, and knowledge bases. These media are so common employers expect that you will understand and be able to use the various types of media.

Writing-Based Social Media

Although there are many different formats of social media, this section focuses on those that involve writing, like social networking sites, blogs, microblogs, wikis, forums, and knowledge bases. Many of these media enable the use of graphics, but the main format of communication is writing, rather than video or photographs.

Depending on the company, you could easily become involved in your company's social media efforts—or even be put in charge of them if your company is small. Although you may already be familiar with social media for personal use, there are many considerations you need to be aware of if you are representing your company. You could be tasked with being a creator, or a curator, depending on how tech-savvy your company is.

Your company might have just one social media presence, like a Facebook page, or might have a comprehensive strategy using many available forms. In fact, many companies integrate their social media, so that Facebook posts and tweets link to blogs, or blogs feature the company's Twitter feed in a sidebar. In order for you to thrive in your new role, you'll need to understand how these technologies work.

Here for instance is the SmartDraw blog page (Figure 1). It contains posts written by company employees explaining ways to conduct (while using SmartDraw applications) effective meetings. Notice that it is part of their website: users open

Figure 1.
SmartDraw web page screen grab

Source: SmartDraw

(Continued)

SmartDraw
@SmartDraw
SmartDraw helps you work smarter through visual communication.
Automatically create flowcharts, org charts, and more. Get a FREE
TRIAL at smartdraw.com
San Diego, CA   http://tiny.cc/SDtwtt

Follow   👤▾

3,130 TWEETS
750 FOLLOWING
1,199 FOLLOWERS

Tweet to SmartDraw

@SmartDraw

Tweets   >
Following   >
Followers   >
Favorites   >
Lists   >
Recent images   >

Tweets

SmartDraw @SmartDraw   1 Jun
RT @AlliCrow: The Visual-Thinker recap is out! bit.ly/hnG7F6 ▸ Top
stories today via @SmartDraw @feistycoach @apelad
Expand

SmartDraw @SmartDraw   31 May
Use a mind map to Get'r Done! bit.ly/JRbZ9N
Expand

SmartDraw @SmartDraw   31 May
@cwestervelt16 Very entertaining and funny video! A more simple
approach would be our "Single Slide" Presentation: bit.ly/JUjqNu
💬 View conversation

SmartDraw @SmartDraw   30 May
@Amrein2 I'll be sure to let our development team know :) Which
templates do you use most often?
💬 View conversation

Figure 2.
@SmartDraw twitter feed
Source: SmartDraw

it by clicking on Blog in the upper right corner. Notice too that below the main menu bar is another bar that connects users to Twitter, Facebook, and other technologies. Clicking the Twitter icon opens the company-run Twitter site (Figure 2), filled with comments by SmartDraw employees and customers. This interlinked system is a common approach to corporate use of social media.

Using Social Media

Before getting to work on any social media projects, make sure you are familiar with your company's social media policy. If your company doesn't have one, you may be asked to help put one together. Many larger companies have addressed the most common social media issues and have comprehensive guides on netiquette (etiquette on the Internet) and online behavior, including Microsoft, the *LA Times*, Cisco, the American Red Cross, General Motors, Harvard Law, and Nordstrom (Dishman). These guidelines are remarkably similar. They encourage employees to use social media, with "good judgment" and in general to act in the social media realms as representatives of the company, whose actions reflect well or poorly on the company.

Because the point of social media is that users can be involved, you need to understand both what social media use involves, and the audience's role before attempting to leverage it. Whether you're familiar with social media or are new to the concept, this section will help you understand your role as a professional social media user. The section explains corporate uses for different social media formats and details best practices for communicating with both external and internal audiences.

Social Networking Sites

As of this writing, Facebook is the dominant social networking site being used on the Internet in the United States. Companies that use it well can spread information that the audience wants without being intrusive. In addition, a presence on Facebook allows readers to engage with your brand and helps foster a sense of community.

If you are part of the social media team working with your company's Facebook page, you will most likely be granted rights as an administrator. Since Facebook is continuously updating how the site functions, the privileges and responsibilities of an administrator change from time to time. To learn the most recent practices you will want to review the help section for administrators to get a handle on your role and your privileges.

Two different methods can be used to promote your organization on Facebook: Pages and Groups. According to Cohen, "Facebook Pages are useful for marketing brands, causes and organizations/businesses. Facebook Groups are useful for having open discussions (privately or publicly) with a small group of fewer than 5,000 users."

Pages

A Facebook page is a website hosted by the Facebook Corporation. "Pages are for businesses, organizations and brands to share their stories and connect with people. Like timelines, you can customize Pages by adding apps, posting stories, hosting events and more." ("What") For example, Porsche has a Facebook Page. On that Page, the company shares photos of their newest cars, links to video premiers of cars, announces upcoming motorsport events that Porsche is participating in, shows photos of cars that are still in development, and even posts fan photos and comments. By "liking" the Facebook page, those who are interested in Porsches get periodic updates about a subject that appeals to them. (Porsche)

Groups

A group is a Facebook site that focuses on a particular subject, which can be anything from swing dancers to Barbie Dolls. Groups can have one or two members or over a million. Groups can control their visibility by being open, closed, or secret. Admission into a Group is determined by administrators or group members, thus controlling privacy and participation. Common uses for Groups are for academic programs at universities, local chapters of a national organization, or social clubs. For instance, the Master of Science in Technical and Professional Communication program at the University of Wisconsin-Stout has a Group page. Groups are also an excellent way for departments within companies to use Facebook for internal social networking, as you'll see later in this section.

How To's for Business Posting

To successfully use Facebook for business, follow these best practices (Ryan) to use when posting on behalf of your company whether you're in charge of a Page or a Group:

(Continued)

- **Ask questions**—When you ask customers or clients questions, you are encouraging your audience to respond and interact, which is the reason for social media.
- **Keep variety in your posts**—Make sure you are posting about different subjects and that your posts add value for customers. It is also a good idea to use various tools besides text like photos, links, and video. Consider posting information that your customers can use (the meaning of sunblock ratings on a sunscreen site), stories about others who use your product, links to related stories perhaps on your website or blog.
- Create **Status** posts, upload **Photos or Video,** or announce **Events or Milestones**.
- **Keep posts brief**—People scan status updates, they don't read them. If you have a lot of information to convey, provide links to longer texts. Try to keep posts under 140 characters.
- **Consider your audience**—There is potential for many people to read your posts. What you say should attract people to your company. This includes editing and revising for your audience.
- **Use proper punctuation**—While you may use shortened text and abbreviations with your friends, your professional Facebook posts are a reflection on your company. Punctuation errors are to be avoided.

Social Networking on the Inside: Internal Social Network Use

Only recently have companies begun to understand the value of internal social networking. Whether using an existing platform or creating one specific to the company, organizations are beginning to tap into the many uses of social networking for their employees. Internal social networking sites use the same strategies for posting as external sites. And as with external sites, internal sites could be a site for the entire company or a group (e.g. the MS in Technical and Professional communication mentioned earlier).

Some companies are resistant to social networking for fear of people misusing it; however, many forward-thinking executives have found that it enhances company culture, and employee happiness which is passed on to customers. According to a study conducted by Google and Millward Brown, internal social networking has been found to, " . . . improve internal efficiencies while deepening relationships among employees and providing customers and stakeholders with the improved goods and services that tend to result from more joined up thinking" (13).

Use an internal social networks to:

1. **Encourage sharing:** Employees can share ideas, input, and opinions about subjects related to the company. In addition, leaders can be provided with constant, real-time updates about the status of projects, research, and development across all departments within a company.
2. **Capture knowledge:** When an employee uses social networking to share knowledge discovered during the course of a project–that knowledge is recorded and can be referenced.
3. **Enable action:** Social networking gives employees a way to engage with one another on projects without having to have everyone in the same

location. Whether in a different department or a different office, employees can work together or communicate on projects.

4. **Empower employees:** This form of communication empowers employees to contribute directly to the enterprise. They can post ideas, links, information, questions, answers, and achievements for anyone in the company to see. (Li)

Microblogs

Microblogs are social media platforms that allow users to share information, but they impose a limit on the amount of content shared. One of the most popular current microblogs is Twitter. With this service, users can post 140 character updates about themselves or their interests. Users of this service "Follow" friends, acquaintances, or celebrities by subscribing to those people's Twitter feeds. They can reply to each others' posts, send direct messages, share something another person wrote by "retweeting" a post, send modified retweets, and index tweets by subject so they are searchable later. Users can also designate "favorite" tweets and link to expanded content like websites or photos (Brockmeier).

Companies have successfully leveraged the microblogging trend to make announcements about the company, products, and employees. Tweets are also a useful way for you to link to press releases, media coverage, or other company social media. For instance, on the Oreo Twitter page, the company announces contests, asks customers' opinions, thanks people for being fans, and acknowledges comments in a friendly way, to foster a sense of community (Oreo).

Twitter has even established best practices for businesses that use the service: "Build your following, reputation, and customer's trust with these simple practices:

1. Share. Share photos and behind the scenes info about your business. Even better, give a glimpse of developing projects and events. Users come to Twitter to get and share the latest, so give it to them!
2. Listen. Regularly monitor the comments about your company, brand, and products.
3. Ask. Ask questions of your followers to glean valuable insights and show that you are listening.
4. Respond. Respond to compliments and feedback in real time.
5. Reward. Tweet updates about special offers, discounts, and time-sensitive deals.
6. Demonstrate wider leadership and know-how. Reference articles and links about the bigger picture as it relates to your business.
7. Champion your stakeholders. Retweet and reply publicly to great tweets posted by your followers and customers.
8. Establish the right voice. Twitter users tend to prefer a direct, genuine, and of course, a likable tone from your business, but think about your voice as you Tweet. How do you want your business to appear to the Twitter community?" (Twitter).

From https://support.twitter.com/articles/68916-following-rules-and-best-practices

Microblogging on the Inside: Internal Corporate Microblogging

There are several different uses for internal microblogging. If you are asked to moderate the company's internal microblog as a part of the social media team, these are the types of posts you may be asked to respond to (Pontefract):

Employee Questions: Using internal microblogs for questions means that management gets a pulse on what employees do and don't know. Answers can be provided and will be shared throughout the company, rather than between two people talking, or within a few people exchanging information in an e-mail.

Sharing: The sharing is usually more far-reaching than an e-mail or other type of communication. Subjects are searchable by subject, since microblog posts are typically indexed. This type of sharing allows information to be more widespread, but less directed and formal than an e-mail.

Opinions: Microblogs provide an opportunity for employees to express their opinions, although some companies may be less open to this sharing than others. Organizational issues can be addressed in short form during chats in microblogging sites, limiting off-topic discussions.

Recognition: Microblogging facilitates recognition. When the company or an individual employee achieves something outstanding, praise and recognition can be shared with the whole company in a brief, widespread way.

Blogs

Blogs are online journals where organizations can share news or information about their companies, services, or products. Blogs are considered interactive media because most of them feature comments sections where readers can offer feedback on a post. You may be asked to write articles for the blog, edit others' articles, or monitor the comments section of the blog to ensure that contributors follow your company's rules for participation.

What Should You Write About?

The content of a blog will vary from company to company; however, use a blog to make sure that your customers get information that is useful to them, and not just marketing about your company's products. Blogs are more about building relationships with customers than they are about telling them how successful and interesting your company is. For example, if your company sells lawn equipment, it is better to give your customers techniques that will help them mow their lawns more quickly, show them how to kill quackgrass, or edge their sidewalks more effectively. If you share useful information on your blog, customers will most likely turn to your products to help them. Smart Draw, for instance, explains how to run better meetings. Their posts do not focus solely on their technology.

Gaining Followers

The only way a blog can help your company is if people read it. One of the best ways to gain readers is to get involved in the blog community surrounding your industry. Read others' blogs, and take time to participate in the conversation by posting in their blog comment sections. You will learn what subjects readers are interested in reading about. Your participation will establish your credibility as an expert in the field, and can lead readers to your blog. Linking your blog from the company website is a simple way to attract readers. The blog can also be advertised on print pieces, Facebook, and Twitter as well.

Blogging on the Inside: Internal Corporate Blogging

Companies often use blogs to keep workers informed about internal information that doesn't need to be seen by outsiders. As a technical writer, you may be asked to contribute to or edit articles being posted to an internal company blog. Common uses for internal blogs are to provide information to employees regarding:

- Details about upcoming products
- Announcements about new projects, services, or products
- Company policies and procedures
- Work-related news like changes to benefits
- Crisis communication policies

Internal blogging can improve a company's ability to keep employees informed and connected. Sharing knowledge and ideas can be useful to help an organization grow and to promote goodwill within the company.

Blogs can sustainably replace company newsletters, and can take the place of impersonal informational e-mail blasts. Unlike those forms of media, blogs invite interactivity. For instance, you can write blog post about an employee inventing a new product, and his or her coworkers can add their congratulations in the comments section of the blog. This interaction turns previously impersonal announcements into a channel for fostering a sense of community.

Wikis

Wikis are pages that users can edit. These pages can be small, such as collaborations to create documents, plans for a family reunion or large such as Microsoft's TechNet Wiki. Collaborators join the wiki and then can create, add to, subtract from, and edit each others' entries. Administrators can limit the collaborators to just a few people (as in a group writing a report) or can open participation to anyone, as Microsoft does with TechNet Wiki.

Typically a corporate wiki contains numerous sections dealing with specific topics. As administrator, you might post information about the topic, but the point is that the users will develop the topic. Microsoft introduces TechNet Wiki with this statement "This is a grassroots effort led by you and us: **we** the community. We don't have all the answers. There will be kinks in the system, lessons to learn with the community, features to hatch and processes to evolve. And there will be content. Lots of great content written by all of us.

Join. Help us. Contribute boldly!" ("Wiki"). If your company website hosts a wiki, you are inviting users to contribute to the body of knowledge and welcoming their input.

(Continued)

Internal Wikis

According to Hobbs, using a wiki in your company involves a culture shift. Contributors need to know that the participation—including adding and rewriting—is welcome.

Those involved can contribute to the original document by editing errors, adding information, eliminating outdated information, categorizing information within the document, providing structure or clarifying information.

You can use a wiki in any of these ways:

- **Collaborative documents:** Allow a group of contributors to work directly on a document, rather than e-mailing proofs and ideas to each other.
- **Project management:** Collect Schedules, notes, progress reports and other types of information used for projects in one accessible space.
- **Brainstorming:** Provide a space for people to put up their suggestions. It both saves on paper and eliminates meetings.
- **Frequently asked questions:** By posting a Q & A section, everyone in the company has access to the answers for questions that are asked the most.
- **Training:** Along with frequently asked questions, wikis can be a repository for training materials for new employees. Unlike an employee handbook, a wiki can be updated along with policy changes, and added to as new training needs arise. (Roderigo)

In order for a wiki to work for your company, you need to make sure you know why the wiki format the best tool for the job. The collaboration that wikis allow also cause many employees to shy away from using the technology. You will have to work with those employees to show them how this "culture shift" in communication is valuable.

Forums

Forums are a kind of message board where users can post and respond to each others' messages. For instance Intel hosts Intel Support Community. An important feature of the site is "Recent Discussion" in which users post a question about a problem and wait for other users to supply an answer. One post was "Is there a way to enable or disable Intel Virtualization Technology with syscfg utility ? four days ago." The user explained his problem and others suggested answers. A popular version of a forum is Ask.com where users post and receive answers to questions such as "What do the colors mean on a mood ring?"

Companies can sponsor forums that allow customers to discuss products and share experiences. Here are five important things to remember if you are in charge of your company's forum. The forum should

1. Have a clear focus related to the company product (woodworking equipment stores has a forum on household projects
2. help your readers do something (share techniques for building bookcases)
3. foster community (allow users to share plans, ideas, photos and other relevant items)
4. have a moderator who keep the forum friendly

Internal Forums

Like those created for customers, internal forums can be a great way to share knowledge. If your company has many divisions, locations, services, or products, having an internal forum can enhance communication and increase a sense of community. Companies can share information, outline policies, make announcements, work on projects, brainstorm, refine ideas, provide instructions and training, or post updates about the company. The interactive component allows coworkers to give feedback to one another, share best practices, and collaborate.

Knowledge Bases

A knowledge base is a collection of questions and answers published online for users to reference. Questions are usually submitted via online form, e-mail, or during customer support calls. Apple, for instance, has a Search Support site in which a user can enter keywords in order to find relevant articles. Entering key words like "iPad camera" connects the users to many articles dealing with the topic ("Advanced Search"). The University of Wisconsin-Stout has a similar site. However, at the bottom of the article you receive is a questionnaire asking whether the article was helpful. Users answers to that survey will help you answer questions more precisely ("Ask5000").

Knowledge bases are less interactive than other media, because the exchange is asynchronous. If a customer or user submits a question, those questions are not seen by general users until they are posted by moderators to a knowledge base.

This format can be used when you just want to make sure that your users get answers to their questions as quickly as possible. Your involvement as a writer may be needed to ensure that answers to technical questions are clear, understandable, and easy to read.

Knowledge bases can be used for the same reasons by both external users and internal users.

Exercises

These exercises assume that they will occur in a computer lab where it is possible to project a site onto many screens or one large one. In that situation, small groups of two or three and oral reporting seem most effective. However, if individual work and written reports work better in the local situation, use that approach.

▶ You Create

1. Create a simple webpage that includes a title, text about yourself, and at least one visual aid. If you do not have access to a program like Dreamweaver, use a free commercial site such as Jimdo.

2. Using the page you created in Exercise 1, create two other versions of it. Keep the content the same, but change the design.

3. Create a series of paragraphs that presents the same information in promotional, concise, and scannable text.

4. Create or download an image. Present it on a webpage in three different sizes. In groups of two or three, discuss how you achieved the differences and the effect of the differences. Report your findings orally to the class.

5. Using the page you created in Exercise 1, make several different backgrounds. In groups of three or four, review the effectiveness of the background (Is it distracting? Does it obscure the text?), and demonstrate both a good and bad version to the class.

▶ You Analyze/Group

6. Go to any website. In groups of two or three, assess the role the reader is asked to assume. Orally report your findings to the class. Alternative: Write a brief analytical report in which you identify the role and present support for your conclusion.

7. Go to any website. In groups of two or three, assess the style of the text. Is it promotional, concise, or scannable? Present your findings orally to the class. Or, as in Exercise 6, write a brief analysis.

8. Go to any website. In groups of two or three, assess the use of visuals at the site. Review for clarity, length of time to load, physical placement on the site. Present an oral report to the class, or write an information analysis.

9. Go to any website. In groups of two or three, assess the template. Are types of information effectively grouped? Is it easy to figure out where the links will take you? Report orally or in writing as your instructor requires.

10. In groups of two or three, critique any of the Examples (pp. 334–339) or use one of the samples in the website that accompanies this text. Judge them in terms of style, screen design, and audience role. Explain where you think the examples are strong and where they could be improved.

Writing Assignment

1. Create an informational website; if possible, load it onto the Web so that others may review it. Determine a purpose and an audience for the site. Create a homepage and documents that carry out the purpose. The site should have at least three nodes. Before you create the site, fill out the planning sheet on page 340. Your instructor will place you in a "review group"; set

up a schedule with the other group members so that they can review your site for effectiveness at several points in your process. To review the site, use the points in the section "Testing" (pp. 328–331) or use the Worksheet for Evaluating a Website (pp. 332–333).

2. In groups of two or three investigate plagiarism on the Web. Write an IMRD detailing your findings. For instance, many types of instructions are verbatim copied from one site to another. In a book like this, examples can't be shown, but they are there. Finding them will take some keyword searching expertise.

Web Exercise

Review two or three websites of major corporations in order to determine how they use the elements of format. Review the homepage, but also review pages that are several layers "in" (e.g., Our Products/Cameras/UltraCompacts) in the Web; typically, pages further "in" look more like printed pages. Write a brief analytical or IMRD report discussing your results.

Social Media Exercises

1. Create a blog using a blog hosting service. Depending on your instructor's directions, create this for a company, for the entire course, or for some aspect of the course (e.g. The Style Blog; the Instructions Blog). OR Create a unified site for your class. Open a Facebook group (closed or secret), create a blog (use a host site like Blogger or Blogdrive), create a Twitter feed. Throughout the semester create and encourage posts about the class and issues related to the topic of the class. At the end of the semester create a Learning Report. See Writing Assignment 8, Chapter 5 (p. 137) for details.

2. In groups of two to four investigate a company or institution whose Facebook, Twitter, Blog, and YouTube sites interact with one another. Create a report that explains the strategies that they use, for example, building customer loyalty, posting events, media, articles of interest but not about the company.

3. Find a blog related to one of your professional interests, analyze five blog posts. Create an analytical report or a brief slide presentation to show your findings. Use this report to help your classmates learn what to post on their business site.

4. If you don't already have one, set up a Twitter account. If you already set one up, find a few technical writing practitioners to follow. After several

weeks report your findings to the class, either in a shared report (via e-mail perhaps) or on the class blog, or as a slide presentation.

5. If you don't have a Facebook or LinkedIn account, create one. If you have a student Facebook account, you may want to consider creating a new one for your professional life.

6. Find an Internet forum concerned with your professional specialty. Choose a thread to read. Create an account and add to the discussion. After several weeks report your findings to the class, in the format your instructor requires.

Works Cited and Consulted

"Advanced Search." Apple. 2012. Web. 2 June 2012. <http://support.apple.com/kb/index?page=search&locale=en_US&q=>.

"Ask5000 HelpDesk." University of Wisconsin-Stout. 2012. Web. 2 June 2012. < http://helpdesk.uwstout.edu/>.

Brockmeier, Joe. "A Beginner's Guide to Twitter." *ReadWriteWeb.com*. ReadWrite Enterprise. 9 January 2012. Web. 1 June 2012 < http://www.readwriteweb.com/enterprise/2012/01/a-beginners-guide-to-twitter.php>.

Bunday, Karl M. "Building Better Web Sites." 2000. Learn in Freedom. Web. 28 May 2012. <http://learninfreedom.org/technical_notes.html>.

Cohen, Andrew. "Table: Facebook Pages vs. Facebook Groups (vs. LinkedIn Groups)." *Forum One*. Forum One. 20 January 2012. Web. 31 May 2012.

Coney, Mary, and Michael Steehouder. "Role Playing on the Web: Guidelines for Designing and Evaluating Personas Online." *Technical Communication* 47.3 (August 2000): 327–340. Print.

Dishman, Lydia. "More Social Media Policies: LA Times, Harvard Law, Microsoft, and Cisco." *Fast Company*. July 2010. Web. 30 May 2012. <http://www.emailinstitute.com/social-media/more-social-media-policies-la-times-harvard-law-microsoft-and-cisco>.

Google & Millward Brown. "How Social Technologies Drive Business Success." *MillwardBrown.com*. 15 May 2012. Google & Millward Brown. Web. 31 May 2012. <www.millwardbrown.com/.../Googe_MillwardBrown_How-Social-Technologies-Drive-Business-Success_201205.pdf>.

Halvorson, Kristina. *Content Strategy for the Web*. Berkeley, CA: New Riders Press, 2009. Print.

Hobbs, Jeff. "HOW TO: Use Wikis for Business Projects." *Mashable.com*. Mashable.com. 1 July 2012. Web. 31 May 2012. <http://mashable.com/2009/07/01/wikis-business-projects/>.

Horton, William. *Designing and Writing Online Documentation: Hypermedia for Self-Supporting Products*. 2nd ed. New York: Wiley, 1994. Print.

Intell Support Community. Intel. 2012. Web. 2 June 2012. <http://communities.intel.com/community/tech/general>.

Kaplan, Andreas M. and Michael Haenlein. "Users of the World, Unite! The Challenges and Opportunities of Social Media." *Business Horizons* 53.1 (2010): 59–68. Print.

Li, Charlene. "Report: Making the Business Case for Enterprise Social Networks." *AltimeterGroup.com.* 22 February 2012. Web. 31 May 2012. < http://www.altimetergroup.com/2012/02/making-the-business-case-for-enterprise-social-networks.html>.

Macromedia Dreamweaver Support Center. "Accounting for text expansion and contraction." 2001. Web. 28 May 2012. <www.adobe.com/support/dreamweaver/manage/localization_design/localization_design03.html>.

Nielsen, Jakob. "F-Shaped Pattern for Reading Web Content." *Useit.com.* 17 April 2006. Web. 20 May 2012. <http://www.useit.com/alertbox/reading_pattern.html>.

Nielsen, Jakob. "Writing Style for Print vs. Web." *Useit.com.* 9 June 2008. Web. 20 May 2012. <http://www.useit.com/alertbox/print-vs-online-content.html>.

Oreo. "Oreo Twitter Site." *Twitter.* 2012. Web. 30 May 2012.

Pearrow, Mark. *Web Site Usability Handbook.* Rockland, MA: Charles River Media, 2000.

Pereira, Arun. "The Rationale of Cultural Customization." Web. 28 May 2012. <www.arunpereira.com/Chapter2.pdf>.

Pfaff-Harris, Kristina. "Copyright Issues on the Web." The Internet TESL Journal II.10 (October 1996). Web. 24 April 2012. <http://iteslj.org/Articles/Harris-Copyright.html>.

Pontefract, Dan. "6 Use Cases for Enterprise Micro-Blogging." *Brave New Org.* Brave New Org, n.d. Web. 31 May 2012. < http://www.danpontefract.com/?p=2041>.

Porsche. "Porsche Facebook Page." *Facebook.* 2012. Web. 2 June 2012.

Rodrigo, Alexis. "25 Ways to Use Wikis in Your Intranet." *Business2Community.com.* Business2Community. 24 April 2012. Web. 31 May 2012. < http://www.vialect.com/ways-to-use-wikis-in-your-intranet>.

Spyridakis, Jan. "Guidelines for Authoring Comprehensible Web Pages and Evaluating Their Success." *Technical Communication* 47.3 (August 2000): 359–382. Print.

Summers, Kathryn, and Michael Summers. *Creating Websites That Work.* Boston: Houghton Mifflin, 2005. Print.

"Text Expansion & Contraction in Translation" The International Business Edge 2012. Web. 28 May 2012. <http://www.globalization-group.com/edge/2010/05/text-expansion-contraction-in-translation/ >

Focus on
HTML

HTML (*Hypertext Markup Language*) is the invisible structure of the Web. Viewers can see a Web document because the browser (e.g., Safari or Internet Explorer) "reads" an HTML document and displays the results on the screen. Actually, HTML is a code, a series of typed orders placed in the document. For instance, to make a word appear boldfaced on the screen, the writer places a "start bold" () and an "end bold" () command in the HTML document:

I want you to read this book.

The browser displays the sentence

I want you to read this book.

HTML code exists for everything that makes a document have a particular appearance on screen. If the item appears on screen, it appears because the code told the browser to display it. Codes exist for paragraphing, fonts, font sizes, color, tables, and all other aspects of a document. Codes tell which visual aid should appear in a particular place in a document. Figures 1 and 2 show the HTML code for a simple document and the document as it displays on a browser.

```
<html>
<head>
<title>
Sample Display Techniques Illustrated          browser
</title></head>                                 title
<body>
<b><h3>Sample Display Techniques Illustrated    title
</b></h3> by Dan Riordan

<p>These pages illustrate several techniques    text
for displaying information. I have illustrated
ways to use lists, the align command, the
anchor command, and escape links.

<p><b>List</b> I like to teach, especially
<ul>
<li>in groups and                               list
<li>using technology.</ul>

<p><b>Align Center</b> My wife and I often
visit a cabin up north. Here is what the cabin  visual
looks like: <center><img src="twridal.gif"      aid
width=100 height=167></center>

<p><b>Anchor</b> This device allows you to
"link" inside a document. I illustrate the
device by letting you read a series of
<a href="twrida5.htm">letters</a> my family's
immigrants wrote in the 1850s.
<p><b>Escape Links</b> Here are "escape links"
to sites connected to this one:
<br><b><i><small>Return to <br><a
href="http://www.uwstout.edu/english/riordan/   "escape
techwrit/techwrit.htm">Technical Writing        links"
</a>|</b>
<b><i> <a href="http//www.uwstout.edu/english/
projects.htm">English Department Student
Projects</a>|</b>
<b><i> <a href="http//www.uwstout.edu/english/
english.htm">English Department</a></p></b></i>
</small>

</body>
</html>
```

© Cengage Learning

Figure 1.
HTML Code for a Web Document

The classic way to develop a site is to create material using an ASCII text editor like Notepad (DOS) or TextEdit (Mac). The method for creating it is easy.

- Open a file in one of these two programs.
- Type in certain HTML commands.
- Type in your text.

But typing in code is time-consuming and susceptible to errors. If, for instance, one of the brackets (>) is omitted in the boldface code, the word will not appear as boldfaced. As a result, most Web authors use a Web authoring program that creates the code as the writer designs the webpage on screen. Many such programs exist. Some of the most frequently used are Sharepoint Designer, DreamWeaver, and SeaMonkey Composer. In addition many sites, such as Jimdo, allow you to open an account and then simply use their templates to build a website. As of this writing this method of creating websites has become very popular because the creation is so easy. Instructions on using such programs are beyond the scope of this book; however, many good instruction books are available, and all the programs have help menus and training tutorials. The best advice is to begin to practice with the programs to learn their features and to develop enough proficiency so that you can achieve the effects that you visualize.

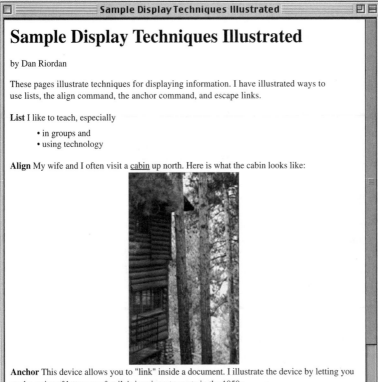

Figure 2.
Browser Display of the Code in Figure 1

Formal Reports

CHAPTER CONTENTS

CHAPTER 12 **IN A NUTSHELL**

Formal format presents documents in a way that makes them seem more "official." Often the format is used with longer (ten or more pages) documents, or else in documents that establish policy, make important proposals, or present the results of significant research.

Formal format requires a title page, a table of contents, a summary, and an introduction, in that order.

The *title page* gives an overview of the report—title, author, date, report number if required, and report recipient if required. Place all these items, separated by white space, at the left margin of the page.

The *table of contents* lists all the main sections and subsections of the report and the page on which each one begins.

The *summary*—often called "executive summary" and sometimes "abstract"—presents the report in brief. The standard method is to write the summary as a "proportional reduction"; each section of the summary has the same main point and the relative length as the original section. After your readers finish the summary, they should know your conclusions and your reasons.

The *introduction* contains all the usual introductory topics but gives each of them a head—background, scope, purpose, method, and recommendations.

Formal reports are those presented in a special way to emphasize the importance of their contents. Writers often use formal reports to present recommendations or results of research. Other reasons for using a formal approach are length (over ten pages), breadth of circulation, perceived importance to the community, and company policy. Although a formal report looks very different from an informal report, the contents can be exactly the same. The difference is in the changed perception caused by the formal presentation. This chapter explains the elements of formal reports and discusses devices for the front, body, and end material.

The Elements of a Formal Report

To produce a formal report, the writer uses several elements that orient readers to the report's topics and organization. Those elements unique to the formal report are the front material and the method of presenting the body. Other elements—appendixes, reference sections, introductions, conclusions, and recommendations—are often associated with the formal report but do not necessarily make the report formal; they could also appear in an informal report.

The formal front material includes the title page, the table of contents, and the list of illustrations. Almost all formal reports contain a summary at the front, and many also have a letter of transmittal. The body is often presented in "chapters," each major section starting at the top of a new page.

Because these reports often present recommendations, they have two organizational patterns: traditional and administrative (Freeman and Bacon; ANSI). The traditional pattern leads the reader through the data to the conclusion (Freeman and Bacon). Thus conclusions and recommendations appear at the end of the report. The administrative pattern presents readers with the information they need to perform their role in the company, so conclusions and recommendations appear early in this report.

Traditional	**Administrative**
Title page	Title page
Table of contents	Table of contents
List of illustrations	List of illustrations
Summary or abstract	Summary or abstract
Introduction	Introduction
Discussion—Body Sections	Conclusions
Conclusions	Recommendations/Rationale
Recommendations/Rationale	Discussion—Body Sections
References	References

Front Material

Transmittal Correspondence

Transmittal correspondence is a document, usually a "that directs the report to someone. The correspondence contains

▶ The title of the report.
▶ A statement of when it was requested.
▶ A very general statement of the report's purpose and scope.
▶ An explanation of problems encountered (for example, some unavailable data).
▶ An acknowledgment of those who were particularly helpful in assembling the report.

	To: Ms. Elena Solomonova, Vice-President, Administrative Affairs From: Rachel A. Jacobson, Human Resources Director Subject: Proposal for the Spousal Employment Assistance Program

Title of report
Cause of
writing
 	
Purpose of
report
 	
Statement of
request
 	
Praise of
coworkers
	Attached is my report "Proposal for the Implementation of a Spousal Employment Assistance Program," which you requested after our March 15 meeting.
	
	    The report presents a solution to the problems identified by our large number of new hires. In brief, those new hires all had spouses who had to leave careers to move to Rochester. This proposal recommends initiating a spousal employment assistance program to deal with relocation problems.
	
	    Compiled by the Human Resources staff, this report owes a significant debt to the employees and their spouses who agreed to be interviewed as part of its preparation.

Figure 12.1   Sample E-mail of Transmittal

© Cengage Learning

Title Page

Well-done title pages (see Figure 12.2, p. 359) give a quick overview of the report, while at the same time making a favorable impression on the reader. Some firms have standard title pages just as they have letterhead stationery for business letters. Here are some guidelines for writing a title page:

**PROPOSAL FOR THE IMPLEMENTATION
OF A SPOUSAL EMPLOYMENT
ASSISTANCE PROGRAM**

By
Rachel A. Jacobson
Director, Human Resources

May 1, 2005

Corporate Proposal
HRD 01-01-2005

Prepared for
Elena Solomonova
Vice-President, Administrative Affairs

© Cengage Learning

Figure 12.2   Title Page for a Formal Report

▶ Place all the elements at the left margin (ANSI). (Center all the elements if local policy insists.)
▶ Name the contents of the report in the title.
▶ Use a 2-inch left margin.
▶ Use either all caps or initial caps and lowercase letters; use boldface when appropriate. Do not use "glitzy" typefaces, such as outlined or cursive fonts.
▶ Include the writer's name and title or department, the date, the recipient's name and title or department, and a report number (if appropriate).

Table of Contents

A table of contents lists the sections of the report and the pages on which they start (see Figure 12.3). Thus it previews the report's organization, depth, and emphasis. Readers with special interests often glance at the table of contents, examine the abstract or summary, and turn to a particular section of the report. Here are some guidelines for writing a table of contents:

▶ Title the page *Table of Contents.*
▶ Present the name of each section in the same wording and format as it appears in the text. If a section title is all caps in the text, place it in all caps in the table of contents.
▶ Do not underline in the table of contents; the lines are so powerful that they overwhelm the words.

<table>
<tr><td colspan="2" align="center">**TABLE OF CONTENTS**</td></tr>
</table>

© Cengage Learning

Figure 12.3 Table of Contents for an Administrative Report

▶ Do not use "page" or "p." before the page numbers.
▶ Use only the page number on which the section starts.
▶ Set margins so that page numbers align on the right.
▶ Present no more than three levels of heads; two is usually best.
▶ Use *leaders*, a series of dots, to connect words to page numbers.

List of Illustrations

Illustrations include both tables and figures. The list of illustrations (Figure 12.4) gives the number, title, and page of each visual aid in the report. Here are guidelines for preparing a list of illustrations:

▶ Use the title *List of Illustrations* if it contains both figures and tables; list figures first, then tables.
▶ If the list contains only figures or only tables, call it *List of Figures* or *List of Tables*.
▶ List the number, title, and page of each visual aid.
▶ Place the list on the most convenient page. If possible, put it on the same page as the table of contents.

<table>
<tr><td colspan="2" align="center">**LIST OF ILLUSTRATIONS**</td></tr>
</table>

© Cengage Learning

Figure 12.4 List of Illustrations for a Formal Report

Summary or Abstract

A summary or an abstract (or executive summary) is a miniature version of the report.

In the summary, present the main points and basic details of the entire report. After reading a summary, the reader should know

- The report's purpose and the problem it addresses.
- The conclusions.
- The major facts on which the conclusions are based.
- The recommendations.

Because the summary "covers" many of the functions of an introduction, recent practice has been to substitute the summary for all or most of the introductory material, placing the conclusions and recommendations last. Used often in shorter (10- to 15-page) reports, this method eliminates the sense of overrepetition that is sometimes present when a writer uses the entire array of introductory elements.

Follow these guidelines to summarize your formal report:

- Concentrate this information into as few words as possible—one page at most.
- Write the summary *after* you have written the rest of the report. (If you write it first, you might be tempted to explain background rather than summarize the contents.)
- Avoid technical terminology (most readers who depend on a summary do not have in-depth technical knowledge).

SUMMARY

Recommendation given first
Background

This report recommends that the company implement a spousal assistance program. Swift expansion of the company has brought many new employees to us, most of whom had spouses who left professional careers. Because no assistance program exists, our employees and their spouses have found themselves involved in costly, time-consuming, and stressful situations that in several instances have affected productivity on the job.

Basic conclusions
Benefits

A spousal assistance program will provide services that include home- and neighborhood-finding assistance, medical practitioner referrals, and employment-seeking assistance. Advantages include increased employee morale, increased job satisfaction, and greater company loyalty.

Cost
Implementation

Cost is approximately $54,000/year. The major benefit is productivity of the management staff. The program will take approximately six months to implement and will require hiring one spousal employment assistance counselor.

© Cengage Learning

Figure 12.5 Summary Example

Introduction

The introduction orients the reader to the report's organization and contents. Formal introductions help readers by describing purpose, scope, procedure, and background. Statements of purpose, scope, procedure, and background orient readers to the report's overall context.

To give readers the gist of the report right away, many writers now place the conclusions/recommendation right after the introduction. Recently, writers have begun to combine the summary and the introductory sections to cut down on repetition. Figure 12.6 (pp. 363–364) illustrates this approach.

Purpose Statement

State the *purpose* in one or two sentences. Follow these guidelines:

▶ State the purpose clearly. Use one of two forms: "The purpose of this report is to present the results of the investigation" or "This report presents the results of my investigation."
▶ Use the *present* tense.
▶ Name the alternatives if necessary. (In the purpose statement in the example later, the author names the problem [lack of a spousal assistance program] and the alternatives that she investigated.)

Scope Statement

A *scope statement* reveals the topics covered in a report. Follow these guidelines:

▶ In feasibility and recommendation reports, name the criteria; include statements explaining the rank order and source of the criteria.
▶ In other kinds of reports, identify the main sections, or topics, of the report.
▶ Specify the boundaries or limits of your investigation.

Procedure Statement

The *procedure statement*—also called the *methodology statement*—names the process followed in investigating the topic of the report. This statement establishes a writer's credibility by showing that he or she took all the proper steps. For some complex projects, a methodology section appears after the introduction and replaces this statement. Follow these guidelines:

▶ Explain all actions you took: the people you interviewed, the research you performed, the sources you consulted.
▶ Write this statement in the *past tense*.

◗ Select heads for each of the subsections. Heads help create manageable chunks, but too many of them on a page look busy. Base your decision on the importance of the statements to the audience.

Brief Problem (or Background) Statement

In this statement, which you can call either the *problem* or *background statement,* your goal is to help the readers understand—and agree with—your solution because they view the problem as you do. You also may need to provide background, especially for secondary or distant readers. Explain the origin of the problem, who initiated action on the problem, and why the writer was chosen. Follow these guidelines:

◗ Give basic facts about the problem.
◗ Specify the causes or origin of the problem.
◗ Explain the significance of the problem (short term and long term) by showing how new facts contradict old ways.
◗ Name the source of your involvement.

In the following example, the problem statement succinctly identifies the basic facts (relocating problems), the cause (out-of-state hires), the significance (decline in productivity), and the source (complaints to Human Resources). Here are the purpose, scope, procedure, and background statements of the proposal for Spousal Employment Assistance:

INTRODUCTION

Purpose

Two-part purpose: to present and to recommend

This proposal presents the results of the Human Resources Department's investigation of spousal employment assistance programs and recommends that XYZ Corp. implement such a program.

Scope

Lists topics covered in the report

This report details the problems caused by the lack of a spousal employment assistance program. It then considers the concerns of establishing such a program here at XYZ. These concerns include a detailed description of the services offered by such an office, the resources necessary to accomplish the task, and an analysis of advantages, costs, and benefits. An implementation schedule is included.

(Continued)

Procedure

Enough
information
given to
establish
credibility

The Human Resources Department gathered all the information for this report. We interviewed all 10 people (8 women and 2 men) hired within the past 12 months and 6 spouses (4 men and 2 women). We gathered information from professional articles on the subject. The human resources office provided all the salary and benefits figures. We also interviewed the director of a similar program operating in Arizona and a management training consultant from McCrumble University.

Problem

Background
(cause)

Basic facts

In the past year, XYZ has expanded swiftly, and this expansion will occur throughout the near future. In the past year, ten new management positions were created and filled. Seven of these people moved here from out of state. Several of these people approached the Human Resources Department for assistance with the problems involved in relocating.

Source of
impetus to
solve problem

Possible
solution

Some of these problems were severe enough that some decline in productivity was noted and was also brought to the attention of Human Resources. Four of the managers left, citing stress as a major reason. That turnover further affected productivity. A spousal employment assistance office is one common way to handle such concerns and offset the potential bad effects of high turnover.

© Cengage Learning

Figure 12.6 Introduction Example

Lengthy Problem (or Background) Statements

Some reports explain both the problem and its context in a longer statement called either *Problem* or *Background*. A *background statement* provides context for the problem and the report. In it you can often combine background and problem in one statement.

Some situations require a lengthier treatment of the context of the report. In that case, the background section replaces the brief problem statement. Often this longer statement is placed first in the introduction, but practices vary. Place it where it best helps your readers.

To write an effective background statement, follow these guidelines:

▶ Explain the general problem.
▶ Explain what has gone wrong.
▶ Give exact facts.
▶ Indicate the significance of the problem.
▶ Specify who is involved and in what capacity.
▶ Tell why you received the assignment.

BACKGROUND

General problem

Management increases have brought many new persons into the XYZ team in the past year. This increase in personnel, while reflecting an excellent trend in a difficult market, has had a marked down side. The new

Data on what is wrong

personnel have all experienced significant levels of stress and some slide in productivity as a result of the move. All ten of the recent hires had spouses who left professional career positions to relocate in Rochester. These people have experienced considerable difficulty finding career opportunities in our smaller urban region, and all the families have reported a certain amount of stress related to everything from finding a home to finding dentists. Four of

Significance

these managers subsequently left our employ, citing stress as the major reason to leave. These departures caused us to undertake costly, time-consuming personnel searches.

Why the author received the assignment

After interviews revealed the existence of such stress, the Executive Committee of Administrative Affairs discussed the issue at length and authorized Human Resources to carry out this study. The Director of Human Resources chaired a committee composed of herself, one manager who did not leave, and a specialist on budget. HR staff conducted the data gathering.

© Cengage Learning

Figure 12.7 Background Example

Conclusions and Recommendations/Rationale

Writers may place these two sections at the beginning of the report or at the end. Choose the beginning if you want to give readers the main points first and if you want to give them a perspective from which to read the data in the report. Choose the end if you want to emphasize the logical flow of the report, leading up to the conclusion. In many formal reports, you present only conclusions because you are not making a recommendation.

Conclusions

The conclusions section emphasizes the report's most significant data and ideas. Base all conclusions only on material presented in the body. Follow these guidelines:

▶ Relate each conclusion to specific data. Don't write conclusions about material you have not discussed in the text.
▶ Use concise, numbered conclusions.
▶ Keep commentary brief.
▶ Add inclusive page numbers to indicate where to find the discussion of the conclusions.

Conclusions presented in same order as in text

> This investigation has led to the following conclusions. (The page numbers in parentheses indicate where supporting discussion may be found.)
>
> 1. The stresses experienced by the new hires are significant and are expected to continue as the company expands (6).
> 2. Stress is not related to job difficulties but instead is related more to difficulties other family members are experiencing as a result of the relocation (6).
> 3. Professionals exist who are able to staff such programs (7).
> 4. The program will result in increased employee morale, increased job satisfaction, and greater company loyalty (9).
> 5. A program could begin for a cost of $54,000 (10).
> 6. The major benefits of the program will be increased productivity of the management staff and decreased turmoil created by frequent turnover (11).
> 7. A program would take six months to initiate (13).

© Cengage Learning

Figure 12.8   Conclusions

Recommendations/Rationale

If the conclusions are clear, the main recommendation is obvious. The main recommendation usually fulfills the purpose of the report, but do not hesitate to make further recommendations. Not all formal reports make a recommendation.

In the rationale, explain your recommendation by showing how the "mix" of the criteria supports your conclusions. Follow these guidelines:

- Number each recommendation.
- Make the solution to the problem the first recommendation.
- If the rationale section is brief, add it to the appropriate recommendation.
- If the rationale section is long, make it a separate section.

Solution to the basic problem

Other recommendations on implementation

> 1. XYZ should implement a spousal employment assistance program. This program is feasible and should eliminate much of the stress that has caused some of the personal anxiety and productivity decreases we have felt with the recent expansion.
> 2. The Executive Committee should authorize Human Resources to begin the procedure of writing position guidelines and hiring an SEA counselor.

© Cengage Learning

Figure 12.7   Recommendations

The Body of the Formal Report

The body of the formal report, like any other report, fills the needs of the reader. Issues of planning and design, covered in other chapters, all apply here. You can use any of the column formats displayed in Chapter 6 for laying out pages. Special concerns in formal reports are paginating and indicating chapter divisions.

Paginating

Be consistent and complete. Follow these guidelines:

▶ Assign a number to each piece of paper in the report, regardless of whether the number actually appears on the page.
▶ Assign a page number to each full-page table or figure.
▶ Place the numbers in the upper right corner of the page with no punctuation, or center them at the bottom of the page either with no punctuation or with a hyphen on each side (-2-).
▶ Consider the title page as page 1. Do not number the title page. Most word processing systems allow you to delete the number from the title page.
▶ In very long reports, use lowercase roman numerals (i, ii, iii) for all the pages before the text of the discussion. In this case, count the title page as page i, but do not put the i on the page. On the next page, place a ii.
▶ Paginate the appendix as discussed in "End Material" (later).
▶ Use headers or footers (phrases in the top and bottom margins) to identify the topic of a page or section.

Indicating Chapter Divisions

To make the report "more formal," begin each new major section at the top of a page (see Example 12.2, which starts on p. 373).

End Material

The end material (glossary and list of symbols, references, and appendixes) is placed after the body of the report.

Glossary and List of Symbols

Traditionally, reports have included glossaries and lists of symbols. However, such lists tend to be difficult to use. Highly technical terminology and symbols should not appear in the body of a report that is aimed at a general or multiple audience. Place such material in the appendix. When you must use technical terms in the body of the report, define them immediately; informed readers can simply skip over the definitions. Treat the glossary as an appendix. If you need a glossary, follow these guidelines:

▶ Place each term at the left margin, and start the definition at a tab (two or three spaces) farther to the right. Start all lines of the definition at this tab.
▶ Alphabetize the terms.

References

The list of references (included when the report contains information from other sources) is discussed along with citation methods in Appendix B.

Appendix

The appendix contains information of a subordinate, supplementary, or highly technical nature that you do not want to place in the body of the report. Follow these guidelines:

▶ Refer to each appendix item at the appropriate place in the body of the report.
▶ Number illustrations in the appendix in the sequence begun in the body of the report.
▶ For short reports, continue page numbers in sequence from the last page of the body.
▶ For long reports, use a separate pagination system. Because the appendixes are often identified as Appendix A, Appendix B, and so on, number the pages starting with the appropriate letter: A-1, A-2, B-1, B-2.

Worksheet for Preparing a Formal Report

☐ **Determine the audience for this report.**

• Who is the primary audience and who the secondary? How much does the audience understand about the origins and progress of this project? How will they use this report? Will it be the basis for a decision?

☐ **Plan the visual aids that will convey the basic information of your report.**

☐ **Construct those visual aids.**

- Follow the guidelines in Chapter 7.

☐ **Prepare a style sheet for up to four levels of heads and for margins, page numbers, and captions to visual aids.**

☐ **Decide whether each new section should start at the top of a new page.**

☐ **Create a title page.**

☐ **Prepare the table of contents.**

- How many levels of heads will you include? (Two is usual.) Will you use periods for leaders?

☐ **Prepare the list of illustrations.**

- Present figures first, then tables.

☐ **Determine the order of statements (purpose, scope, procedure, and so forth) in the introduction.**

- In particular, where will you place the problem and background statements? In the introduction? In a section in the body?

☐ **Prepare a glossary if you use key terms unfamiliar to the audience.**

☐ **List conclusions.**

☐ **List recommendations, with most important first.**

☐ **Write the rationale to explain how the mix of conclusions supports the recommendations.**

☐ **Write the summary.**

☐ **Prepare appendixes of technical material.**

- Use an appendix if the primary audience is nontechnical or if you have extensive tabular or support material.

Examples

Example 12.1 is the body of the report whose introduction is explained in this chapter. Example 12.2 is a brief formal report.

Example 12.1

Formal Report Body

DISCUSSION

In this section, I will describe spousal employment assistance (SEA), discuss the advantages and benefits of it, and develop a time schedule for the implementation of it.

NATURE OF THE PROBLEM

Many complex issues arise when relocating a dual-career family. Issues such as a new home, a new mortgage, two new jobs in the family, a new and reliable child care service—to name only a few. These issues, if not dealt with in an efficient manner, can create tremendous stress in the new employee—stress that dramatically affects productivity on the job.

Productivity and protection of our company's human resources investment are the key issues we are dealing with in this program. The intention is that the more quickly the employee can be productive and settled in a new area, the less costly it will be for our company.

DESCRIPTION OF THE PROGRAM

I am proposing a separate office within the company for the SEA program. It would be staffed by a consultant who would research and develop the following areas:

- Home-finding counseling
- Neighborhood finding
- Mortgage counseling
- Spouse and family counseling
- Spouse employment assistance
- Child care referrals
- School counseling
- Cost-of-living differences
- Doctor and dentist referrals

All counseling services would be handled by our SEA office employee except for formal employment assistance, which would be contracted with a third-party employment firm. A third-party firm can provide the advantage of objectivity as well as a proper level of current employment information.

ADVANTAGES OF THE PROGRAM

The program is a service provided by us, and paid for by us, that is for the sole purpose of assisting the new employee. The advantages are increased employee morale, increased job satisfaction, and greater company loyalty. The employee feels that the company is concerned with the problems he or she is facing in the relocation process. The assistance the employee receives makes the move easier, so adjusting to the new job is quicker. The result is a more productive employee.

WHAT ARE THE COSTS VERSUS THE BENEFITS?

Costs

The comprehensive program will cost the company approximately $54,500 per year. As illustrated in Table 1, this includes $2,500 for research and development, $27,000 for the SEA consultant, and $25,000 for the third-party employment firm (ten contracts at $2,500 each). This figure doesn't include the cost of office completion, which would run about $1,100 to finish the first-floor office space (room 120), which isn't currently occupied.

Also in Table 1, I have estimated the dollar amount our company invests yearly on new relocating managers. $54,500 is a drop in the bucket when you realize that we spend at least $290,000 yearly on new hires alone.

TABLE 1
Cost/Employee Investment Comparison

	Estimated Yearly Cost of Program*	Estimated Value of Human Resource Investment	
Research/development	$ 2,500	New relocating managers	
SEA consultant	27,000	Approx. 10 @ $29,000 each	
			$290,000
Employment firm contracts 10 @ approx. $2,500 each	25,000		
TOTAL	$ 54,500	TOTAL	$290,000

*Doesn't include the one-time cost of office completion (about $1,100).

Benefits

The benefit to our company is the increase in productivity of the management staff. The cost to our company shouldn't be considered a luxury or frill expense, but a way to protect and enhance the company's human resources investment. The yearly cost of the program ($54,500) compared with the estimated yearly

(Continued)

cost of new employees who would use it ($290,000) shows that the expense is far outweighed by the investment we've made in new management hires.

WHAT ABOUT IMPLEMENTATION?

Implementation time is estimated at six to seven months depending on when the SEA consultant is hired. This is because, after a three-month hiring and selection period, the new consultant would be given three months to begin the research and development of the program. After these three months, research would continue, but client consultation would also begin (refer to Figure 1).

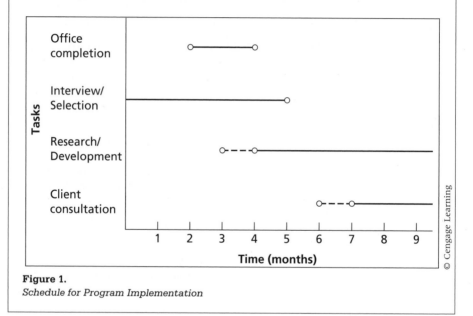

© Cengage Learning

Figure 1.
Schedule for Program Implementation

Example 12.2

Formal Report
Excerpts

This sample reproduces only the executive summary, the introduction, and one section of the report. The title page, table of contents, list of figures and tables are deleted. The first section of the Findings section reproduced here illustrates the way that the author presented the entire body. Each subsection of Findings is treated identically.

EXECUTIVE SUMMARY

The purpose of this field study project was to investigate the usability of the University of Wisconsin-Stout's new website that was launched in the fall of 2011.

Methodology

The study involved twelve participants who met to partake in the usability study. The methods used in during the testing included the following:

- Participants were briefed about the study using the preamble (**Appendix 1**) and asked to sign the consent form should they agree to participate in the study (**Appendix 2**).
- Then participants were asked to complete the tasks on the task list (**Appendix 3, 4, 5**).
- After completing the task list participants were then directed through a series of post-test questions (**Appendix 6, 7, 8**).

Findings

- Six out of the twelve participants felt that the videos on the homepage should be placed in a different location. Ex: Placing it in its own tab or on the top-level links.
- All eight of the eight participants said that locating D2L was difficult.
- Six of the twelve participants said that the color, tabs, pictures, and videos presented on the website needed design improvements.
- Four of the four participants noted that finding the degree requirements for a major was difficult.
- All twelve of the twelve participants felt that the website needs navigational changes in order to function effectively.

Recommendations

The study generated the following recommendations:

- Place videos in a different area of the website.
- Label the Logins link Personal Accounts.
- Have an introduction video detailing commonly used features of the site.
- Ensure that the video portion of the website contains only videos.
- Narrow down redundant content on the website like ways to give for example.
- Fix tabs so that they do not dropdown behind other content on the website.
- Make sure links go to the right webpage.
- Change the Freshman link to First Year Students.
- Label the Class Search link Access Stout.
- Consider using a different color besides orange for the website.
- Logins link should be placed under the Current Students tab.
- Place Undergraduate Majors under the Future Students tab.

(Continued)

INTRODUCTION

This study investigated the usability of the new website of the University of Wisconsin-Stout. The study focused primarily on navigation and visuals in order to identify usability issues and make recommendations.

This study involved twelve participants from three most typical target users of the website:

Group 1: Four *prospective* students
Group 2: Four *current* students
Group 3: Four *faculty* members

The three groups of participants were asked to complete the following tasks respectively:

Prospective Students:

1. Explore the website for five minutes
2. Find and play a video on the homepage
3. Find a program that interests you and locate its degree requirements
4. Find a list of faculty members in a program that interests you
5. For the program that interest you find its program director's name
6. Locate the Applied Arts Building on the campus map using the homepage

Current Students:

1. Explore the website for five minutes
2. Find and play a video on the homepage
3. Log into Student e-mail
4. Locate Desire to Learn
5. Locate Access Stout
6. Find a list of faculty members in the Department of English and Philosophy
7. Locate the Applied Arts Building on the campus map using the homepage

Faculty:

1. Explore the website for five minutes
2. Find and play a video on the homepage
3. Locate Access Stout
4. Locate Desire to Learn
5. Find the webpage of the Human Resources Department
6. Find a list of faculty members in the Department of English and Philosophy
7. Locate the Applied Arts Building on the campus map using the homepage

Methodology and Procedure

The study was conducted in the Technical Communication lab on the UW-Stout campus.

- First subjects were introduced to the study using the *preamble* (**Appendix 1**), and asked to sign the *consent form* should they agree to participate in the study (**Appendix 2**).
- Then subjects were asked to sit in front of the laptop and load the university homepage.
- Next participants set out to complete six tasks in Group 1 and seven tasks in Group 2 and Group 3 as I observed them. They were asked to think aloud. (**Appendix 3, 4, 5**).
- Each participant had 45 minutes to complete the entire study.
- Participants were asked questions from the *post-test questionnaire* (**Appendix 6, 7, 8**).

(Continued)

FINDINGS

The results can be classified into two categories. The first were my notes on any problems participants had while they completed the tasks and the second were the problematic responses participants expressed to me when asked the post-test questions.

Observational Notes

This section discusses what participants expressed verbally to be problematic as they completed the tasks. The two different issues stated by participants were with design and locating key functionalities.

Design Related Issues
Group 1: Prospective Students

"Why do the tabs get stuck behind other information on the screen when clicked?" Figure 1 shows the tab clicked being stuck behind content on the website. [Figure 1 not shown here. It is a screen shot of the website.]

Example 12.3

Executive
Summary of a
20-Page Report

General Education Senior Level Assessment, Fall 2011

EXECUTIVE SUMMARY

The following is a summary of the main findings from the fall 2011 General Education Senior Level Assessment survey. The senior students who completed this survey were part of the seventh cohort to receive laptop computers as part of the e-scholar program. The survey was completed by 97 senior students for a response rate of 14%.

The authors believe the data support the following general conclusions:

General Education Categories

1. The survey was divided into nine General Education course categories of writing effectively, speaking effectively, listening effectively, analytic reasoning, health and physical education, humanities and arts, social and behavioral science, natural science, and technology. Average category ratings ranged from 3.16 to 3.49 on a 5-point scale where 1 = *None* and 5 = *Strong*. For all categories, the 95% confidence intervals for each skill overlap with the rest of the intervals for each skill, which suggests that all questions were rated similarly.

   a. *Write Effectively* (3.45) and *Speak and Present Ideas Effectively* (3.49) had the highest mean ratings.
   b. *Natural Science* had the lowest average (3.16).

2. In comparison, mean category ratings for fall 2010 ranged from 3.23 to 3.50.
3. Each of the nine course categories consisted of one or more questions. Six categories had at least four questions. Listed below is the range of averages (95% confidence intervals) for the categories with four or more questions. For all categories, the 95% confidence intervals for each skill overlap with the rest of the intervals for each skill, which suggests that all questions were rated similarly.

   a. Write effectively average ratings ranged from 3.13 to 3.60.
   b. Speak or present ideas effectively average ratings ranged 3.39 to 3.55.
   c. Analytic reasoning average ratings ranged from 3.28 to 3.44.
   d. Humanities and arts average ratings ranged from 3.24 to 3.36.
   e. Social and behavior sciences average ratings ranged from 3.25 to 3.42.
   f. Technology average ratings ranged from 3.25 to 3.41.

Individual Question Ratings

4. The average ratings on all questions were above 3.0 on the 5-point scale, and ranged from 3.16 to 3.60. For all categories, the 95% confidence intervals for each skill overlap with the rest of the intervals for each skill, which suggests that all questions were rated similarly.

5. In comparison fall 2010 mean ratings ranged from 3.15 to 3.60.
6. *To reason and to explain reason* was the skill with the highest mean rating (3.60).
7. *Understand how the natural and physical sciences affect daily life and how daily life affects the natural and physical sciences* was the skill with the lowest mean rating (3.16).

This sample provides the introductory material from an engineering firm. The title page, table of contents, and executive summary show you how this is handled. Notice that the executive summary presents the important conclusion in bold in a solitary paragraph.

Phase I Environmental Assessment
Proposed Shopko Property
Junction Avenue
Stanley, WI

June 2012

Prepared for (Users):
City of Stanley
116 E. 3rd Avenue
Stanley, WI 54767

Prepared by:
Cedar Corporation
604 Wilson Avenue
Menomonie, WI 54751
CC #S0133-009-302-01

(Continued)

Example 12.4

Introductory
Material and
Executive
Summary

APPENDICES

Appendix A - Environmental Records Search Data
Appendix B - Sanborn7 Map Report
Appendix C - Aerial Photographs
Appendix D - Proposed CSM I.................................EXECUTIVE SUMMARY

Cedar Corporation has performed a Phase 1 Environmental Site Assessment in conformance with the scope and limitations of ASTM Practice E1527-05 of the proposed Shopko property located along the proposed extension of Junction Avenue in Stanley, WI, the property. Any exceptions to, or deletions from this practice are described in Sections II.C. of this report.

This assessment has revealed no evidence of recognized environmental conditions in connection with the property.

The results of this study are based upon the professional interpretation of information available to Cedar Corporation during the time and budget constraints of this assessment. Cedar Corporation has considered that the information provided by the cited references is complete and correct. Cedar Corporation does not warrant that this report represents an exhaustive study of all possible environmental concerns at the Property. The items investigated as part of this study represent the most likely sources of environmental concern, and are consequently believed to address your needs at this time.

Exercises

▶ You Create

1. For Exercise 10, Chapter 6 (p. 168), create one or all of the following: a title page, table of contents, summary, conclusion, and recommendation.
2. For Exercise 10, Chapter 6 (p. 168), create a page layout.

▶ You Revise

3. Redo this table of contents to make it more readable.

4. Edit the following selection, which is taken from the discussion section of a formal report. Add at least two levels of heads, and construct one appropriate visual aid. Write a one- or two-sentence summary of the section.

> The cost of renting a space will not exceed $150. Our budget allows $140 per month investment at this time for money available from renting space. The cost of renting a space is $175 per month for 100 square feet of space at Midtown Antique Mall. This exceeds the criteria by a total of $25 per month. The cost for renting 100 square feet of space at Antique Emporium is $100 per month, which is well within our criteria based on our budget. Antique Emporium is the only alternative that meets this criterion. The length of the contract cannot exceed 6 months because this is what we have established as a reasonable trial period for the business. Within this time, we will be able to calculate average net profit (with a turnover time no longer than 3 months) and determine if it is worth the time invested in the business. We will also be able to determine if we may want to continue the business as it is or on a larger scale by renting more floor space. The Midtown Antique Mall requires a 6-month contract. The Antique Emporium requires an initial 6-month contract which continues on a month-to-month basis after the contract is fulfilled. Both locations fulfill the contract length desired in the criteria. The possibility of continuing monthly at the Antique Emporium is an attractive option compared with renewing contracts bi-yearly.

▶ Group

5. In groups of two to four, discuss whether the conclusions and recommendations in Examples 13.1 and 13.2 (pp. 396–401) should have the recommendation section at the beginning or end of the reports. Be prepared to give an oral report to the class. If your instructor requires, rewrite the introductory material so it follows the pattern of the human resources proposal (pp. 363–364) Alternate: Rewrite the human resources introduction by using the executive summary method.

6. Your instructor will hand out a sample report from the *Instructor's Manual*. In groups of two to three, edit it into a formal report. Change introductory material as necessary.

7. If you are working on formal format elements, bring a draft of them to class. In groups of two or three, evaluate each other's material. Use the guidelines (for Title Page, Table of Contents, Summary, Introduction, Conclusion, Recommendation) in this chapter as your criteria. Rewrite your material as necessary.

Writing Assignments

1. Create a formal report that fulfills a recommendation, feasibility, proposal, or research assignment, as given in other chapters of this book.

   a. Create a template for your formal report. Review Chapter 6, page 160, and Chapter 15, pages 457–459.

   b. Choose an introductory combination and write it.

   c. Write the conclusions and recommendations/rationale sections.

   d. Divide into pairs. Read each other's draft from the point of view of a manager. Assess whether you get all the essential information quickly. If not, suggest ways to clarify the material.

2. Write a learning report for the writing assignment you just completed. See Chapter 5, Writing Assignment 8, page 137, for details of the assignment.

Web Exercise

Review two or three websites of major corporations in order to determine how they use the elements of format. Review the homepage, but also review pages that are several layers into the site (e.g., Our Products/Cameras/Ultra-Compacts); typically, pages further "in" look more like hard copy. Write a brief analytical or IMRD report discussing your results.

Works Cited

American National Standards Institute (ANSI). *Scientific and Technical Reports—Preparation, Presentation, and Preservation.* ANSI/NISO Z39.18-2005. New York: ANSI, 2005. Print.

Freeman, Lawrence H., and Terry R. Bacon. *Writing Style Guide.* Rev. ed. GM1620. Warren, MI: General Motors, 1991. Print.

Recommendation and Feasibility Reports

CHAPTER CONTENTS

CHAPTER 13 **IN A NUTSHELL**

Feasibility studies and recommendations present a position based on credible criteria and facts. *Feasibility studies* use criteria to investigate an item in order to tell the reader whether or not to accept the item. *Recommendations* use criteria to compare item A to item B in order to tell the reader which one to choose. To decide whether or not to air condition your house is a feasibility issue; to decide which air conditioning system to purchase is a recommendation issue.

Report strategy. In the introduction, *set the context:* tell the background of the situation, explain the methods you used to collect data, and state why you chose these criteria. In the body, *deal with one criterion per section.* A helpful outline for a section is

▶ Brief introduction to set the scene

▶ Discussion of data, often subdivided by alternative

▶ A helpful visual aid

▶ A brief, clear conclusion

Based on criteria. Criteria are the framework through which you and the reader look at the subject.

▶ Select topics that an expert would use to judge the situation. (For the air conditioner, a criterion is cost.)

▶ Select a standard, to limit the criterion. (The limitation is "the system may not cost more than $6000.")

▶ Apply the criteria. (Look at the sales materials of two reputable systems.)

▶ Present the data and conclusion clearly. Report the appropriate facts from your investigation, create a useful visual aid, and use heads and chunks to guide the reader through the subsections.

Professionals in all areas make recommendations. Someone must investigate alternatives and say "choose A" or "choose B." The "A" or "B" can be anything: which type of investment to make, which machine to purchase, whether to make a part or buy it, whether to have a sale, or whether to relocate a department. The decision maker makes a recommendation based on *criteria:* standards against which the alternatives are judged.

For professionals, these choices often take the form of *recommendation reports* or *feasibility reports.* Although both present a solution after alternatives have been investigated, the two reports are slightly different. Recommendation reports indicate a choice between two or more clear alternatives: this distributor or that distributor, this brand of computer or that brand of computer (Markel). Feasibility reports investigate one option and decide whether it should be pursued. Should the client start a health club? Should the company form a captive insurance company? Should the company develop this prototype? (Bradford; Hofstrand and Holz-Clause; "Professional"). This chapter explains how to plan and write both types of reports.

Planning the Recommendation Report

In planning a recommendation, you must consider the audience, choose criteria for making your recommendations, use visual aids, and select a format and an organizational principle.

Consider the Audience

In general, many different people with varying degrees of knowledge (a multiple audience) read these reports. A recommendation almost always travels up the organizational hierarchy to a group—a committee or board—that makes the decision. These people may or may not know much about the topic or the criteria used as the basis for the recommendation. Usually, however, most readers will know a lot about at least one aspect of the report—the part that affects them or their department. They will read the report from their own point of view. The human resources manager will look closely at how the recommendation affects workers, the safety manager will judge the effect on safety, and so on. All readers will be concerned about cost. To satisfy such readers, the writer must present a report that enables them all to find and glean the information they need.

Choose Criteria

To make data meaningful, analyze or evaluate them according to criteria. Selecting logical criteria is crucial to the entire recommendation report because you will make your recommendation on the basis of those criteria and because your choice of the "right" criteria establishes your credibility.

The Three Elements of a Criterion

A *criterion* has three elements: a name, a standard, and a rank (Holcombe and Stein). The *name* of the criterion, such as "cost," identifies some area relevant to the situation. The *standard* is a statement that establishes the limit of the criterion—for instance, "not to exceed $500.00." The standard heavily influences the final decision. Consider two very different standards that are possible for cost:

1. The cost of the water heater will not exceed $500.

2. The cheapest water heater will be purchased.

If the second standard is in effect, the writer cannot recommend the more expensive machine even if it has more desirable features.

The *rank* of the criterion is its weight in the decision relative to the other criteria. "Cost" is often first, but it might be last, depending on the situation.

Discovering Criteria

Criteria vary according to the type of problem. In some situations, a group or individual will have set up all the criteria in rank order. In that case, you show how the relevant data for the various alternatives measure up to these criteria.

When criteria have not been set up, you need to discover them by using your professional expertise and the information you have about needs and alternatives in the situation. One helpful way of collecting relevant data is to investigate appropriate categories: technical, management/maintenance, and financial criteria (Markel).

Technical criteria apply to operating characteristics such as the necessary heat and humidity levels in an air-moving system. *Management/maintenance criteria* deal with concerns of day-to-day operation, such as how long it will take to install a new air-moving system. *Financial criteria* deal with cost and budget. How much money is available, and how big a system will it purchase?

Applying Criteria

Suppose you were to investigate which of two jointers to place in a high school woods lab. To make the decision, you need to find the relevant data and create the relevant standards. To find relevant data, answer questions derived from the three categories.

▶ Technical—Does the jointer have appropriate fence size? Table length? Cutting capacity?

▶ Financial—How much does each jointer cost? How much do optional features cost? How much money is available? What is the standard?

▶ Management/Maintenance—Which one is safer? Will we need to reconfigure the lab or its electrical service? Will the jointers be available by the start of school in August?

To create standards, you must formulate statements that turn these questions into bases for judgment. You derive these standards from your experience, from an expert authority (such as another teacher), or from policy. For instance, because you know from your own experience the length of your typical stock, your standard will read "Must be able to handle up to 52 inches." Another teacher who has worked with these machines can tell you which features must be present for safety. School policy dictates how you should phrase the cost standard.

Use Visual Aids

Although you might use many kinds of visuals—maps of demographic statistics, drawings of key features, flow charts for procedures—you will usually use tables and graphs. With these visuals you can present complicated information easily (such as costs or a comparison of features). For many sections in your report, you will construct the table or figure first and then write the section to explain the data in it. Visual aids help overcome the problem of multiple audiences. Consider using a visual with each section in your report.

In the following example, the author first collected the data, then made the visual aids, and *then* wrote the section. Note that the table combines data from several criteria; this technique avoids many small, one-line tables. (The entire report appears as Example 13.1 at the end of the chapter.)

The fence serves as a guide for planing face and edge surfaces. The size of the fence, width and length, is directly related to cutting efficiency. The fence size of the jointer currently in operation is 3" × 28". In purchasing a new jointer the fence size should be increased for improved accuracy and squaring efficiency.

- Delta DJ-20. As Table 1 shows, the Delta fence size is 5" × 36". These dimensions represent a 2" × 8" increase, which will result in more efficient operations.
- Powermatic-60. The Powermatic fence is 4" × 34½", a 1" × 6½" increase over that of the existing fence (see Table 1).

Conclusion. Both machines exceed the fence size criterion of 3" × 28". Delta DJ-20 has the greatest increase, 2" × 8", and will result in greater squaring accuracy and longitudinal control when jointing edge surfaces.

(Continued)

TABLE 1
8" Jointer Capabilities Comparison

Criteria	Standard	Delta DJ-20	Powermatic-60
Fence size	Minimum of 3" wide × 28" long	5" × 36"	4" × 34½"
Table length	Minimum of 52"	76½"	64"
Cutting capacity	Minimum depth of $3/8$"	$5/8$"	½"
Cost	Not to exceed $2300	$2128.00	$2092.00

© Cengage Learning

© Cengage Learning

Figure 13.1   Sample Section with Illustrative Visual Aid

Select a Format and an Organizational Principle

As you plan your report, you must select a format, an organizational principle for the entire report, and an organizational principle for each section.

Select a Format

Your choice of format depends on the situation. If the audience is a small group that is familiar with the situation, an informal report will probably do. If your audience is more distant from you and the situation, a formal format is preferable. The informal format is explained in Chapter 10, the formal format in Chapter 12.

In addition to selecting format type, create a style sheet of heads and margins. Review Chapter 6 and the examples in Chapters 10 and 12. Your style sheet should help your audience find what they need to do their job.

Organize the Discussion by Criteria

Organize the discussion section according to criteria, with each criterion receiving a major heading. Review Examples 13.1 and 13.2; each major section is the discussion of one criterion. Your goal is to present comparable data that readers can evaluate easily.

Organize Each Section Logically

Each section deals with one criterion and evaluates the alternatives in terms of that criterion. Each of these sections should contain three parts: an introduction, a body, and a conclusion. In the introduction, define the criterion and discuss its standard, rank, and source, if necessary. (If you discuss the standard, rank, and source somewhere else in the report, perhaps in the introduction, do not repeat that information.) In the body, explain the relevant facts about each alternative in terms of the criterion; in the conclusion, state the judgment you have made as a result of applying this criterion to the facts. You will find a sample section on page 394.

Drafting the Recommendation Report

As you draft the recommendation report, carefully develop the introduction, conclusions, recommendations/rationale, and discussion sections.

Introduction

After you have gathered and interpreted the data, develop an introduction that orients the readers to the problem and to the organization of the report. Your goal is to make readers confident enough to accept your recommendation. In recommendation reports, as in all reports, you can mix the elements of the introduction in many ways. Always include a purpose statement and add the other statements as needed by the audience. Four common elements in the introduction are

▶ Statement of purpose.
▶ Explanation of method of investigation.
▶ Statement of scope.
▶ Explanation of the problem.

Purpose

Begin a recommendation report with a straightforward statement, such as "The purpose of this report is . . . " or, more simply, "This report recommends. . . . " You can generally cover the purpose, which is to choose an alternative, in one sentence.

Method of Gathering Information

State your method of gathering information. As explained in Chapter 5, the four major methods of gathering data are observing, testing, interviewing, and reading. Stating your methodology not only gives credit where it is due but also lends authority to your data and thus to your report.

In the introduction, a general statement of your model of investigation is generally sufficient: "Using lab and catalog resources here at the university and after discussion with other Industrial Arts teachers in this area, I have narrowed my choices to two: Delta Model DJ-20 and Powermatic Model 60."

Scope

In the scope statement, cite the criteria you used to judge the data. You can explain their source or their rankings here, especially if the same reasons apply to all of them. Name the criteria in the order in which they appear in your report. If you have not included a particular criterion because data are unavailable or unreliable, acknowledge this omission in the section on scope so that your readers will know you have not overlooked that criterion. Here is an example:

Each machine has been evaluated using the following criteria, in descending order of importance:

1. Fence size
2. Table length
3. Cutting capacity
4. Cost

Background

In the background, discuss the problem, the situation, or both. To explain the problem, you must define its nature and significance: "Considering that the machine has been under continuous student use for 27 years and has reduced accuracy because of the small table and fence size, I indicated I would contact you regarding a new jointer." Depending on the audience's familiarity with the situation, you may have to elaborate, explaining the causes of various effects (Why does it have reduced accuracy? Just what is the relationship of the table and the fence? What *are* tables and fences?).

To explain the situation, you may need to outline the history of the project, indicate who assigned you to write the report, or identify your position in the corporation or organization.

The following informal introduction effectively orients the reader to the problem and to the method of investigating it. The report was sent as an attachment to an "e-mail cover letter." Because of the relative formality of the situation, the e-mail had this text:

E-mail head	Date: December 2, 2014 To: Joseph P. White, Superintendent of Schools From: David Ayers Subject: Purchase recommendation for 8" jointer Dr. White, attached is the purchase recommendation for a new 8" jointer for the high school woodworking lab. I consulted various catalogs and talked with representatives of Delta and Powermatic. I recommend that we purchase the Delta Model DJ-20. If you have questions or concerns, I am happy to meet with and other interested parties at your convenience. David Ayers
Situation and background	Recently, Jim DeLallo and I discussed at length the serious problems he was having in operating the jointer at the high school. Considering that the machine has been under continuous student use for 27 years and has reduced accuracy because of the small table and fence size, I indicated I would contact you regarding a new jointer. You asked that I forward 2014–2015 budget requests by December 15.

<table>
<tr><td>Cause of writing</td><td>

Therefore, I have prepared this recommendation report for choosing a new 8" jointer. Based on lab and catalog resources available on the Web and after dis-</td></tr>
</table>

Cause of writing

Method

Therefore, I have prepared this recommendation report for choosing a new 8" jointer. Based on lab and catalog resources available on the Web and after discussion with other Industrial Arts teachers, I have narrowed my choices to two: Delta Model DJ-20 and Powermatic Model 60. Before completing this evaluation I also met with representatives of Delta and Powermatic. Each machine has been evaluated using the following criteria, in descending order of importance:

Scope

1. Fence size
2. Table length
3. Cutting capacity
4. Cost

Preview

The remainder of the report will compare both machines to the criteria.

© Cengage Learning 2014

Figure 13.2   Sample Informal Introduction

Conclusions

Your conclusions section should summarize the most significant information about each criterion covered in the report. One or two sentences about each criterion are usually enough to prepare the reader for your recommendation. Writers of recommendation and feasibility reports almost always place these sections in the front of the report. Remember, readers want the essential information quickly.

All elements in the criteria have been met. The slightly higher cost of the Delta, $36.00, is more than offset by increased efficiency and capacity as noted below:

1. A larger fence size—for better control of stock when squaring
2. A larger table size—resulting in more efficient planning
3. A greater depth of cutting capacity—for improved softwood removal and increased rabbeting capacity

Recommendations/Rationale Section

The recommendation resolves the problem that occasioned the report. For short reports like the samples presented here, one to four sentences should suffice. For complex reports involving many aspects of a problem, a longer paragraph (or even several paragraphs) may be necessary.

Recommendation

Possible negative factor explained

It is recommended that the district budget for capital purchase of the Delta Model DJ-20 8" jointer in 2014. Selection of the Delta jointer is a departure from the practice of purchasing Powermatic equipment for the woodworking shop. It is my feeling that the Delta jointer is best suited for the current and future needs of the woodworking program. Service and repair will not be a problem in changing equipment manufacturers, since N. H. Bragg services both lines of equipment.

Discussion Section

As previously noted, you should organize the discussion section by criteria, from most to least important. Each criterion should have an introduction, a body discussing each alternative, and a conclusion. Here is another part of the discussion section from the recommendation report on jointers. Table 1, which is mentioned in this section, appears earlier on page 390. The entire report is shown in Example 13.1.

In-feed and out-feed tables are combined and referred to as table length. Increased table length improves accuracy when jointing and provides greater stability when planing face surfaces. On the existing machine, the table length is 52". When planing and jointing stock over 40", it is difficult to maintain accuracy. To realize improved handling and accuracy on a new jointer, the table length should be above 52".

- Delta DJ-20. As Table 1 shows, the table length of the Delta is 76½", a 24½" increase. This increased size will allow for greater efficiency when planing stock to approximately 60".
- Powermatic-60. As Table 1 shows, the table length of this jointer exceeds the minimum length by 12". Improved planing can be increased to approximately 50".

Conclusion. Both jointers exceed the 52" minimum table length size. The significant increase in the Delta jointer table length will offer improved planing and jointer accuracy and increased handling capacity.

© Cengage Learning 2014

Figure 13.3   Sample Discussion Section

Planning the Feasibility Report

Feasibility reports investigate whether to undertake a project. They "size up a project before it is undertaken and identify those projects that are not worth pursuing" (Ramige 48). The project can be anything: place a golf course at a particular site, start a capital campaign drive, or accept a proposal to install milling machines. The scope of these reports varies widely, from analyses of projects costing millions of dollars to informal reviews of in-house proposals. Your goal is to investigate all relevant factors to determine whether any one factor will prevent the project from continuing. Basically you ask, "Can we perform the project?" and you provide the rationale for answering yes or no. Follow the same steps as for planning recommendation reports. In addition, consider the following guidelines:

- Consider the audience.
- Determine the criteria.
- Determine the standards.
- Structure by criteria.

Consider the Audience

Generally, the audience is familiar with the situation in broad outline. Your job is to give specific information. They know, for example, that in any project a certain time frame is allowed for cost recovery, but they do not know how much time this project needs. Your goal is to make them confident enough of you and the situation to accept your decision.

Determine the Criteria

Criteria are established either by a management committee or by "prevailing practice." Either a group directs investigators to consider criteria such as cost and competition level, or "prevailing practice"—the way knowledgeable experts investigate this type of proposed activity—sets the topics. For instance, cost recovery is always considered in the evaluation of a capital investment project.

If you have to discover the topics yourself, as you often do with small projects, use the three categories described on page 388—technical, management/maintenance, and financial criteria. The criteria you choose will affect the audience's sense of your credibility.

Determine the Standards

To determine standards is to state the limits of the criteria. If the topic is reimbursement for acceptable expenses, you must determine the standards to use to judge whether the stated expenses fall within the acceptable limit. These standards require expert advice unless they exist as policy. If the policy is that a new machine purchase must show a return in investment of 20 percent, and if the machine under consideration will return 22 percent, buying the machine is feasible.

Structure by Criteria

The discussion section of a feasibility report is structured by criteria. The reimbursement report could include sections on allowable growth, time of recovery of investment, and disposal costs.

Writing the Feasibility Report

To write the feasibility report, choose a format and write the introduction and the body.

Choose a Format

The situation helps you determine whether to use a formal or an informal format for your feasibility studies. As a rule of thumb, use the formal format for a lengthy report intended for a group of clients. The informal format is suitable for a brief report intended to determine the feasibility of an internal suggestion.

Write the Introduction and Body

In the introduction, present appropriate background, conclusions, and recommendations. Treat this introduction the same as a recommendation introduction. In the discussion, present the details for each topic. As in the recommendation report, you should present the topic, the standard, relevant details, and your conclusion. Organize the material in the discussion section from most to least important. As with all reports, use appropriate visual aids, including tables, graphs, and even maps, to enhance your readers' comprehension. A large collection of feasibility reports on line (Witt) indicates that there is no single format accepted by all writers. Many writers place the Conclusions and Recommendation sections in the executive summary and at the end of the report.

The following section from an informal internal feasibility report presents all four discussion elements succinctly:

WHOLESALE COST

Introduction

This section examines the wholesale costs associated with producing the Heaven 'n Nature Sticker Christmas Card Kit. The standard set for this criterion was, "The wholesale cost of the product should not exceed $6.00 per set."

Analysis

A list of potential suppliers was provided by Illuminated Ink. These sources were then reviewed by searching through wholesale supply catalogs, visiting local vendors, and searching online sources. After an acceptable supplier had been located for each component of the kit, the total price of the kit was calculated and compared to the standard set for this criterion. See Table 1.

CONCLUSION

This research reveals that it is possible for Illuminated Ink to produce this kit for $5.13 per kit, a price that is lower than the standard required.

TABLE 1
Breakdown of Wholesale Component Costs

Component	Description	Qty	Supplier	Cost/Unit
Card base	Cranberry Red, Forest Green, and Midnight Blue cardstock	3	Picture This (1)	$0.54
Card insert	white 20# bond, green ink	3	Picture This (1)	$0.12
Winter friends stickers	Frances Meyer	3	Picture This (1)	$2.64
Snowflake stickers	silver, gold, & white	24	Picture This (1)	$0.48
Brads	gold minibrads	6	Picture This (1)	$0.24
Cotton ball	small, white	3	Wal-Mart (2)	$0.03
Envelope	white fiber A2 envelope	3	Impact Images (3)	$0.72
Soft fold box	6" × 9"	1	Impact Images (3)	$0.28
Instruction sheet	white 20# bond paper, black ink	1	Illuminated Ink (4)	$0.04
Mailing labels	1" × 25/8", white	8	Sam's Club (5)	$0.04
			Total Cost Per Kit	**$5.13**

© Cengage Learning

Figure 13.4   Sample Feasibility Section

Several brief informal feasibility reports appear in the examples and exercises of this chapter. In addition, you can find a wide range of examples by searching Google with the keyword phrases "feasibility report" or "feasibility study."

Worksheet for Preparing a Recommendation/Feasibility Report

☐ **Analyze the audience.**

- Who will receive this report?
- Who will authorize the recommendation in this report?
- How much do they know about the topic?
- What is your purpose in writing to them? How will they use the report?
- What will make you credible in their estimation?

(Continued)

☐ **Name the two alternatives or name the course of action that you must decide whether to take.**

☐ **Determine criteria.**

☐ **Ask technical, management/maintenance, and financial questions.**

☐ **For each criterion, provide a name, a standard, and a rank.**

☐ **Rank the criteria.**

☐ **Prepare background for the report.**

- Who requested the recommendation report? Name the purpose of the report. Name the method of investigation. Name the scope. Explain the problem. What is the basic opposition (such as need for profit versus declining sales)?
- What are the causes or effects of the facts in the problem?

☐ **Select a format—formal or informal.**

☐ **Prepare a style sheet including treatment of margins, headings, page numbers, and visual aid captions.**

☐ **Select or prepare visual aids that illustrate the basic data for each criterion.**

☐ **Select an organizational pattern for each section, such as introduction, alternative A, alternative B, visual aid, and conclusion.**

Worksheet for Evaluating Your Report

☐ **Evaluate the introduction.**

- Does the introduction give you the gist of the report?
- Does the introduction give you the context (situation, criteria, reason for writing) of the report?
- Do you know the recommendation after reading five to ten lines?

☐ **Evaluate the criteria.**

- Do they seem appropriate?
- Are all the appropriate ones included? If not, which should be added?
- Can you find a statement of the standard for each one? If no, which ones?

- Can you really evaluate the data on the statement of standard?
- Do you understand why each criterion is part of the discussion?
- Do you understand the rank of each criterion?

☐ **Evaluate the discussion.**

- Is the standard given so you can evaluate?
- Are there enough data so you can evaluate?
- Do you agree with the evaluation?
- Do you understand where the data came from?

☐ **Evaluate the visual aid and the paper's format.**

- Are the two levels of heads different enough? See pages 145–147.
- Is the discussion called "discussion"?
- Does a visual appear in each spot where one would help communicate the point?
- Are any of the visuals more or less useless; that is, they really do not interact with any points in the text?
- Is the visual clearly titled and numbered?
- Is the visual on the same page as the text that describes it?
- Does the text tell you what to see in the visual?

Worksheet for Evaluating a Peer's Report

☐ **Interview a peer. Ask these questions:**

1. Why did you include each sentence in the introduction? (Your partner should explain the reason for each one.)
2. Why did you use the head format you used?
3. Why did you choose each criterion?
4. Why did you write the first sentence you wrote in each criterion section?
5. Why did you organize each section the way you did? Do you think a reader would like to read it this way?
6. What one point have you made with the visual aid? Why did you construct it the way you did and place it where you did?
7. If you had to send this paper to someone who paid you money regularly for doing a good job, would you? If not, what would you do differently? Why don't you do that for the final paper?
8. Are you happy with the level of writing in this paper? Do you think these sentences are appropriately professional, the kind of thing you could bring forward as support for your promotion? If not, how will you fix them? If you try to fix them, do you know what you're doing?

Examples

Examples 13.1–13.3 illustrate informal recommendation and feasibility reports.

Example 13.1

Informal Recommendation Report

Purchase Recommendation for 8" Jointer
Prepared by David Ayers
For Joseph P. White, Superintendent

INTRODUCTION

Recently, Jim DeLallo and I discussed at length the serious problems he was having in operating the jointer at the high school. Considering that the machine has been under continuous student use for 27 years and has reduced accuracy because of the small table and fence size, I indicated I would contact you regarding a new jointer. You asked that I forward 2014–2015 budget requests by December 15.

Therefore, I have prepared this recommendation report for choosing a new 8" jointer. Based on lab and catalog resources available on the Web and after discussion with other Industrial Arts teachers in this area, I have narrowed my choices to two: Delta Model DJ-20 and Powermatic Model 60. Before completing this evaluation I also met with representatives of Delta and Powermatic. Each machine has been evaluated using the following criteria, in descending order of importance:

1. Fence size
2. Table length
3. Cutting capacity
4. Cost

RECOMMENDATION

It is recommended that the district budget for capital purchase of the Delta Model DJ-20 8" jointer in 2014. Selection of the Delta jointer is a departure from the practice of purchasing Powermatic equipment for the woodworking shop. It is my feeling that the Delta jointer is best suited for the current and future needs of the woodworking program. Service and repair will not be a problem in changing equipment manufacturers, since N. H. Bragg services both lines of equipment.

All elements in the criteria have been met. The slightly higher cost of the Delta, $36.00, is more than offset by increased efficiency and capacity, as noted below.

1. A larger fence size—for better control of stock when squaring.
2. A larger table size—resulting in more efficient planing.
3. A greater depth of cutting capacity—for improved softwood removal and increased rabbeting capacity.

The remainder of the report will compare both machines to the criteria.

CRITERIA

Fence Size

The fence serves as a guide for planing face and edge surfaces. The size of the fence, width and length, is directly related to cutting efficiency. The fence size of the jointer currently in operation is 3" × 28". In purchasing a new jointer the fence size should be increased for improved accuracy and squaring efficiency.

- Delta DJ-20. As Table 1 shows, the Delta fence size is 5" × 36". This represents a 2" × 8" increase, which will result in more efficient operations.
- Powermatic-60. The Powermatic fence is 4" × 34½", a 1" × 6½" increase over that of the existing fence (see Table 1).

Table 1
8" Jointer Capabilities Comparison

Criteria	Standard	Delta DJ-20	Powermatic-60
Fence size	Minimum of 3" wide × 28" long	5" × 36"	4" × 34½"
Table length	Minimum of 52"	76½"	64"
Cutting capacity	Minimum depth of ⅜"	⅝"	½"
Cost	Not to exceed $2300	$2128.00	$2092.00

Conclusion. Both machines exceed the fence size criterion of 3" × 28". Delta DJ-20 has the greatest increase, 2" × 8", and will result in greater squaring accuracy and longitudinal control when jointing edge surfaces.

Table Length

In-feed and out-feed tables are combined and referred to as table length. Increased table length improves accuracy when jointing and provides greater stability when planing face surfaces. On the existing machine, the table length is 52". When planing and jointing stock over 40", it is difficult to maintain accuracy. To realize improved handling and accuracy on a new jointer, the table length should be above 52".

- Delta DJ-20. As Table 1 shows, the table length of the Delta is 76½", a 24½" increase. This increased size will allow for greater efficiency when planing stock to approximately 60".
- Powermatic-60. As Table 1 shows, the table length of this jointer exceeds the minimum length by 12". Improved planing can be increased to approximately 50".

(Continued)

Conclusion. Both jointers exceed the 52" minimum table length size. The significant increase in the Delta jointer table length will offer improved planing and jointer accuracy and increased handling capacity.

Cutting Capacity (Depth of Cut)

Jointer cutting capacity is determined by the maximum depth of cut. This depth of cut is created when the in-feed table is lowered. For production work with softwoods and edge rabbeting, a large depth of cut is desired. The existing jointer has a ⅜" maximum depth of cut. This is a limiting factor when doing softwood production work and constructing edge rabbets over ⅜". When purchasing a new machine, the depth of cut should be at least ⅜".

- Delta DJ-20. As Table 1 shows, the depth of cut on this machine is ⅝", ¼" above the minimum standard. This will be an important feature when edge rabbeting and doing softwood production work.
- Powermatic-60. As Table 1 shows, the depth of cut for this machine is ½", a ⅛" increase above the minimum standard.

Conclusion. Both machines exceed the ⅜" minimum criterion set. The Delta jointer has the greatest depth of cut, ⅝", which will allow for greater softwood removal and maximum rabbeting.

Cost

The jointer is a capital equipment item and the cost cannot be department budgeted if in excess of $2300, unless prior approval is granted by the secondary committee. Costs (including shipping, stand, and three-phase conversion) are

- Delta DJ-20: $2128.00
- Powermatic-60: $2092.00

Conclusion. Both machines meet the fourth criterion. The Powermatic is slightly lower in cost, but does not have all the capacity and features of the Delta model. The additional cost of the Delta jointer ($6.80 on a 20-year depreciation schedule) is more than offset by the increase in table and fence size and improved cutting depth.

Introductory E-mail "Cover Letter"

Date: December 3, 2013
To: Steve Zubek, President/Widget Printing
From: Mark Jezierski, Production Manager
Subject: Purchase of New ABDick 9840 Printing Press

Steve, attached is my report recommending that we purchase the ABDick 9840 press. I am available to meet with you and other officers at any time.
Mark.

Example 13.2

Informal Recommendation Report

Purchase of New Printing Press
Mark Jezierski, Production Manager

I am writing in response to your recent request to determine which of the proposed printing presses, the Multigraphic 1860 or the AB Dick 9840, would best satisfy our needs. The age and inefficiency of the current equipment used in production have prompted this report. I have thoroughly researched these machines through trade journals and with other printers who currently use these machines to determine which of these machines best fit the criteria set by upper management.

My recommendation to you is that the Widget Printing Company purchase the AB Dick 9840. The criteria I have used to determine the feasibility of this purchase according to their rank and importance consist of:

1. Machine must be able to print a minimum of 8000 impressions per hour.
2. The press must be able to print an area image up to 11.5" × 17.25".
3. The press must be able to print a paper size of 12" × 17.5".
4. The cost to purchase the new machine must be under $25,000.

DISCUSSION

A. Impressions Per Hour

Definition. For a press to be economically efficient it should be able to print a minimum of 8000 impressions per hour. Most of the presses on the market today are capable of this speed, but some produce at even faster speeds.

Comparison

Press	Impressions Per Hour
AB Dick 9840	10,000
Multigraphic 1860	8000

As you can see, both of the proposed presses are rated as acceptable according to the criterion set in this area. As the comparison shows, the AB Dick 9840 is

(Continued)

capable of printing 2000 more impressions per hour than the Multigraphic 1860. With impressions per hour being the most important criterion, the AB Dick would be the best choice.

B. Maximum Image Area

Definition. The press purchased must be able to print an image up to 11.5" × 17.25". This image area will allow the printing of 8.5" × 11" and 11" × 17" jobs that have from one side to four sides that bleed.

Comparison

Press	Maximum Image Area
Multigraphic 1860	13.19" × 17.50"
AB Dick 9840	12.50" × 17.25"

Again you can see both of the presses meet the suggested criterion of 11.5" × 17.25". However, the Multigraphic 1860 is capable of the largest maximum image area (13.19" × 17.50"). This gives a considerable excess amount of image area to be used to print jobs that require larger image areas than the specified maximum of the criterion. The AB Dick 9840 is capable of an image area of 12.5" × 17.25", which is acceptable according to the standards we have set. When considering only image area, the Multigraphic 1860 would be the best choice.

C. Paper Size

Definition. The press purchased must be able to print a paper size of 12" × 17.5". This paper size will allow the printing of 8.5" × 11" and 11" × 17" jobs that have from one side to four sides that bleed.

Comparison

Press	Maximum Paper Area
Multigraphic 1860	15.00" × 18.00"
AB Dick 9840	13.50" × 17.25"

Again, as you can see, both of the proposed presses meet the criterion established for this area. The Multigraphic 1860 is capable of printing on the largest size paper. It can print on paper up to 15" × 18", which by far exceeds the paper size requirement needed by Widget Printing. This capability, however, will allow us a greater flexibility in projects that require larger than specified paper sizes. The AB Dick 9840 also exceeds the established criterion, although by a smaller margin than the Multigraphic 1860, but it will also allow us a degree of flexibility in the undertaking of larger projects. When considering only paper size, the Multigraphic 1860 would be the best choice.

D. Price

Definition. The price of the press purchased is not a major criterion. The capabilities of the press are more important factors. However, a price range has been set and cannot be overlooked when deciding on which press to purchase. We have established a price ceiling of $25,000.

Comparison

Press	Price
AB Dick 9840	$17,395
Multigraphic 1860	$20,000

Once again both of the proposed machines meet the established criterion. Both of these presses are well within the price range set. Based on only price as a criterion, the AB Dick would be the best choice.

RECOMMENDATION

After comparing the AB Dick 9840 and the Multigraphic 1860 presses, I feel the AB Dick 9840 is the press that the Widget Printing Company should purchase. The AB Dick meets all of the established criteria and has the largest capabilities in the area of impressions per hour, which is the most important criterion established. It also is the best press per dollar of purchase available. Therefore I feel that it is in our best interest to purchase the AB Dick 9840.

Example 13.3

Professional Feasibility Report

The following sections indicate a completely different "look" than proposed elsewhere in this text. The firm had to answer prewritten questions. To save space on the first two pages of the report are reproduced here. There is no introduction. The Alternatives section requires the authors to explain why they selected and rejected options. They are following the "feasibility way of thinking," which asks writers to select an option and defend their selection. The paragraphs are numbered in order to ease referring to sections in discussions.

PROJECT NARRATIVE: East Business Park Wetland Water Quality Certification

Location: Township 29 North, Range 4 West, Section 31
Lat 44°57'19"N, Long 90°54'55"W

Schedule: The work is proposed to occur in July 2012 to August 2013.

Site Wetlands: A wetland delineation was completed by Mark Iverson, Evergreen Irrigation, Inc. (a copy of the Wetland Delineation Report is provided in

(Continued)

Appendix E of the Storm Water Management Plan Narrative). Four wetlands were found on the current project site. Wetland #1 and Area #4 will be not be impacted by the proposed project. Wetland #2 will be impacted by a proposed 0.48 acre wetland fill. Wetland #3 will be impacted by a proposed 0.19 acre wetland fill.

INFORMATIONAL REQUIREMENTS FOR PRACTICABLE ALTERNATIVES ANALYSIS

I. Background/Description of Project.

a. **Describe the purpose and need for the project:** The City of Stanley has proposed to construct the East Business Park to attract development in the City of Stanley.

b. **Is your project an expansion of existing work or is it new construction? Explain:** Extension of existing Pine Street and widening of existing Junction Avenue

c. **When did you start to develop a plan for your project?** Phillip Epping and Denzine Land Surveying were retained by the City of Stanley to prepare a feasibility study in February 2012. Cedar Corporation was retained by the City of Stanley to prepare construction plans in March 2012.

d. **Explain why the project must be located in or adjacent to wetlands:** The proposed wetland fill to Wetland #2 is required to extend Pine Street along its desired alignment (that is, roughly centered between CTH X and STH 29). The proposed wetland fill to Wetland #3 is required to improve the intersection of CTH X and Junction Avenue and to widen Junction Avenue on its existing alignment.

II. Alternatives.

a. **How could you redesign or reduce your project to avoid the wetland, and still meet your basic project purpose?** The City of Stanley has determined that the selected alternatives are the only options that would meet the basic project purpose at a reasonable cost.

b. **Other sites**

 i. **What geographical area(s) was searched for alternative sites?** The City considered lands owned by the City and the adjacent hospital.

 ii. **Were other sites considered?** Yes, see below.

 iii. **Have you sold any lands in recent years that are located within the vicinity of the project? If so, why were they unsuitable for the project?** No.

c. **For each of the alternatives you identified, explain why you eliminated the alternative from consideration (include cost comparisons, logistical, technological, and any other reasons).**

i. Wetland 2: Wetland 2, which appears to be under state jurisdiction, is located in the center of the site, extending from the northeast corner of the Hospital property roughly north by northwest. Wetland 2 has historically been farmed. The preliminary layouts of Pine Street showed the roadway extending through the delineated wetland on its way to Junction Avenue. The City of Stanley has reviewed the following alternatives with regard to these wetland impacts. A sketch can be found in **Appendix A**:

1. Option A: Construct Pine Street as a 37'-wide roadway through the wetland. This option would require a wetland fill of approximately 0.64 acres. This option would provide east–west access through the park.

2. >>>**SELECTED ALTERNATIVE**<<< Option B: Construct Pine Street as a 29'-wide roadway (two 12'-wide lanes and two 30"-wide curbs, no shoulders or on-street parking) through the wetland. This option would require a wetland fill of approximately 0.48 acres. We estimate the reduced cost of this reduced width roadway to be $21,000. This option would provide east–west access through the park.

3. Option C: Construct Pine Street around the north side of the wetland. This option would still require wetland disturbance in order to extend the sanitary sewer and water main. This option would result in three sharp curves in the roadway which may not be compatible with the City's intent to develop the adjacent land as a business park. This option would require an additional 1,320 feet of roadway as compared to Option A. We estimate the additional cost of this additional roadway to be $302,000. This option would provide east–west access through the park.

4. Option D: Construct Pine Street around the south side of the wetland. This option would not require any wetland disturbance. This option would result in three sharp curves in the roadway which may not be compatible with the City's intent to develop the adjacent land as a business park. This option would require an additional 110 feet of roadway as compared to Option A. We estimate the additional cost of this additional roadway to be $45,000. In addition, this option would require purchasing approximately 1.1 acres from the Hospital. The Hospital is unwilling to sell any land to the City of Stanley at this time. This option would provide east–west access through the park.

5. Option E: Construct Pine Street with two cul-de-sacs, one on either side of the wetland. This option would still require wetland disturbance in order to extend the sanitary sewer and water main. This option would allow a reduction of 410 feet of roadway as compared to Option A. We estimate the reduce cost of the reduced roadway to be $98,000. This option would <u>not</u> provide east–west access through the park.

Exercises

▶ Group

1. In groups of two to four, analyze the community attitudes that are addressed by the authors of Examples 15.1–15.3 or of the examples in Exercises 2 and 3 later. What factors have the writers obviously tried to accommodate? What kind of report is expected? What length? Do they desire to prove conclusively that the material is accurate? Or is there an informal understanding that only a few words are necessary?

   Alternate: In the groups, role-play the sender and receiver of the reports. Receivers interview the senders to decide whether to implement the recommendation.

▶ You Analyze

2. Analyze this sample for organization, format, depth of detail, and persuasiveness. If necessary, rewrite the report to eliminate your criticisms. Create the visual aid that the author mentions at the end of the report. Alternate: Rewrite the report as a much "crisper," less chatty document. Alternate: Construct a table that summarizes the data in the report.

   > The purpose of this report is to determine from which insurance company I should purchase liability insurance for my 2014 Chevrolet Malibu. Data for this report were gathered from personal interviews with agents representing their companies. After comparing different companies, I narrowed my choice to decide which one I should buy. I evaluated Ever Safe and Urban Insurance using the following criteria, which are ranked in importance:
   >
   > 1. Cost—Could annual insurance of liability be less than $250?
   > 2. Payments—Could it be paid semiannually?
   > 3. Service—Is the agent easily accessible?
   >
   > After this evaluation, I concluded that Ever Safe was the best company to purchase my liability insurance. First, this insurance company costs $245, which is less than the $250 limit that I proposed to spend. Second, it can be paid semiannually. And, third, Ever Safe offers toll-free claim service 24 hours a day.
   >
   > Ever Safe costs $245 a year, with Urban Insurance costing $240 a year, which both met my required criteria of purchasing liability insurance for under $250. Urban Insurance is $5 less; however, Ever Safe does have other options that are worth the extra money in means of purchasing.
   >
   > Ever Safe and Urban Insurance both offer semiannual payments. In terms of this aspect, they are both weighted the same.

Ever Safe offers toll-free claim service 24 hours a day. Urban Insurance is long distance with limited working hours. They are available after working hours but only through an answering machine that will record your message for the agent to get in touch with you on the following day.

In the decision of an insurance company, it is plain to see that Ever Safe meets the requirements of my criteria and that Urban Insurance does not. Urban Insurance is cheaper, allows semiannual payments, but does not fulfill the service that I was looking for. For the extra dollars of payment, the service in Ever Safe is worth it.

3. Analyze this section for organization, format, depth of detail, and persuasiveness. Rewrite the memo, if necessary, to eliminate your criticisms. Create a visual aid that the author mentions at the end of the report. Summarize the data that support the recommendation. Alternate: Rewrite the report as a much "crisper," less chatty document. Alternate: Construct a table or figure that effectively summarizes the data in the report.

COST

INTRODUCTION

Since the 49'R Pulling Team does not have any sponsors, they can only spend money on parts that are necessary and feasible. It was proposed that a new supercharger should not cost more than $5000 after the trade-in of the 49'R Pulling Teams' current supercharger. Just the purchase of the supercharger minus the estimated value of the current supercharger would have been fine. But there would have been a cost associated with the equipment criterion that would push the total cost too high.

RESEARCH

The 49'R Pulling Team wanted to trade in their existing supercharger, so a sales representative would have to be contacted to negotiate a price. Since the 49'R Pulling Team has been in this sport for many years, they provided some information about who to contact about purchasing a new supercharger. The one person they have done business with is John Knox, who is with Sassy Engines in New Hampshire. It was estimated the value of the existing supercharger would be $1500.

Research from the Internet provided only a few companies that offer the style and size supercharger that the 49'R Pulling Team was looking for. The companies included Littlefield, SSI, SCS, Kobelco, and Kuhl. The cost for a new Kobelco 14-71 hi-helix supercharger was $5800. For a SSI 14-71 hi-helix supercharger, the outright cost was $5800 dollars. The cost for a Kuhl 14-71 hi-helix retro-fit supercharger was $5250. Those prices did not include any money from a trade-in or sale.

So far, all three superchargers were under $5000 and this criterion would have had a positive recommendation. But after looking into the equipment criterion and finding out there would be an additional $3500 cost, all three superchargers exceed the $5000 limit. The reason for the equipment cost can be found under the equipment link.

www.kuhlsuperchargers.com/cat_p02.htm
www.kocoa.com/2005_pricing_schedules.htm
www.sassyengines.com/Blowerdriveparts.html

CONCLUSION

This table gives the results that were accumulated and the cost after a trade-in or possible sale estimated at $1500 and with the $3500 equipment cost.

Supercharger Brand	Cost for New	Cost After Trade-In	Final Cost
Kobelco	$5800	$4300	$7800
Kuhl	$5250	$3750	$7250
SSI	$5800	$4300	$7800

After the estimated value of the existing supercharger was subtracted from the cost of a new 14-71 supercharger from three different companies and the addition of the equipment cost, all three were more than the $5000 that was set in this criterion. Under the restrictions of this criterion, the purchase of a new supercharger would not be recommended to the 49′R Pulling Team.

▶ You Revise

4. Rewrite this brief section. Create a table that illustrates the data.

INTRODUCTION

The University of Wisconsin–Stout has experienced a series of large budget cuts in recent years. The athletic department has been granted an estimated budget of $250,000 for updates of current facilities. For a new floor installation to be feasible, the total cost must not reach over $250,000.

FIGURES

According to Connors Flooring, the total installation cost of a maple sports floor is $81,300 with over 38 years of life expectancy. This is based on a 10,000-square-foot floor.

The Maple Floor Manufacturers Association has concluded that wood floors also require regular cleaning, sanding, lines repainted, and floor refinishing approximately every three years. This is a cost of approximately $8,000.00 per three years, equaling $2,666.00 p/year. With a 38 year life expectancy, the total for cleaning, sanding, painting, etc. = 101,308.

CONCLUSION

The total cost of the floor is under $200,000, which meets the first criteria of not exceeding $250,000.

5. Analyze this sample for organization, format, depth of detail, and convincingness. If necessary, rewrite the memo to eliminate your criticisms. Create the visual aid that the author mentions at the end of the report. Alternate: rewrite the memo as a much "crisper," less chatty document.

Feasibility of building a new receiving dock

This report is in reference to a proposal I received about building a new receiving dock at our Wheeler facility. I would like to point out certain criteria that must be met before any plans can be made up. First, the space needed for shipping trucks to maneuver around must be large enough. Second, the dock would need to be located somewhere in the middle of the plant. And third, the cost of this project must be within the budget constraints.

During the past few months I have done some research on this matter. In doing so I have talked to three other Hankedo and Cokeby plants about these three criteria. From that I have determined that the space needed for shipping trucks and the location of the dock are acceptable, but the cost criterion is beyond our budget. The only feasible alternative would be to allocate funds from the next five years' budgets. From the data I have gathered my recommendation for this proposal would be to go ahead with the construction as soon as possible.

The criteria used were based on data I received from three other Hankedo and Cokeby plants: St. Paul Hankedo and Cokeby plant, Cottage Grove Hankedo and Cokeby plant, and Austin, Texas, Hankedo and Cokeby plant. From this information I determined that the dock should be located in the middle of the plant because it would be as close as possible to the production areas. This will help increase fork truck efficiency and lower maintenance due to less travel time.

Also from those data, I concluded that with the size of the Wheeler plant compared to the other three, our trucking space should be at least 1000 square yards. This would give the truckers ample room to maneuver as they see fit. In my opinion it is better to have a little too much room than not enough.

The cost criterion did not meet the standards allotted to the Wheeler Hankedo and Cokeby company. I have talked to the budgeting committee, and they suggested that it would be feasible to take out money from budgets of the next five years. This is due to the fact that there are no large renovations planned during that time. The lowest cost estimate that we received was $2.5 million. The budget is set at $2 million. As you can see, we are over budget by $.5 million. But, as stated earlier, we have solved this problem.

▶ You Create

6. As your instructor requires, perform the following exercises in conjunction with one of this chapter's Writing Assignments.

   a. Perform the actions required by the worksheet.
   b. Write a discussion section. Construct a visual aid that depicts data for each criterion. Write an introduction for the section: define the criterion and tell its significance, rank, and source. Point out the relevant data for each section. Write a one-sentence conclusion. Word it positively. (Say X is cheaper than Y, not Y is more expensive than X.)
   c. Write an introduction that orients the reader to the situation and to your recommendation. Choose one of the several methods shown in this chapter and Chapter 10.
   d. In groups of two or three, review each other's problem statements and the criteria derived from them. Make suggestions for improvement.
   e. In groups of two or three, read a body section from each other's reports. Assess whether it presents the data that support the conclusion.
   f. In groups of two or three, compare conclusions to the recommendations. Do the conclusions support the recommendations?
   g. In groups of two or three, assess each other's introductions. Do they contain enough information to orient the reader to the situation and the recommendation?
   h. In groups of two or three, read the near-final reports for consistency of format. Are all the heads at the appropriate level? Are all the heads really informative? Is the style sheet applied consistently? Does it help make the contents easy to group? Do the visual aids effectively communicate key points?

Writing Assignments

1. In groups of two to four recommend a collaboration software to your instructor for use in his or her classes. Interview the instructor to determine his or her criteria, then investigate the software currently available. It is critical that your final recommendation meet not just his or her criteria.

2. Assume that you are working for a local firm and have been asked to evaluate two kinds, brands, or models of equipment. Select a limited topic (for instance, two specific models of ten-inch table saws, the Black and Decker model 123 and the Craftsman model ABC), and evaluate the alternatives in detail. Write a report recommending that one of the alternatives be purchased to solve a problem. Be sure to explain the problem. Both alternatives should be workable; your report must recommend the one that will work better.

Gather data about the alternatives just as you would when working in industry—from sales literature, dealers, your own experience, and the experience of others who have worked with the equipment. Select a maximum of four criteria by which to judge the alternatives and use a minimum of one visual aid in the report. Aim your report at someone not familiar with the equipment. Fill out the worksheet in this chapter, and perform the parts of Exercise 6 that your instructor requires.

2. Assume that you are working for a local firm that wants to expand to a site within 50 miles. Pick an actual site in your area. Then write a feasibility report on the site. Devise criteria based on the situation. Do all the research necessary to discover land values, transportation systems, governing agencies, costs, and any other relevant factors. Your instructor will provide you with guidance about how to deal with the local authorities and how to discover the facts about these topics. Use this chapter's worksheet, and perform those parts of Exercise 6 that your instructor requires.

3. Assume that you have been asked to decide on the feasibility of a proposed course of action. Name and describe the proposal. Then establish the relevant criteria to determine feasibility. Apply the criteria and write an informal report. Use this chapter's worksheet, and perform those parts of Exercise 6 that your instructor requires.

4. Find a firm or an agency in your locale that has a problem that it will allow you to solve. Research the problem, and present the solution in a report. The report may be either formal or informal, recommendation or feasibility. Your instructor will help you schedule the project. This project should not be an exercise in format and organization, but a solution that people need in order to perform well on their jobs. Use this chapter's worksheet, and perform those parts of Exercise 6 that your instructor requires.

5. Assume that your manager wants to create a webpage. Investigate the situation, and write a report explaining the feasibility of creating and maintaining a website.

6. Write a learning report for the writing assignment you just completed. See Chapter 5, Writing Assignment 8, page 167, for details of the assignment.

Web Exercises

1. Assume that your manager wants to create a social media presence for the company. Investigate the situation and write a report explaining the feasibility of creating and maintaining a social media presence.

2. Write a report on whether or not the Web is a feasible source of information that you can use to perform your duties as a professional in your field. For instance, is the Web a more feasible source than hard copy of OSHA regulations or ASTM standards?

Works Cited

Bradford, Michael. "Four Types of Feasibility Studies Can Be Used." *Business Insurance* (19 June 1989): 16. Print.

Hofstrand, Don and Mary Holz-Clause. "Feasibility Study Outline." Iowa State University Extension and Outreach. 2009. Web. 22 May 2012. <http://www.extension.iastate.edu/agdm/wholefarm/html/c5-66.html>.

Holcombe, Marya W., and Judith K. Stein. *Writing for Decision Makers: Memos and Reports with a Competitive Edge.* Belmont, CA: Lifelong, 1981. Print.

Markel, Mike. "Criteria Development and the Myth of Objectivity." *The Technical Writing Teacher* 18.1 (1991): 37–47.

"Professional and Technical Writing/Feasibility." Wikibooks. 2012. Web. 22 May 2012. <http://en.wikibooks.org/wiki/Professional_and_Technical_Writing/Feasibility>.

Ramige, Robert K. "Packaging Equipment, Twelve Steps for Project Management." *IOPP Technical Journal* X.3 (1992): 47–50. Print.

Witt, Will. "300 Links to Feasibility Study Examples and Samples." Will Witt Limited. Web. 2 June 2012.

Proposals

CHAPTER CONTENTS

CHAPTER 14 **IN A NUTSHELL**

The goal of a proposal is to persuade readers to accept a course of action as an acceptable way to solve a problem or fill a need. Internal proposals show that the situation is bad and your way will clearly make it better. External proposals show that your way is the best.

Basic proposal issues. Four issues for you to discuss convincingly in a proposal are:

▶ The *problem*—how some fact negatively affects positive expectations (high absenteeism on manufacturing line 1 is causing a failure to meet production goals) and that you know the cause (workers are calling in sick because of sore backs).

▶ The *solution*—actions that will neutralize the cause (eliminate bending by reconfiguring the work tables and automating one material transfer point).

▶ The *benefits* of the solution—what desirable outcome each person or group in the situation will obtain.

▶ The *implementation*—who will do it and how, how long it will take.

Develop credibility. To accept your solution, your readers must feel you are credible. Your methods must be clear and sound—an expert's assessment of the situation. Your analyses of the problem, the cause, the benefits, how long it will take, the cost, etc., must show a reasoned regard of each concern, one that will not cause surprises later on.

Basic guidelines. Follow these guidelines:

▶ Use a top-down strategy.

▶ Describe the situation and use visual aids.

▶ Provide context in the introduction.

▶ Provide a summary that clearly states the proposed solution.

Aproposal persuades its readers to accept the writer's idea. This chapter explains two types of proposals: the grant proposal and the internal proposal. Not discussed in this chapter is the long external proposal in which one firm proposes to perform work for another firm, for example, to create a guidance system for a type of plane owned by an airline. In a grant proposal a person, or more likely, a non-profit institution asks funding agency or foundation to support an idea or program created by the institution. In an *internal proposal,* the writer urges someone else in the company to accept an idea or to fund equipment purchases or research.

Grant Proposals for Non-Profit Organizations

Background of Non-Profits

Non-profit institutions abound in the United States. Over 1.5 million of them ("Frequently") operate everything from historical buildings to major art institutions to neighborhood social programs. A non-profit organization is one that the United States Internal Revenue Service has designated a "501(c)3 organization." To receive the designation the organization's earnings, unlike those of a for-profit organization, may not be used for the benefit of shareholders or directors of the organization ("Exemption"). The designation allows the non-profit to be exempt from taxes on the money it receives and the designation allows donors to use their contributions as a tax-deduction on their yearly tax obligations.

Typically non-profits do not take in enough money to cover their operating costs; thus they have to find money elsewhere to cover those costs. Suppose a community theater group decides to run a year-long theatrical program for disadvantaged fourth-graders. The program, let's say, will cost $2000 for space rental and visiting artists. In addition the group has $10,000 in other expenses. As a result the theater group needs $12,000 to cover its normal expenses plus host the fourth-grader program. However the group can only raise $10,000 from its ticket sales. It has to raise the other $2000 from some other sources. To do so, the theater group, like many non-profits, writes a grant proposal to a foundation, whose mission is to give money to non-profit groups. For instance, a regional bank may have a foundation whose mission is to support arts organizations in its operating area. The theater company would send a grant proposal to that foundation asking for financial support for the fourth-grader program. If the grant proposal is persuasive, the foundation probably will fund the program. In other words, successful grant proposals are a major method by which non-profits are able to provide programs in their community or region.

How Foundations Announce That They Support Non-Profits

In order to attract non-profit groups to apply for available funding, Foundations issue an RFP (Request for Proposal) or they announce their mission and criteria for funding on a website. If you have never read an RFP for non-profits, you probably should enter those terms into a search engine to find and review several. One excellent place to begin (at the time of this writing) is "Requests for Proposals" (http://foundationcenter.org/pnd/rfp/) at The Foundation Center's site PND: Philanthropy News Digest. This site lists RFPs that have recently been posted. Reading any of these RFPs will give you a good sense of whether or not to apply for funds to that foundation and which sections and documents you must submit in order for them to agree to fund your proposal.

Here's an example: Southern Arts, "founded in 1975 to build on the South's unique heritage and enhance the public value of the arts," offered grants up to $2500 so that a local art organization could present performing artists from outside the state. In their material Southern Arts explains their conditions for considering any proposal and they explain the process to submit a proposal. Any organization that wishes to apply to them must conform to their guidelines or Southern Arts will reject the proposal. Southern Arts indicates that they will pay 50 percent of an artist's fee, up to $2500. The requesting organization must have a budget of $150,000 or less, exist in the nine-state region that Southern Arts supports, prove that they are a 501(c)3 organization, host the artist during a certain time period, and submit the proposal online at the Southern Arts website. Submitting online has become the most common way to apply to foundations.

Southern Arts also lists all the sections and documents that applicants must include: an application, a letter of intent or contract between the artist and the organization, and proof of their 501(c)3 status. The application must contain sections that present or explain all of the following:

- the mission of the organization,
- the budget of the event,
- the project description in one sentence,
- a timeline of the project,
- samples of the artist's work,
- why the project is important to the community,
- the target audience,
- the accessibility of venue,
- the organization's ability to carry out the project,
- how the project fits one of Southern Arts priorities.

The application instructions also indicate how long each section can be. The site includes sample application that was funded in a different year ("Grant Programs").

Here's a second example: The Otto Bremer Foundation accepts grants to support events in the region it serves (Minnesota, North Dakota, Wisconsin). Like Southern Arts, the Bremer Foundation issues comprehensive guidelines for applicants on its website. They explain what type of program they will and will not fund, dates for submitting proposals, documents that must be included (cover sheet, narrative, and attachments including proof of tax-exempt status, and various financial information forms). The cover sheet is online. The applicant simply fills in the blanks. The narrative must contain seven different statements, including a description of the project and how it will be evaluated. The application may be mailed or e-mailed to them ("Guide").

As you can see, foundations have generally the same approach. They announce that they have money available; they explain the conditions for receiving money from them; and they explain how to apply.

Planning the Proposal

This section explains how to plan a grant proposal to submit to a foundation. To do so, you must

▶ Read the foundation's guidelines carefully
▶ Collect all relevant data
▶ Write or adapt usual elements of grant proposals

Read the Foundation's Guidelines Carefully

As the scenarios just mentioned make clear, your first step in grant writing is to determine all the requirements or constraints set by the foundation. You need to determine the submission date in order to set a schedule for completing all the work. In addition you must clarify the method of submitting the proposal. If the organization requires online submission, they may refuse to consider a proposal that arrives in the mail. You also must determine whether the foundation will fund your type of program and your type of organization. For instance the Otto Bremer foundation specifically states that it will not fund artistic performances, though it might fund a community-improvement program submitted by a theater group. Questions you must answer include the following: Will they fund a project with your size budget? Do they restrict applications geographically or some other way? Exactly which sections and documents do they require? Do they have a length that sections may not exceed? If they say "a total of five pages" they might refuse to consider a 12-page application.

Be sure to clarify the length and form of the application. Almost always the funding source will clearly state maximum lengths. Southern Arts, for instance, has this guideline for its schedule section:

"Schedule. (Maximum 1,250 characters including spaces, approximately ¼ page.)"

The Levin Family Foundation states that the executive summary may only be one page long. The foundation also limits the amount of space that any organization may use to answer the questions.

In addition to length the funding source will clarify how to send them the application and how many to send. They might say "six copies on paper" or "e-mail all pages to Ms. Y as an MS Word attachment to an e-mail; send only one copy."

Collect All the Relevant Data

As you can see from the list of required sections in the Southern Arts guidelines, you have to collect quite a bit of data before you can begin to write. You must have a budget for your project, but you must also have a number of other sections, including some that might be difficult to write, such as why the project is important to the community or the manner in which your project fulfills one of their priorities. Other sections that you might have to clarify are the accessibility of the venue, and samples of the artist's work.

For almost all grants you have to collect data ahead of time in order to meet their requirements. You must develop your concept, research costs, speak with people who will be involved (for instance to set dates and fees). To help yourself make a check list of everything that has to be included with the application.

To see the types of information that one grant proposal writer had to collect, turn to Figure 14.1 (pp. 418–424), a proposal from the East End Community Services Corporation, a non-profit institution, to the Levin Family Foundation. The Levin Foundation requires applicants to use the Levin application form. Notice that the writer had to know the yearly budget and other funding sources, and had to describe the project in a very small amount of space, all on page 1. Other relevant data includes the demographic details of population the project will serve (p. 4) why the project is a "best practice" (pp. 5–6) and other collaborating organizations (p. 6).

Writing the Non-Profit Grant Proposal

Various experts (Carlson and O'Neal-McElrath; Davis; Fritz; "Proposal"; "Writing") usually list five elements common to grant proposals. This section will explain those five. Not all grant applications can be written exactly as the general guidelines suggest. In many ways the best advice is write clearly, use informative and persuasive details, and follow the constraints that the funder requires. The five elements are:

▶ Executive Summary
▶ Need
▶ Project Description
▶ Budget
▶ Organizational Information

Executive Summary

The executive summary presents the reader with an overview of your organization and project. Often funding officers read this section in order to decide whether to read the rest of the application. This section should briefly (one page or so) explain the need; describe the project, including what will happen, when, and for whom; list the amount requested; and explain the organization mission and history. Even though you present this section first, write it last after you have detailed your project in the other sections.

In Figure 14.1 (pp. 418–424) the executive summary (pp. 419–423) is the longest section in the proposal. It explains the problem ("trailing the district"), the program ("expansion of our academic enrichment", using "increase in the number of tutors"), and result ("striking success"). The last paragraph explains the impact in persuasive detail, noting that graduation rates have increased and students have met with the governor. Note that the budget and organization are not dealt with in this executive summary because they had been outlined on page 1, under "Request Data."

Need

The need statement gives evidence and support for why this program will solve whatever problem you are dealing with. Need statements must avoid circular reasoning, which means that the absence of your solution is the problem. An illustrative example is "Grade school children do not have a mentoring system available. This proposal will solve that problem by creating one." Those two statements are circular. The question the writer has to answer is—Why is the lack of a mentoring program a problem? If grade school children enter the system scoring below state standards and if they only fall further behind without mentoring, then mentoring will be a solution to the problem.

To write a need statement

▶ explain the situation.
▶ tell why the situation is important.

Explain the situation. Don't assume that the reader understands the issue, whatever it is.

Tell why the situation is important. Do so by explaining both factual and emotional evidence. The need statement of the proposal in Figure 14.1 clearly explains the problem. The youth test scores indicate that students enter school lacking in skills and without intervention they will never catch up. The need statement also gives evidence that the program will help the students catch up as indicated by test scores (55 percent were proficient by third grade).

Project Description

The project description is the heart of the grant proposal. Explain these factors:

- what you will do
- who will do it
- how you will evaluate whether your actions had the impact or outcome you wish

In Figure 14.1 the project description is broken into two parts: section four and section five. In section four, the writer fills in details of what will happen, including its goals and a table showing who will be affected by those goals, the actions that make it a "best practice," its duration. She uses a Gantt chart to illustrate the schedule. The objectives section is a list of quantifiable outcomes. The writer has already listed broad goals, but objective section shows that the program is committed to a high level of quality when she says that 88% of the participants will show a 10% increase in their scores.

The evaluation section explains what impact you hope to achieve and how you will know that you have achieved it. A common question for evaluations is—What does success look like? As can be seen in Figure 14.1 other questions include who will fill out evaluation forms and how will results be disseminated. The evaluation section shows that the East End organization has completely thought through evaluating their program. The writer details how they define success, the range of stakeholders (staff, parents, youth) who will be asked to evaluate and the ways in which East End will disseminate information.

Budget

The budget is the amount of money you are requesting. You should break the figures into sections so that the reader understands how the money will be used.

The East End proposal does not present a detailed budget. In the Request Data section on page 1 the writer asks for $5000. In the evaluation section (p. 7) she notes that a number of other organizations have committed large amounts to this program.

Organizational Information

If the organization does not have a statement about itself you will have to write one. This statement should explain

- The mission of the organization
- A brief history including when it was started, other projects it is currently undertaking, and other projects it has completed.
- A list of who is on the organization's board and what the board's responsibilities are.

In Figure 14.1, one page is devoted to describing the East End organization, including its history, vision, current programs, and whom it serves. These statements are all provided succinctly in the space allowed by the application.

Version 01/05 Page 1

LEVIN FAMILY FOUNDATION GRANT APPLICATION

ALL SPACES ON THIS PAGE MUST BE FILLED AND ALL SIGNATURES ARE REQUIRED

Section One

Organization Data

Applicant Organization (Legal Name): East End Community Services Corporation
Doing Business As: East End Community Services
Previous Name, if changed: Not Applicable
Street Address: 624 Xenia Ave.
City: Dayton State: _Ohio_____ Zip: _45410_____ County: Montgomery County
E-mail: jlepore-jentleson@east-end.org Web site: www.east-end.org
Phone: (937) 259-1898 Fax: (937) 259-1897
IRS Name, as listed on 501(c) (3) letter: East End Community Services Corporation
IRS letter date: / / Tax Exempt ID number (EIN): 31-1508554
Executive Director: Jan Lepore-Jentleson Direct Phone: (937.) 259-1898 ext 16
Organization's Budget: $1,558,737 Endowment Size: $ NA
Organization's Major Funding Sources: Mathile Family Foundation, Montgomery County Dept of Job and Family Services, Ohio Department of Education, Human Services Levy, private foundations, United Way

Organization's Affiliation and/or accreditation body (check all that apply)

United Way ___ X__ Fine Arts Fund _____ Better Business Bureau __X___
Chapter of national or regional organization _____
Other (Specify) License from the Ohio Department of Job and Family Services for Child Care.

Request Data

Program/Project Title: Success for Youth in East Dayton
Amount of this request: $5,000
Proposal contact person information: Name: Amy Jomantas
Title Associate Executive Director Phone (937) 251-1066 Fax (937) 251-1067
E-mail ajomantas@east-end.org
Community/Counties served by this Program/Project: Inner East Dayton, specifically the area surrounding Ruskin School. However, we serve students who attend Ruskin but live in other neighborhoods.

Projected number of people to be served by Program/Project. This grant will serve approximately 244 students and is part of matching funding to other foundation and governmental funds.

Brief demographic description of population served by this Program/Project: This project serves students attending Ruskin PreK-8 Neighborhood School Center. Of the children that attend Ruskin School, 100% are eligible for free or reduced lunch, more than 60% are white, primarily Appalachian, 15% are African American and 15% are Latino and the remainder are mixed race. This is an extremely impoverished neighborhood, where the youth are at high risk for exposure to family and street violence, alcohol, and drugs.

Type of request (check all that apply)
Capital _____ Program/Project ___ X____ Endowment _____
Operating _____ Technical Assistance _____ Start-up _____
Signature of Executive Director: /
Printed Name _Janice Lepore-Jentleson _ (*date*)
Signature of Board President: /
Printed Name_Gordon Heller ___ (*date*) **Version 01/05 Page 2**

EXECUTIVE SUMMARY—1 Page Only

This proposal builds upon the success that we have achieved serving K-8th grade students at the new Ruskin PreK-8 Neighborhood School Center as well as older teens from the neighborhood. The proposal is consistent with the increasing numbers of children that we serve at Ruskin and from area high schools. The proposal also continues the expansion of our academic enrichment, wellness, sports, music/arts curriculum, including tutoring in reading, math, and science that occurred starting in the Fall of 2008 and continued through June 2009. One of the largest improvements that occurred in 2009 is a tremendous increase in the number of tutors matched to students and the provision of a new online test preparation and remediation services called "Study Island." The increase in tutors led to striking results! Following the merger of the, East End Community (charter) School with Dayton Public Schools (DPS) at Ruskin in August 2008) the October 2008 Reading Ohio Achievement Test (OAT) results indicated that that most of the third grade was scoring in the limited range and was trailing the district. Thanks to the increased tutoring, the third graders are scoring much higher in Reading than the overall DPS district. In fact, results show Ruskin School and the afterschool students scored higher than the overall DPS district in all Ohio Achievement Test results. **In presentation to the East End Board of Trustees, Dr. Stanic, Superintendent of DPS praised East End's efforts and stated, "if we want to improve Dayton Public Schools, this is the way to do it."** In addition, the after-school program was one of 200 after-school programs throughout the State that participated in a survey of parents, teachers, faculty, community stakeholders, volunteers and school

(Continued)

administrators conducted by The Ohio State University in January of 2009. In this study, East End Community Services' after-school program scored above the state mean on all indicators in the Ohio-Quality Assessment Rubric. With this proposal, we will continue to offer high quality services to students and supports to parents with the goal of increasing academic outcomes and building personal resiliency.

The proposal also supports our overall programming for 140 teens, including academic tutoring, service-learning projects, prevention programming, athletics, jobs skills training, job placement, drivers education, financial aid/ scholarship assistance and connection to post secondary education to youth between the ages of 12 and 22 years of age. The program is divided into sections for middle school youth 6th, 7th and 8th grades, high school youth 9th through 12th grades and also young adults—college age. This program has a history of helping youth increase developmental assets, avoid teen pregnancy/parenting, reduce juvenile crime involvement, remain engaged in school, improve grades and prepare for college or vocational careers. Many of the youth have gone on to attend college at Sinclair Community College or Wright State University. Of all the youth surveyed, 95% report that they want to attend college. The program targets youth who are living in east Dayton in some of the most impoverished neighborhoods in Montgomery County. The programs not only address academic achievement, but help create youth that make responsible choices about the health and are prepared to become leaders in the community. We have been successful in greatly improving the graduation rate and 100% of teens remain engaged in school. 98% of the teens do not become pregnant that engaged in teen pregnancy prevention programming. Last year, our youth were involved in making presentations at Liquor License hearing, City Commission and School Board meetings. They also participated in running a Neighborhood Investment Bank that reached out to area organizations including businesses, labor unions, civic organizations and churches to sponsor study teams that later completed neighborhood beautification projects. Two of the youth traveled to Columbus to speak with Governor Strickland about educational funding. The funds requested from the Levin Family Foundation leverages funding from government, other foundations and private donations, filling a gap created by the state budget crisis that affected funding for 2009 and 2010, **continuing a much needed, proven program that is consistent with the mission as well as the priorities of the Levin Family Foundation. Version 01/05 Page 3**

Section Two—Profile of Organization

1) Brief summary of organization's history
East End Community Services has provided more than 11 years of services to the neighborhoods that comprise inner east Dayton. East End has grown to

serve more than 2,000 individuals, families and children annually, providing a wide range of services that include afterschool, youth programming, employment services, housing development, specialized services for immigrants and most recently, economic development. In 2005, the Better Business Bureau awarded East End Community Services the Eclipse Integrity Award. In 2006, Advocates for Basic Legal Equality gave the agency the Access to Justice Award. We were recognized and quoted in the Working Harder for Working Families Hunger Report for 2008, a national report completed by the Bread for the World Institute.

2) Brief statement of organization's vision/mission*

Our mission is to listen to and work with the residents of Twin Towers and surrounding neighborhoods in bringing about a prosperous, healthy, and caring community. Consistent with this mission, our goal is to help parents and children improve their lives through access to education, services, and employment. Increasing the incomes of the parents and other adults who live in these neighborhoods is essential to fulfilling our mission and ties directly back to success for children.

3) Brief description of current programs/projects and activities

We provide the following services which are consistent with the priorities of our strategic plan.

- More than 240 children and teens are provided with academic enrichment, one to one tutoring, mentoring, skill development, and service learning opportunities each year at Ruskin School.
- 700 individuals, including more than 150 Latino individuals, are provided with case management, advocacy services, assistance in applying for benefits through Ohio Benefit Bank, career assessment, job readiness skill training, job placement, and retention services.
- We are developing housing and were just awarded more than a $8 million in Low Income Housing Tax Credit (LIHTC) dollars to create 40 new units and will be breaking ground this fall.
- Outreach activities reach the 1,300 households that comprise Twin Towers in Dayton each year.

4) Description of organization's constituency and geographic region

As of July 2009, Dayton has a current unemployment rate of 13.7% and the University of Dayton Business Research Group believes the rate in inner east Dayton is much higher. (1) Since 2001, there has been a loss of 30,000 manufacturing jobs in the Dayton area and more than 64,000 jobs in all sectors, including manufacturing between 2000 and 2006. (2) Of the persons specifically seeking help from our agency, 39% have less than a high school education and experience many barriers to employment, including problems with childcare,

(Continued)

transportation, low self esteem, depression, and other behavioral health issues, including alcohol and drug abuse. Many of the individuals are reading on fourth grade level of less. Twin Towers and the surrounding neighborhoods in inner east Dayton not only have a disproportionate rate of poverty, they have a disproportionate number of foreclosures fueled by subprime mortgages. (3) In Twin Towers more than 60% of all the loans are subprime. Additionally, crime and victimization issues, including family violence, interfere with success for families. The Dayton Police Department statistics indicate that in the first quarter of 2009, murder and nonneglient manslaughter was up 200% from last quarter, across the city, although East End has been very successful in lowering crime within Twin Towers. (4) Despite these considerable problems in the geographic area, the community has many strengths. More than 2,500 hours of volunteerism are provided by neighborhood residents annually including attending planning meetings, working at cultural events like the Appalachian Festival, conducting book, clothing, and homework supply giveways, mowing vacant lawns, collecting trash, working on crime prevention activities, and meeting the needs of families in crisis. The neighbors have been active in planning the governance and physical design of the new Ruskin School and have been an integral part of the housing development planning process.

1. Ohio Labor Market Information, Unemployment Rates by City, July 2009
2. 22 County Business Patterns, (NAICS) Montgomery County Major Industry 2000 to 2006.
3. Ohio Supreme Court, U.S. Census Bureau, Policy Matters Ohio review of filings in U.S. district courts. **Version 01/05 Page 4**

Dayton Police Department statistics indicate that in the first quarter of 2009, murder and nonnegligent manslaughter was up 200% from last quarter, across the city, although East End has been very successful in lowering crime within Twin Towers. (4) Despite these considerable problems in the geographic area, the community has many strengths. More than 2,500 hours of volunteerism are provided by neighborhood residents annually including attending planning meetings, working at cultural events like the Appalachian Festival, conducting book, clothing and homework supply giveways, mowing vacant lawns, collecting trash, working on crime prevention activities, and meeting the needs of families in crisis. The neighbors have been active in planning the governance and physical design of the new Ruskin School and have been an integral part of the housing development planning process.

4. City of Dayton Police Department COMPSTAT Citywide Profile of Crime Statistics updated April 19, 2009.
5. East End Community School. Statistics on Free and Reduced Lunches 2006–2007 School Year.

6. Montgomery County Children's Services Board 2003 Reports of Abuse and Neglect by Zip Code.
7. Dayton Police Department Domestic Violence Calls Dispatched to Sectors in Zip Code Areas 2002–2004.
8. Ohio Hospital Association. Persons treated in Emergency Rooms but not admitted for alcohol/drug diagnoses.
9. Posner, J. K. and D. L. Vandell. Low-income children's after-school care: Are there beneficial Effects of after-school programs? *Child Development,* 1994. 65: pp. 440–456.

Section Three—Statement of Need

1) Statement of need project is attempting to meet and evidence of that need.
After-school time becomes particularly critical when children enter school unprepared. Dayton Public Schools Kindergarten Readiness Assessment-Literacy (KRA-L) scores indicate that only 21.18% scored in the highest level, in the Fall of 2008. Nearly 41% of students tested needed targeted instruction, and another 38% needed intense instruction. Thus, youth are entering school academically behind, and without intervention they are likely to have difficulty catching up. The three and one half hours children spend at Ruskin after the formal school day provides the opportunity for such intervention/remediation.

At the first testing in the Fall of 2008, 62% of Ruskin third graders scored in the lowest level ("limited"), another 24% scored in the basic range, and only 2% were proficient or above. However, following tutoring, academic enrichment and use of Study Island, **55% of third graders scored proficient in the Spring of 2009**—a significant increase. At Ruskin, 100% of the students are eligible for free or reduced lunches, a school indicator of poverty. (5) They come from neighborhoods primarily in 45410 and 45403 ZIP codes that have disproportionately high rates of child abuse, domestic violence and alcohol/drug abuse. Our program is based upon research regarding afterschool needs.

Section Four—Program/Project Description & Methodology

1) Description of program/project, including:

a) Activities to accomplish programs/project (Is this a new or ongoing activity?)
This is a current service. We have provided after-school services to inner East Dayton children for the past 11 years. However, this project is an expansion and addresses a shortage created with state budget cuts.

(Continued)

b) Goals/objectives Our goal is to provide children and teens with academic enrichment and prevention programming that builds school success, promotes character development and improves achievement test results.	244 children/teens of which 150 attend regularly, at least 30 days per year.

Figure 14.1 Proposal from the East End Community Services Corporation

The Internal Proposal

The internal proposal persuades someone to accept an idea—usually to change something, or to fund something, or both. Covering a wide range of subjects, internal proposals may request new pieces of lab equipment, defend major capital expenditures, or recommend revised production control standards. The rest of this chapter explains the internal proposal's audiences, visual aids, and design.

Planning the Internal Proposal

The goal of a proposal is to convince the person or group in authority to allow the writer to implement his or her idea. To achieve this goal, the writer must consider the audience, use visual aids, organize the proposal well, and design an appropriate format.

Consider the Audience

The audience profile for a proposal focuses on the audience's involvement, their knowledge, and their authority.

How Involved Is the Audience?

In most cases, readers of a proposal either have assigned the proposal and are aware of the problem or have not assigned the proposal and are unaware of the problem. For example, suppose a problem develops with a particular assembly line. The production engineer in charge might assign a subordinate to investigate the situation and recommend a solution. In this assigned proposal, the writer does not have to establish that a problem exists, but he or she does have to show how the proposal will solve the problem.

More often, however, the audience does not assign the proposal. For instance, a manager could become aware that a new arrangement of her floor space

Ethics and Proposals

Proposals are an attempt to persuade an audience to approve whatever it is that is being proposed. Whether the proposal is internal or external, solicited or unsolicited, it is a kind of contract between the technical writer (or company) and the audience. Because proposals often deal with time and money, your trustworthiness and accountability are at stake. Consider your audience's needs and write sympathetically and knowledgeably for them. The ethical writer considers the audience's requirements, not what he or she can get out of the situation.

could create better sales potential. If she decides to propose a rearrangement, she must first convince her audience—her supervisor—that a problem exists. Only then can she go on to offer a convincing solution.

How Knowledgeable Is the Audience?

The audience may or may not know the concepts and facts involved in either the problem or the solution. Estimate your audience's level of knowledge. If the audience is less knowledgeable, take care to define terms, give background, and use common examples or analogies.

How Much Authority Does the Audience Have?

The audience may or may not be able to order the implementation of your proposed solution. A manager might assign the writer to investigate problems with the material flow of a particular product line, but the manager will probably have to take the proposal to a higher authority before it is approved. So the writer must bear in mind that several readers may see and approve (or reject) the proposal.

Use Visual Aids

Because the proposal is likely to have multiple audiences, visual aids are important. Visuals can support any part of the proposal—the description of the problem, the solution, the implementation, and the benefits. In addition to the tables and graphs described in Chapter 7, Gantt charts (see Chapter 15) and diagrams can be very helpful.

Gantt Charts

As described in Chapter 7, Gantt charts visually depict a schedule of implementation. A Gantt chart has an X axis and a Y axis. The horizontal axis displays time periods; the vertical axis, individual processes. Lines inside the chart show when a process starts and stops. By glancing at the chart, the reader can see the project's entire schedule. Figure 14.2 is an example of a Gantt chart.

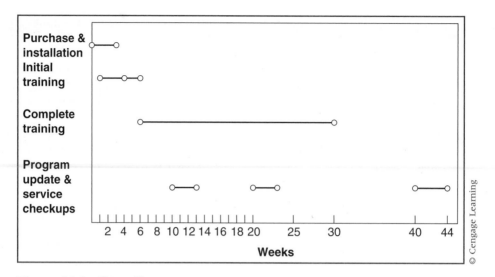

Figure 14.2   Gantt Chart

Diagrams

Many kinds of diagrams, such as flow charts, block diagrams, organization charts, and decision trees, can enhance a proposal. Layouts, for instance, are effective for proposals that suggest rearranging space.

Organize the Proposal

The writer should organize the proposal around four questions:

1. What is the problem?

2. What is the solution?

3. Can the solution be implemented?

4. Should the solution be implemented?

What Is the Problem?

Describing the problem is a key part of many proposals. You must establish three things about the problem:

▶ The data
▶ The significance
▶ The cause

The *data* are the actual facts that a person can perceive. The *significance* is the way the facts fail to meet the standard you hope to maintain. To explain the significance of the problem, you show that the current situation negatively affects productivity or puts you in an undesirable position. The *cause* is the problem itself. If you can eliminate the cause, you will eliminate the negative effects. Of course, almost every researcher soon discovers that there are chains of causes. You must carry your analysis back to the most reasonable cause. If the problem is ultimately the personality of the CEO, you might want to stop the chain before you say that. To be credible, you must show that you have investigated the problem thoroughly by talking to the right people, looking at the right records, making the right inspection, showing the appropriate data, or whatever. In the following section from a proposal, the writers describe a problem:

Significance	Table 1 shows a big jump in the number of parking tickets given out in 2006–2007, an increase of over 3000 tickets. We feel that the increase occurred because of the
Cause Data	inadequate parking lot signs. The current signs are old, plain, vague, and not very sensible. They only state that a permit is required, and one often does not know what kind of permit is needed. The signs don't specify whether they are for faculty, students, or commuters. In addition, the current signs are only 12 inches by 18–24 inches and can be overlooked if people are unaware of them.
Significance	In our survey of some West Central University students, we found that many students who received tickets either did not know that they could not park in the specific lot, were unsure of which lot they were able to park in, or did not see any specific signs suggesting that they could not park there.

TABLE 1
Tickets Given Out per 2500 Parking Stalls at West Central University

Year	No. of Tickets
2004–2005	13,202
2005–2006	13,764
2006–2007	16,867

© Cengage Learning

Figure 14.3   Text and Table Explain Problem

What Is the Solution?

To present an effective solution, explain how it will eliminate the cause, thus eliminating whatever is out of step with the standard you hope to maintain. If the problem is causing an undesirable condition, the solution must show how that condition can be eliminated. If the old signage for parking lots gives insufficient information, explain how the solution gives better information. A helpful approach is to analyze the solution in terms of its impact on the technical, management/maintenance, and financial aspects of the situation.

Solution named

Our solution is to create new permanent signs to be installed at the entrance of each parking lot. The new signs (in their entirety) will measure three feet by four feet so they will be visible to anyone entering the lot. Each sign will include the name of the lot; a letter to designate if the lot is for students (S), faculty (F), or administration (A); a color code for the particular permit needed; and the time and the days that the lot is monitored. The signs will not only present the proper information but will also look nice, making the campus more appealing. See Figure 1 for the proposed design.

Details show how the solution solves the problem.

Benefits

Figure 1:
Entrance Sign

© Cengage Learning

Figure 14.4 Text and Visual Aid Explain Solution

Can the Solution Be Implemented?

The writer must show that all the systems involved in the proposal can be put into effect. To make this clear to the audience, you would explain

- The cost
- The effect on personnel
- The schedule for implementing the changes

This section may be difficult to write because it is hard to tell exactly what the audience needs to know.

IMPLEMENTATION

Agents involved in implementation

The businesses we suggest that you deal with are Fulweil Structures, CE Signs, and University Grounds Services. The reason for choosing these businesses is that you will please the community of Menomonie by doing business in town and these businesses have good prices for a project like this. Also, these companies can provide services over the summer.

Schedule

Implementation of the new signs will take approximately one summer.

A suggested schedule is

1. Order signs from Fulweil.	one week
2. Fulweil constructs signs.	one week
3. CE paints signs.	one week
4. Grounds crew erects signs.	two weeks

Schedule explained

If you compare this schedule to the estimates below, you will see that we have built in some time for delays. The project can be easily finished in a month. We suggest June because it has the fewest students for the most weeks; our second suggestion is August, but then you will have to finish by about the 20th or risk much confusion when school starts on the 25th.

Cost

Cost background

Below is a list of supplies and approximate costs from Fulweil Structures and CE Signs. The total project cost is $13,892.16. Fulweil Structures asked us to inform you that these prices are not binding quotes.

Table 2

Table presents all costfigures

List of Supplies and Approximate Costs for New Entrance Signs

Fulweil Structures (each sign)		
6 ' × 2" × 2" solid bar aluminum (2 in quantity)	$118.91	
3 ' 4" × 3 ' 4" aluminum sheet (1 in quantity)	64.93	
3 hours of labor at $25/hour	75.00	
Total cost		$258.84
CE Signs (each sign)		
3 ' × 3 ' Reflective Scotchguard	$ 50.00	
10–15 letters painted	125.00	
2½ hours labor at $34/hour	85.00	
Total cost		$260.00
Total cost of each sign		518.84
Total cost of 24 signs		12,452.16

(Continued)

Projected cost of erecting signs		
2 hours/sign @ 25.00/hr (24 signs)	$1200.00	
Materials/sign @ 10.00 (24 signs)	240.00	
Total cost of erecting signs		1440.00
Total cost of project (24 signs)		$13,892.16
© Cengage Learning		

© Cengage Learning

Figure 14.5   Implementation Sample

Should the Solution Be Implemented?

Just because you can implement the solution does not mean that you should. To convince someone that you should be allowed to implement your solution, you must demonstrate that the solution has benefits that make it desirable, that it meets the established criteria in the situation, or both.

List of people who benefit

Discussion of each area of benefit

THE BENEFITS OF THIS PROJECT

The benefits of the signs will be felt by you, the students, the faculty, and the administration. You will see the number of appeals decline because the restrictions will be clearly visible, saving much bookwork and time for appeals. You will also answer fewer phone calls from persons needing to know where to park and you will write fewer tickets, thus saving much processing time.

The students, faculty, and administration will be happier because they will know exactly where and when they can and cannot park. Students will not receive as many parking tickets and will save money. Faculty and administration will also benefit by not having students park in their reserved parking spots (or at least not as often).

Figure 14.6   Benefits Section

Design the Proposal

To design a proposal, select an appropriate format, either *formal* or *informal*. A formal proposal has a title page, table of contents, and summary (see Chapter 14). An informal proposal can be a report or some kind of preprinted form (see Chapter 12). The format depends on company policy and on the distance the proposal must travel in the hierarchy. Usually the shorter the distance, the more informal the format. Also, the less significant the proposal, the more informal the format. For instance, you would not send an elaborately formatted proposal to your immediate superior to suggest a $50 solution to a layout problem in a workspace.

Writing the Internal Proposal

Use the Introduction to Orient the Reader

The introduction to a proposal demands careful thought because it must orient the reader to the writer, the problem, and the solution. The introduction can contain one paragraph or several. You should clarify the following important points:

- Why is the writer writing? Is the proposal assigned or unsolicited?
- Why is the writer credible?
- What is the problem?
- What is the background of the problem?
- What is the significance of the problem?
- What is the solution?
- What are the parts of the report?

An effective way to provide all these points is in a two-part introduction that includes a context-setting paragraph and a summary. The context-setting paragraph usually explains the purpose of the proposal and, if necessary, gives evidence of the writer's credibility. The summary is a one-to-one miniaturization of the body. (Be careful not to make the summary a background; background belongs in a separate section.) If the body contains sections on the solution, benefits, cost, implementation, and rejected alternatives, the summary should cover the same points.

A sample introduction follows.

DATE: April 8, 2006
TO: Jennifer Williamson
FROM: Steve Vinz
            Mike Vivoda
            Michele Welsh
            Marya Wilson
SUBJECT: Installing new parking lot signs

Reason for writing: sets context

Parking on campus has been a topic of many discussions here at West Central University and one of much concern. The topics on parking include what lots students are able to park in, when students can park in the lots, and the availability of parking on campus. We believe that students do not know exactly when and where they can park in the campus lots because of the vague and confusing signs.

Summary

We feel that the school should post at each entrance new, more informative, and more readable signs containing all the rules and regulations. These signs

(Continued)

---

would say exactly who can and cannot park in the lot, the times when the lots are patrolled, and what type of permit is needed. The project could be completed in five weeks and would cost $13,892.16. Major benefits include fewer administrative hassles and happier university community members. This report will first discuss the problem, then the solution, implementation, and the benefits.

Figure 14.7   Two-part Introduction

Use the Discussion to Convince Your Audience

The discussion section contains all the detailed information that you must present to convince the audience. A common approach functions this way:

The problem

▶ Explanation of the problem
▶ Causes of the problem

The solution

▶ Details of the solution
▶ Benefits of the solution
▶ Ways in which the solution satisfies criteria

The context

▶ Schedule for implementing the solution
▶ Personnel involved
▶ Solutions rejected

In each section, present the material clearly, introduce visual aids whenever possible, and use headings and subheadings to enhance page layout.

Which sections to use depends on the situation. Sometimes you need an elaborate implementation section; sometimes you don't. Sometimes you should discuss causes, sometimes not. If the audience needs the information in the section, include it; otherwise, don't.

The section presented in Figures 14.3–14.6 (pp. 427–430) illustrates one approach to the body. Other examples appear in the examples.

Worksheet for Preparing a Proposal

☐ **Determine the audience for the proposal.**

- Will one person or group receive this proposal?
- Will the primary audience decide on the recommendations in this proposal?

- How much do they know about the topic?
- What information do you need to present in order to be credible?

☐ **Prepare background.**

- Why did the proposal project come into existence?

☐ **Select a format—formal or informal.**

☐ **Prepare a style sheet of margins, headings, page numbers, and visual aid captions.**

Non-Profit Proposal

☐ **Closely review the granting agency's requirements, including length of proposal and due dates.**

☐ **Prepare a two-column list of the agency's requirements and the ways your proposal meets those needs.**

☐ **Create sections that explain, as required, your needs and project description.**

☐ **Clearly explain the budget.**

☐ **Use statements of information about your organization to explain why you have the expertise to implement and complete the project.**

☐ **Create visual aids that will enable readers to understand and support your organization's request.**

Internal Proposal

☐ **Define the problem.**

- Tell the basic standard that you must uphold (we must make a profit). Cite the data that indicate that the standard is not being upheld (we lost $5 million last quarter). Explain the data's causes (we lost three large sales to competitors) and significance (we cannot sustain this level of loss for another year).

☐ **Construct a visual aid that illustrates the problem or the solution.**

- Write a paragraph that explains this visual aid.

☐ **List all the parameters within which your proposal must stay.**

- Examples include cost restrictions, personnel restrictions (can you hire more people?), and space restrictions.

(Continued)

☐ **Outline your methodology for investigating the situation.**

☐ **Prepare a list of the dimensions of the problem, and show how your proposed solution eliminates each item.**

- (This list is the basis for your benefits section.)

☐ **Write the solution section.**

- Explain the solution in enough detail so that a reader can fully understand what it entails in terms of technical aspects, management/maintenance, and finances. Also clearly show how it eliminates the causes of the undesirable condition.

☐ **Construct the benefits section.**

- Clearly relate each benefit to some aspect of the problem. A benefit eliminates causes of the problem (the bottleneck is eliminated) or causes the solution to affect something else positively (worker morale rises).

☐ **Prepare a schedule for implementation.**

- Assess any inconveniences.

☐ **List rejected alternatives, and in one sentence tell why you rejected them.**

Worksheet for Evaluating a Proposal

☐ **Answer these questions about your paper or a peer's. You should be able to answer "yes" to all of the following questions. If you receive a "no" answer, you must revise that section.**

a. Is the problem clear?
b. Is the solution clear?
c. Do you understand (and believe) the benefits?
d. Does the implementation schedule deal with all aspects of the situation?
e. Does the introduction give you the basics of the problem, the solution, and the situation?
f. Is the style sheet applied consistently? Does it help make the contents clear?
g. Do the visual aids communicate key ideas effectively?

Examples

Examples 14.1, 14.2, and 14.3 illustrate three different methods of handling internal proposals.

Example 14.1

Internal
Proposal

Date: November 7, 2013
To: George Schmidt, Chief Engineer
From: Greg Fritsch, Assistant Engineer
Subject: Unnecessary shearing from joint welds

After talking to you on the phone last week, I mentioned that the Block Corporation is having difficulties with shearing on their engine mount supports. I contacted Mr. Jackson, a research expert, who said the stress from the weight of the engine causes the weld to shear. The shearing then causes the motor to collapse onto the engine mount supports. He advised me to purchase a higher-tensile-strength weld. The new weld I propose will reduce the defect rate from 10% to 0%. This report includes the following information: weld shearing, weld constraints, and shearing solution.

WELD SHEARING

Unnecessary weld shearing of the engine mount supports has been a problem for the Block Corporation since 2010. The company is suffering a 10% defective rate on every 100 engine mounts welded.

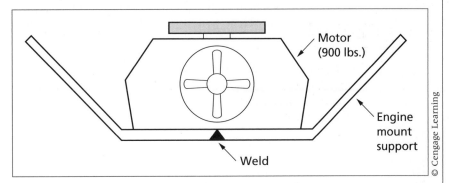

Figure 1:
Engine Mount Weld

As seen in Figure 1, the weld must hold together when 900 lbs. of force are applied to the motor mount supports. A quality weld with a high tensile strength should withstand temperature fluctuation without shearing.

(Continued)

WELD CONSTRAINT

The Block Corporation listed the following constraints for implementing a new weld:

1. Material costs must increase by less than .01¢ per engine mount support welded.
2. Welding machines must not exceed 240 volts.
3. Current welding machines have to be used.
4. Each electrical outlet has to have a separate transformer.

SHEARING SOLUTION

The solution to the company's problem is to implement a higher-tensile-strength weld. The weld is projected to increase material and electrical costs, but is not expected to exceed the company's 1% budget increase for the 2013 fiscal year.

Cost

New welding wire with a higher tensile strength will increase 2¢ for every 100 yards of wire. All engine mount welds require three yards of wire to secure a solid weld. The overall cost increase per engine mount welded will be only .006¢.

Voltage

There will be an increase in the amount of electricity used in the new welding process. The welding machines will be required to switch from 120 to 240 volts.

Use of Current Machines

The welders will use the same welding machines as in the past. The welding machines are compatible with the new welds and do not need to be replaced.

Separate Transformers

An electrical hookup from 120 to 240 volts will be needed at each electrical outlet. A transformer will be required at each individual box to ensure an increase in voltage flow.

Example 14.2

Internal
Proposal

REPLACEMENT OF PRESENT SINGLE-PHASE VENTILATION MOTOR WITH A NEW THREE-PHASE INDUCTION MOTOR

INTRODUCTION

The purpose of this report is to inform you of the inadequacies of the present ventilation system, and the benefits of replacing the current motor. In this report

I will first give you a quick summary of the proposal, followed by the necessary background information required. I will then discuss in detail the following: the problems with the present system, the proposed solution to correct the problem, the implementation of the new system, the rationale behind the decision, followed by the conclusion.

Summary of Proposal

The problem in the ventilation system came to light during a routine inspection. I noticed the following problems with the present system:

1. Insufficient air flow at the southern end of plant
2. Current motor wastes too much electricity

The combination of these two problems creates both an unsafe and an inefficient system. Fortunately, the solution is quite simple and inexpensive. To correct the problem, the present single-phase motor in the system must be removed and replaced with a new three-phase induction motor. This new motor will not only correct the problems of the present system, it will also produce the following benefits:

1. Longer life.
2. Decreased power factor.
3. Expandability.
4. Minimal downtime at installation.

Background Information

One must know the difference between a single-phase and a three-phase motor. A single-phase motor runs on only one electrical phase, but requires additional starting circuitry. Three-phase motors, on the other hand, require all three electrical phases, but do not need any starting circuitry. It is also important to know that the amount of air flow in a system is measured in cubic yards per hour.

DISCUSSION

This section covers problems with the present system, a proposed solution, implementation of the new system, rationale, and benefits of the system.

Problems with the Present System

Air Flow   The main problem with the present ventilation system is that it is unable to produce enough air flow to the southern end of the plant. During my inspection, I took various measurements of air flow throughout the plant using an air flow meter. I noticed that the southern end is receiving only 1800 cubic yards ventilation an hour. OSHA standards require that 2000 cubic yards must be replaced every hour. If this situation is not corrected, we may be endangering the well-being of our employees, not to mention being slapped with a possible fine from OSHA. After closer examination of the ventilation system, I discovered

(Continued)

that the only thing wrong with the system is the motor driving the fan. The motor is old and worn out, and therefore unable to produce the necessary air flow.

Electrical Consumption   The other problem with the present system is its abnormally high power-consumption, which I discovered while taking measurements on the present motor with a digital VOM meter. With these measurements, I calculated that the motor is running at only 50% peak efficiency. An average three-phase induction motor runs at approximately 90% peak efficiency. Over the course of a year, the company loses about $900 from the inefficiency of the present motor.

Proposed Solution

After careful analysis of all information, I have come to the following conclusion: replace the present single-phase motor in the ventilation system with a new three-phase induction motor. A new three-phase motor will not only increase air flow, it will also do it more efficiently.

Air Flow   If a three-phase induction motor were installed in place of the single-phase motor, it would increase air flow by almost 20%. I calculated this by using the torque and speed characteristics of a three-phase motor. This would boost the air flow to the southern end of the plant from 1800 cubic yards to over 2100 cubic yards per hour. This is well within OSHA standards.

Electrical Consumption   One of the biggest advantages of a three-phase motor over a single-phase motor is the efficiency. An average new three-phase motor can run at up to 90% efficiency. An average single-phase motor of the same horsepower could achieve only 80% efficiency at best. The more efficient a device, the less expensive it is to run. See Figure 1.

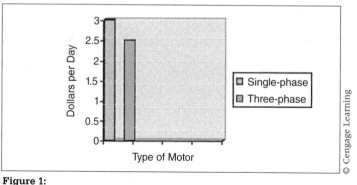

Figure 1:
Electrical Running Cost of a Single-Phase Motor versus a Three-Phase Motor

As you can see from this figure, a three-phase motor requires much less electricity to run per day than the single-phase motor. The reason a three-phase is more efficient than the single-phase motor is that it needs no additional starting circuitry. The addition of this starting circuitry in a single-phase motor is what robs it of maximum efficiency.

Implementation of the New System

The installation of a new three-phase induction motor should pose no problems. In this section, I will concentrate on the main aspects of installation: cost, time, and inconvenience.

Cost   The overall cost of replacing and installing the new motor should not exceed $500. The motor and control box together are $300. The wiring must be done by a certified electrician and overall labor cost should not exceed $150. The remaining $50 will buy new motor mountings, brackets, and wire. A three-phase junction box is within 20 feet of the ventilation system and should pose no installation difficulties for the electrician.

Time   The installation time from start to finish should be no more than five hours. It will take one hour for us to remove the old single-phase motor. Installing the three-phase motor should take no more than an hour and a half. The remaining hour would be used for cleanup work and initial start-up of the system. I received all of these time and cost figures from a certified electrician.

Inconveniences   The new three-phase motor could be installed with only a few slight inconveniences, the most obvious of which is the shutdown of the ventilation system. This cannot be done during working hours, so it will have to be done on a Saturday. The labor costs I have stated earlier reflect the electrician's time-and-a-half rate imposed by working on the weekend. There is also the minor inconvenience of having someone here that Saturday to let the electrician into the building.

Rationale/Benefits

A three-phase induction motor in a ventilation system will provide three main benefits: longer life, decreased power factor, and expandability.

Longer Life   If a three-phase induction motor were installed into the ventilation system, it would provide much longer life than an equivalent single-phase motor. This point is made clear in Figure 2.

As you will notice from Figure 2, a three-phase motor will last much longer than will a single-phase motor doing the same task. The reason for this is the simplicity of operation of a three-phase motor; the single-phase requires starting circuitry, which has a tendency to break down more quickly.

(Continued)

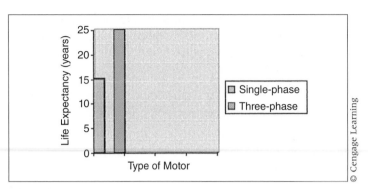

Figure 2:
Expected Life of a Single-Phase Motor versus a Three-Phase Motor

Decreased Power Factor   Power factor is a confusing factor involved with the use of any inductive device, including motors. Power factor, if left unchecked, alters the electricity we receive. Although both single-phase and three-phase motors have some amount of power factor associated with them, a three-phase motor has less. A three-phase motor will reduce the amount we are charged for power factor.

Expandability   Another advantage of a three-phase motor is the expandability we would receive in our ventilation system. With the increase in air flow, we could easily add on to the ventilation system.

CONCLUSION

We cannot afford to let this problem continue. A three-phase motor in the ventilation system will best suit our needs both now and in the future.

Example 14.3

Non-Profit
Grant Proposal

Description of Organization
Based in the Seward neighborhood, the Alliance for Metropolitan Stability is a coalition of organizations advocating for public policies that promote equity in land use. Our mission is to engage communities in eliminating racial, environmental, and economic disparities in growth and development patterns in the Twin Cities metropolitan area.

The Alliance was formed in 1994, in the realization that true stability for the Twin Cities could only be achieved through a comprehensive approach to regional problems. We have 20 member groups and dozens of allied organizations that unite under the recognition that our regional problems are linked through significant intersections of issues, people, and places. Our partners are representatives of:

- Geography-based organizations, such as neighborhood groups.
- Low-income communities and communities of color.
- Faith-based organizations.
- Issue-based organizations, such as environmental, energy, land-use, affordable housing, and transit groups.
- The organized labor community.

We serve the seven-county Twin Cities metro area, with a specific interest in serving low-income people and communities of color.

How a Grant from Seward Co-op Would Be Used

The Alliance's goal is to build strategic, cross-sector partnerships that advance sustainable and equitable approaches to Twin Cities land use and development. This requires that we work closely with large groups of community organizers, leaders, citizens and public officials to build consensus and move collectively toward more equitable public policies for the community and the environment. As a coalition of grassroots organizations, a significant part of the Alliance's work is to host community meetings, forums, organizer roundtables and trainings.

The Alliance likes to purchase food and beverages for these meetings from cooperatives and restaurants that are committed to using locally sourced and organic foods whenever possible. We further demonstrate our commitment to promoting environmental sustainability and local economic development by purchasing from local businesses, such as the Seward Co-op Grocery & Deli.

Our forums and community meetings not only support local businesses and sustainable agriculture, they focus on topics that build community and contribute to the overall sustainability of the Twin Cities and the Seward Neighborhood. Recent topics have included:

- Communicating the value of grassroots organizing.
- Wealth creation in communities of color.
- Blue/green jobs (linking the labor community to the environmental community).
- Environmentally sustainable affordable housing.
- Connecting the racial and environmental justice movements.
- Legislative report card on racial justice.

We expect to host even more events in late 2008 and early 2009, as we are organizing a series of meetings that will bring together all the organizations associated with the regional equity movement in the Twin Cities, including neighborhood, environmental, racial justice, transit, affordable housing, economic justice, labor, and faith-based organizations. A contribution from the Seward

(Continued)

Co-op Grocery & Deli would support these important events that bring together grassroots organizations together to build power around issues of racial, environmental and economic justice.

How a Grant Would Support Seward Co-op's Mission

The Alliance believes that all socially responsible businesses and organizations have a responsibility to play a role in promoting sustainability and local economic development—not only through their missions, but through their actions. That's why we are committed to buying local and organic whenever possible.

In addition, we regularly bring groups of community-minded individuals into the Seward neighborhood for in-house meetings and meetings in community spaces. Each year, hundreds of our project participants learn about local community and economic development initiatives, grassroots organizing for social change, and environmentally sound lifestyle choices through participation in our projects and meetings.

We know that leading by example is the most practical and useful way to inspire our partners to make commitments to buying sustainable products from local businesses.

Dollar Amount of the Alliance's Request

The Alliance was very grateful to receive a grant of $500 from the Seward Co-op in 2007 to support our commitment to providing local and organic foods at our meetings. So far in 2008, we have already spent more than $1,400 toward this purpose.

We request a grant of $1,000 from the Seward Co-op to support these efforts in 2008. A grant from the co-op would be matched dollar-for-dollar through a matching grant challenge from the McKnight Foundation.

Thank you for your support of the Alliance for Metropolitan Stability.

Exercises

▶ You Create

1. Create a visual aid that demonstrates that a problem exists.
2. With the visual aid from Exercise 1, write a paragraph that includes the data, the significance, and the cause of the problem, and write a second paragraph that suggests a way to eliminate the problem.
3. Make a Gantt chart of a series of implementation actions. Write a paragraph that explains the actions.

4. Create two different page designs for the proposal about parking signs on campus (p. 432).

▶ You Revise

5. Rewrite the following paragraphs. The writer is a recreation area supervisor who has discovered the problem; the reader is the finance director of a school district. Shorten the document. Make the tone less personal. Make a new section if necessary. Adopt the table or create new visual aids. If your instructor requires, also add an introduction and a summary.

DISCUSSION OF TRENDS

I have data that establish trends in the building's use (see Table 1). These data show peak adult and student use during the winter months. When school is out (June–August), we have more students and children using the building. Our slow months are in the Spring (April and May) and in the Fall (September and October). These trends coincide with what we know to be true about revenue loss. I have a more difficult time controlling the adult and student population using the building during the winter months. This results in a higher (25%) revenue loss for these months. On the other end, the children and students using the building during the summer are easier to control. This results in a lower (10%) revenue loss.

Table 1
Building Use for Open Recreation

	Adults	Students (Grades 7–12)	Children (Grade 6 & Under)	Total
January	621	583	412	1616
February	645	571	407	1623
March	597	545	393	1535
April	428	372	279	1070
May	210	330	239	779
June	365	701	587	1653
July	276	823	650	1749
August	327	859	718	1904
September	189	268	225	682
October	226	314	275	815
November	398	292	412	1102
December	589	494	384	1467

© Cengage Learning

PROBLEM

The problem of revenue loss really involves two issues. The loss of revenue leads directly to a secondary issue, which is loss of control. When I cannot control the people entering the building, we lose revenue. When these people assume they can get in free, they also assume I cannot control their actions thereafter.

As a supervisor I have many duties. During open rec. I am expected to be at the office window collecting fees. This is all well and fine *if* I could stay there the entire time! Unfortunately, I must occasionally check activities in the weight room, the field house, the pool, and the locker rooms. At these times I am out of the office and cannot control people from just walking in. Even answering the phone causes problems. I must cross the office to the desk, and then I lose direct eye contact to the front entrance.

I really have no good explanation for why I have more problems with the adult-student users during the winter months as compared to the student-children users in the summer. All I know is that when these winter "bucket shooters" start pouring in for open rec., control goes right out the window. The only answer is a barrier to contain them in the lobby area until they have paid.

SOLUTION

The solution is a barrier that extends from the entrance door into the lobby, to the office wall. This is a length (open space) of about 14 feet. I suggest a chain as a temporary solution. Attached at the entrance door frame, it should extend 5½ feet to a stationary post (nonpermanent support), feed through an opening at the top of this post, and continue on another 5½ feet to another post. The remaining 3 feet to the office wall will be the entrance area. This will be chained off as well, and passage will be allowed only after paying the open rec. fee. A sign that reads "DO NOT ENTER" should be attached to the chain at the entrance area by the office. When I am out of the office, people may think twice and remain in the lobby until I return.

The cost in hardware for this barrier will be minimal, and I suggest it only as a temporary measure. I would like to establish the effectiveness of a simple barrier before considering a more permanent structure. There will always be some who ignore the barrier. There is never a perfect solution.

▶ **You Analyze**

6. Analyze Examples 16.1 and 16.2. Follow the instructions for Exercise 8. Alternate: If your instructor requires, rewrite and redesign one of the examples.

❱ Group

7. In groups of two to four, discuss one of the proposals given in this chapter. What do you like? Dislike? Would you agree to implement the solution? Report your results to the class.

8. In groups of three or four, analyze any of the three examples in this chapter. Prepare a report to your class that pinpoints its weaknesses and strengths. Focus on depth of detail, appropriateness for audience, and unnecessarily included items.

Writing Assignments

For each of the following assignments, first perform the activities required by the worksheet (pp. 432–434).

1. Write a proposal in which you suggest a solution to a problem. Topics for the assignment could include a problem that you have worked on (and perhaps solved) at a job or a problem that has arisen on campus, perhaps involving a student organization, or at your workplace. Explain the problem and the solution. Show how the solution meets established criteria or how it eliminates the causes of the problem. Explain cost and implementation. If necessary, describe the personnel who will carry out the proposal. Explain why you rejected other solutions. Use at least two visual aids in your text. Your instructor will assign either an informal or a formal format. Fill out the worksheet from this chapter, and perform the exercises that your instructor requires.

    Your instructor may make this a group assignment. If so, follow the instructions for developing a writing team (Chapter 2), and then analyze your situation and assign duties and deadlines.

2. In groups of three or four, write a simple request for a proposal (RFP). Ask for a common item that other people in your class could write about. (If you've all taken a class in computerized statistics, for example, ask for a statistics software program.) Try to find a real need in your current situation. Interview affected people (such as the statistics instructor) to find out what they need. Then trade your RFP with that of another group. Your group will write a proposal for the RFP you receive. Your instructor will help you with the day-to-day scheduling of this assignment.

3. Write a learning report for the writing assignment you just completed. See Chapter 5, Writing Assignment 8, page 137, for details of the assignment.

Web Exercise

Write a proposal suggesting that you create a website for a campus club or a company division (including a "special interest" site, such as for the company yoga club). Explain how you will do it, why you are credible, the cost, the benefits to the company, and the schedule for production.

Works Cited

Bacon, Terry. "Selling the Sizzle, Not the Steak: Writing Customer-Oriented Proposals." *Proceedings of the First National Conference on Effective Communication Skills for Technical Professionals.* Greenville, SC: Continuing Engineering Education, Clemson University, November 15–16, 1988. Print.

"Exemption Requirements—Section 501(c)3 Organizations" IRS. Web. 17 May 2012. <http://www.irs.gov/charities/charitable/article/0,,id=96099,00.html>.

"Frequently Asked Questions: How Many Nonprofit Organizations Are There in the United States?" The Foundation Center. Web. 16 May 2012. <http://foundationcenter.org/getstarted/faqs/html/howmany.html>.

"Grant Programs: Overview." Southern Arts. Web. 16 May 2012. <http://www.southarts.org/site/c.guIYLaMRJxE/b.1303445/k.E308/Overview.htm>.

"Guide to Grant making at the Otto Bremer Foundation." Otto Bremer Foundation. Web. 17 May 2012. <http://www.ottobremer.org/guidelines.php>.

Carlson, Mim and Tori O'Neal-McElrath. *Winning Grants, Step by Step.* 3rd ed San Francisco: Jossey-Bass, 2008. Print.

Davis, Barbara. *Writing a Successful Grant Proposal. Minnesota Council on Foundations.* 2005. Web. 18 May 2012. <www.mcf.org/system/article-resources/0000/0325/writingagrantproposal.pdf>.

Fritz, Joanne. "How to Write a Grant Proposal for a Foundation". About.com. Nonprofit Charitable Orgs. Web. 18 May 2012. <http://nonprofit.about.com/od/foundationfundinggrants/tp/grantproposalhub.htm>.

Proposal Writing Short Class. *The Foundation Center.* Web. 16 May 2012. <http://foundationcenter.org/getstarted/tutorials/shortcourse/index.html>.

Writing a Successful Grant Proposal. Minnesota Council on Foundations. Web. 16 May 2012. <http://www.mcf.org/nonprofits/successful-grant-proposal>.

User Manuals

CHAPTER CONTENTS

CHAPTER 15 **IN A NUTSHELL**

A manual should be written and designed so that readers are comfortable enough with the machine or object to confidently interact with it. Effective manuals teach readers that machines are objects that require humans to use and control them. Your readers can achieve this position as you help them relate to the machine.

Supply context. Help them see the machine from the designer's point of view. What does this machine or this part do, and why, and what kinds of concerns does that function imply? Once readers get the big picture, they will usually try to use the item for its intended purpose.

Explain what the parts do. List all the visible parts, and explain what they cause, how to stop or undo what they cause, what other parts work in sequence with them.

Explain how to perform the sequences. Think of readers as users or doers. What actions will they perform? Think of common ones like turning the machine on and off. Spend time working on the machine yourself so you can clearly explain how to work it.

Use visual logic. One major section should discuss each of the three areas mentioned earlier. Divide each section into as many subsections as needed. Use heads and white space so readers can easily find sections and subsections. Use clear text and visual aids so readers figure out how to do the actions confidently.

Develop credibility. Give brief introductions that tell the end goal of a series of steps; give warnings before you explain the step; state the results of actions or give clear visual aids so that readers can decide if they are progressing logically through the steps.

Companies sell not only their products but also knowledge of how to use those products properly. This knowledge is contained in manuals. Both the manufacturer and the buyer want a manual that will allow users safely and successfully to assemble, operate, maintain, and repair the product.

Very complex mechanisms have separate multivolume manuals for different procedures such as installation and operation. The most common kind of manual, however, is the user's manual, which accompanies almost every product.

User's manuals have two basic sections: descriptions of the functions of the parts and sets of instructions for performing the machine's various processes. In addition, the manual gives information on theory of operation, warranty, specifications, parts lists, and locations of dealers to contact for advice on parts. This chapter explains how to plan and write an operator's manual.

Planning the Manual

Your goal is for the manual to help readers make your product work. To plan effectively, determine your purpose, consider the audience, schedule the review process, discover sequences, analyze the steps, analyze the parts, select visual aids, and format the pages.

Determine Your Purpose

The purpose of a manual is to enable its readers to perform certain actions. But manuals cannot include every detail about any system or machine. Decide which topics your readers will need, or can deal with. For example, you would choose to explain simple send and receive commands for e-mail beginners, but not complicated directory searching.

Decide the level of detail. Will you provide a sketchy outline, or will you "hand-hold," giving lots of background and explanation? To see the results of a decision to "hand-hold," review Example 9.1 (pp. 262–264) and examples 9.3 and 9.4 (pp. 265–266). Example 9.2, page 264. Making these key decisions will focus your sense of purpose, allowing you to make the other planning decisions detailed in this chapter.

Consider the Audience

Who is your audience? Create an audience profile. Characterize your readers and their situation so that you can include text, visuals, and page design that give them the easiest access to the product. First, determine how much they know about general terms and concepts. Readers who are learning their first word processing program know nothing about "save," "cut," "paste," "open," "close," and "print." Readers learning their fifth program, however, already understand these basic word processing concepts. Early in the planning process, make a list of all the words the readers must understand.

Second, consider your goal for your readers. What should they be able to do as a result of reading the manual? A common answer, of course, is to be able to operate the product, but what are those key abilities they *must* have to do so? Those abilities will help you decide what sections to include and how to write them.

Third, consider how your readers will read the manual. Both beginning and expert audiences usually are "active learners." They do not want to read; instead, they want to accomplish something relevant quickly (Redish). When they do read, they do not read the manual like a story, first page to last. Instead, they go directly to the section they need. To accommodate these active learners who differ widely in knowledge and experience, use format devices—such as heads and tables of contents—that make information accessible and easy to find. This type of thinking will help you with the layout decisions you must make later and will help you decide what information to include in the text.

Fourth, consider where the audience will use the manual. This knowledge will help you with page design. For instance, manuals used in poor lighting might need big pages and typefaces, whereas manuals used in constricted spaces or enclosed in small packages need small pages and typefaces.

Fifth, consider your audience's emotional state. For various reasons, many, if not most, users do not like, or even trust, manuals (Cooper). Further, users are often fearful, hassled, or both. Your goal is to both allay their fears and develop their confidence. The presentation of your manual—its sequence and format— and of your identity as a trustworthy guide will develop a positive relationship.

Determine a Schedule

Early in your planning process, set up a schedule of the entire project. Typically, a manual project includes not just you, the writer, but also other people who will review it for various types of accuracy—technical, legal, and design. In industrial situations, this person might be the engineer who designed the machine. If you write for a client, it will be the client or some group designated by the client.

Think of each draft as a cycle. You write, and then someone reviews, and as a result of their review you rewrite or redesign. At the outset of the project, set dates for each of these reviews and decide who will be part of the review team. In addition, agree with your reviewers on when you expect them to return the draft and on what types of comments they are to make. You can handle the actual schedule in several ways, perhaps write in the actions you will perform during various weeks on a calendar. Or you could make a Gantt chart.

Suppose your tasks are to interview an engineer, create a design, write a draft, have a reader's review, write a second draft, have a second review, and print the manual. Suppose also that your schedule allows you eight weeks. Your Gantt chart might look like Figure 15.1 (p. 450).

Discover Sequences

Discovering all the sequences means that you learn what the product does and what people do as they use it (Cohen and Cunningham). To learn what the

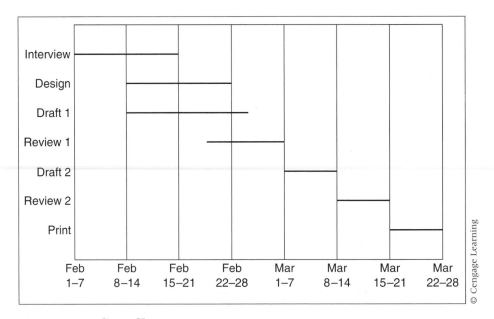

© Cengage Learning

Figure 15.1 Gantt Chart

product does, learn the product so thoroughly that you are expert enough to talk to an engineer about it. Because this process takes a good deal of time, you need to plan the steps you will take to gain all this knowledge. Schedule times to use the product. Talk to knowledgeable people—either users or designers, or both. Your goal is to learn all the procedures the product can perform, all the ways it performs them, and all the steps users take as they interact with the product.

For example, the writer of a manual for a piston filler, a machine that inserts liquids into bottles, must grasp how the machine causes the bottle to reach the filling point, and how the machine injects the liquid into the bottle. Gaining this knowledge requires observing the machine in action, interviewing engineers, and assembling and disassembling sections.

But the writer must also know what people do to make the machine work. The most practical way to gain this knowledge is to practice with the product. These acts become the basis for the sections in the procedures section. As you practice, make flow charts and decision trees. In your flow charts, list each action and show how it fits into a sequence with other actions (see Figure 15.2).

The sequences your manual must teach the user typically include

▶ How to assemble it.
▶ How to start it.
▶ How to stop it.
▶ How to load it.
▶ How it produces its end product.
▶ How each part contributes to producing the end product.

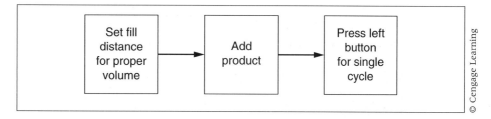

Figure 15.2   Flow Chart

> ▶ How to adjust parts for effective performance.
> ▶ How to change it to perform slightly different tasks.

Analyze the Steps

To analyze the steps in each sequence means to name each individual action that a user performs. This analysis is exactly the same as that for writing a set of instructions (review Chapter 11). In brief, determine both the end goal and the starting point of the sequence, and then provide all the intermediary steps to guide the users from start to finish. Try constructing a decision tree. Make a flow chart for the entire sequence, and then convert the chart into a decision tree.

For an example of such a conversion, compare Figures 15.2 and 15.3. In these steps, taken from a piston filler manual, the writer wants to explain how to insert a specified amount of liquid into a bottle. Figure 15.2 shows the flow chart; Figure 15.3 (p. 452) shows a decision tree based on the flow chart.

Here is the text developed from the two figures:

1. Set the fill distance for the proper volume.
   a. Check specifications for bottle volumes (p. 10).
   b. To determine this distance, find out the diameter of your piston.
   c. Go to the volume chart on page 11.
   d. Find the piston diameter in the left column.
   e. Read across to the volume you need.
   f. Read up to determine the length you need.
   g. Adjust the distance from *A* to *B* (Figure 6 [not shown]) to the length you need.
2. Add the product to the hopper. If you are unsure of the product type, see specifications (p. 11).
3. Press the left button (*A* on Figure 6 [not shown]) for single cycle.

Analyze the Parts

To analyze the parts, list each important part and explain what it does. Then convert these notes into a sentence. If you look at a few common user manuals, say, for a DVD player, you will always find this section in the front of the manual. A helpful method is to make a three-part row for each part. Name the

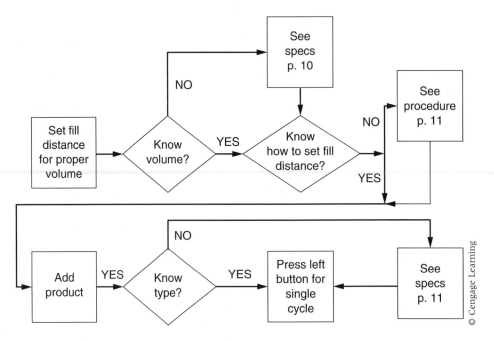

© Cengage Learning

Figure 15.3   Decision Tree Based on Flow Chart

part, write the appropriate verb, and write the effect of the verb. Then turn that list into a comprehensible sentence. Here is the list for a stop button:

Name of Part	*Verb*	*Effect*
stop button	stops/ends	all functions stop

Here is the sentence for the button:

The red emergency stop button immediately stops all functions of the machine.

Select Visual Aids

Visual aids—photographs, drawings, flow charts, and troubleshooting charts—all help the reader learn about the product. In recent years, with the advent of desktop publishing and many graphics software and hardware programs, the use of visual aids has proliferated. Including many visuals is now the norm. Many manuals have at least one visual aid per page; many provide one per step.

Your goal is to create a text–visual interaction that conveys knowledge both visually and textually. Consider this aspect of your planning carefully. If you can use a visual aid to eliminate text, do so. Notice one key use of visual aids in manuals. Visuals give permission. Although many visual aids are logically unnecessary because the text and the product supply all the knowledge, they are still useful. Consider, for instance, Figure 5.3 in Figure 15.7 on page 460 of the video camera manual or the various visuals on pages 466–473 of the baby stroller manual. Neither of these is strictly needed because the user could read the text and look at the machine and see what is described in the text. But the visual reassures the readers that they are "in the right place." Use visuals liberally in this manner. Your readers will appreciate it.

You must decide whether each step needs a visual aid. Most manual writers now repeat visual aids. As a result, the reader does not have to flip back and forth through pages. To plan the visual image needed to illustrate a step, decide which image to include and from which angle users will view it. If they will see the part from the front, present a picture of it from the front. Use a storyboard (Riney), such as the one shown in Figure 15.4, to plan the visual aid. Storyboards are discussed in Chapter 16.

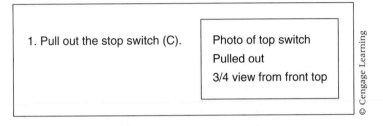

Figure 15.4   Storyboard

Format the Pages

The pages of a manual must be designed to be easy to read. Create a style sheet with a visual logic (see Chapters 6 and 7) that associates a particular look or space with a particular kind of information (all figure captions italic, all page numbers in the upper outside corner, and all notes in a different typeface). You must also design a page that moves readers from left to right and top to bottom. (Review Chapter 6 for format decisions.) This process is more complex than you might think, so carefully consider your options. You might review several consumer manuals that accompany software products or common home appliances.

To produce effectively laid out pages, use a grid and a template. A *grid* is a group of imaginary lines that divide a page into rectangles (see Figure 15.5, p. 454). Designers use a grid to ensure that similar elements appear on pages in the same relative position and proportion. One common grid divides the page into two unequal columns. Writers place text in the left column and visual aids in the right column, as shown on page 454.

A *template* is an arrangement of all the elements that will appear on each page, including page numbers, headers, footers, rules, blocks of text, headings, and visual aids. Figure 15.6 (p. 455) is a template of a page. The arrows indicate all the spots at which the author made a deliberate format decision. Create a tentative template before you have gone very far with your writing because your visual logic is part of your overall strategy (see p. 455) and will influence your word choice dramatically.

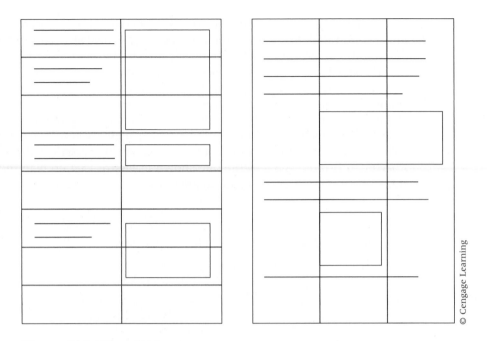

© Cengage Learning

Figure 15.5   Page Grids

The following notes list all the format decisions that the author made for Figure 15.6:

1. Type, font, size, position of header text
2. Position, width, length of header rule
3. Size, type, font, position, and grammatical form (*-ing* word) of level 1 heads
4. Size, type, font, and position of instructional text; space between head and next line of text (leading)
5. Size, type, font, and position of numeral for instructional step
6. Punctuation following the numeral
7. Position of second line of text
8. Space between individual instructions
9. Punctuation and wording of reference to figure
10. Size, type, font, position of notes or warnings
11. Space between text and visual aid
12. Width (in points) and position of frame for visual aid
13. Size, type, font, and position of figure number and brief explanatory text
14. Position, width, length of footer rule
15. Type, font, size, position of footer text
16. Type, font, size, position of page number

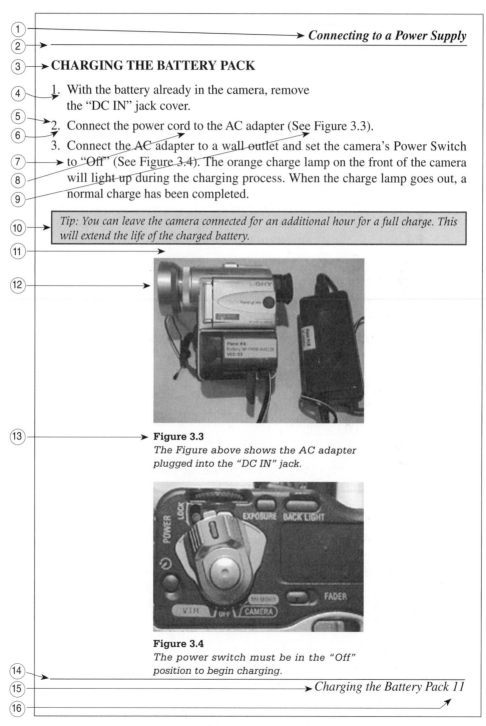

(1)

Connecting to a Power Supply

(2)

(3) **CHARGING THE BATTERY PACK**

(4) 1. With the battery already in the camera, remove the "DC IN" jack cover.

(5)
(6) 2. Connect the power cord to the AC adapter (See Figure 3.3).

3. Connect the AC adapter to a wall outlet and set the camera's Power Switch
(7) to "Off" (See Figure 3.4). The orange charge lamp on the front of the camera
(8) will light up during the charging process. When the charge lamp goes out, a
(9) normal charge has been completed.

(10) *Tip: You can leave the camera connected for an additional hour for a full charge. This will extend the life of the charged battery.*

(11)

(12)

(13) **Figure 3.3**
The Figure above shows the AC adapter plugged into the "DC IN" jack.

Figure 3.4
The power switch must be in the "Off" position to begin charging.

(14)

(15) *Charging the Battery Pack 11*

(16)

© Cengage Learning

Figure 15.6 Page Template with Decision Points

Writing the Manual

The student sample shown in Figure 15.7 is taken from a user's manual for a video camera. This manual uses a one-column page; instructions are aligned on the left margin and visuals are centered. All the visual aids are clear digital photographs.

Introduction

In the introduction explain the manual's purpose and whatever else the reader needs to become familiar with the product: how to use the manual, the appropriate background, and the level of training needed to use the mechanism.

The introduction tells the purpose of the machine, states the purpose and divisions of the manual, and explains why a user needs the machine. In Figure 15.7, the introduction is entitled "Preface."

Arrange the Sections

A manual has two major sections:

▶ Description of the parts (see Figure 15.7, "Identifying Parts of the Camera")
▶ Instructions for all the sequences (see Figure 15.7, "Recording Video")

The Parts Section

The parts and functions section provides a drawing of the machine with each part clearly labeled. The description explains the function of each item. This section answers the question: What does this part do? or What happens when I do this? An easy, effective way to organize the parts description is to key the text to a visual aid of the product. Most appliance manuals have such a section.

The Sequences Section

The sequences section enables users to master the product. Arrange this section by operations, not parts. Present a section for each task. Usually the best order is chronological, the order in which readers will encounter the procedures. Tell first how to assemble, then how to check out, then how to start, to perform various operations, and to maintain. However, be aware that readers seldom read manuals from beginning to end, so you must enable them to find the information they need. Make the information easy to locate and use by cross-referencing to earlier sections. Never assume that readers will have read an earlier section.

In addition, as mentioned earlier, repeat key instructions or visuals. Do not make readers flip back and forth between pages. Rather, place the appropriate information where readers will use it (Rubens).

Table of Contents

TABLE OF CONTENTS

(Continued)

Identifying Parts of the Camera

IDENTIFYING PARTS OF THE CAMERA

LCD Screen   Projects camera image for recording or camcorder image for review. Swivels to any position.

LCD Open Button   Opens LCD screen when pushed

Lens Cap   Covers lens; protects lens from scratches

Microphone   Records sound during recording

Viewfinder   Shows image as it will be recorded

Open/Eject Button   Opens tape cover and ejects tape after being pushed

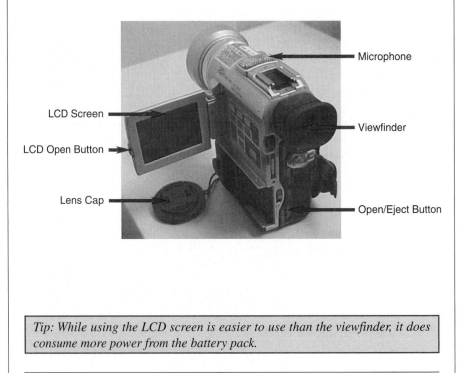

Microphone

LCD Screen

LCD Open Button

Lens Cap

Viewfinder

Open/Eject Button

Tip: While using the LCD screen is easier to use than the viewfinder, it does consume more power from the battery pack.

2

Preparing the Camera

RECORDING VIDEO

PREPARING THE CAMERA

This camera uses DV Mini Video Cassette Tape. They can be purchased in two-packs from either the campus bookstore or local department stores.

1. Remove the lens cap.
2. Connect the power source.
3. Insert a video cassette.
4. Press the small green button on the Power Switch and slide it to the "CAMERA" position (see Figure 5.1).
5. Press "OPEN" on the LCD panel and open the LCD screen to 90 degrees (see Figures 5.2 and 5.3).

Figure 5.1
The power switch is in the "CAMERA" position.
The start/stop button is also circled.

Figure 5.2
The "OPEN" button for the LCD screen is located
just below the Power Switch.

16

(Continued)

Recording Video

SHOOTING FOOTAGE

1. Press the round Start/Stop button with the red dot in the center of the Power Switch to start recording.

> *Tip: As soon as you press the Start/Stop button, the red recording lamp on the front of the camera will light up to indicate that the camera is recording (see Figure 5.4).*

2. Press the red Start/Stop button again to stop recording (see Figure 5.1).

Figure 5.3
The image above shows the LCD screen fully opened.

Figure 5.4
The recording lamp is located next to the A/V jack cover. The lamp turns red while recording.

17

© Cengage Learning

Figure 15.7   Excerpts from an Operator's Manual for a Digital Mini Video Camera

Assume Responsibility

All manual writers have an ethical responsibility to be aware of the dangers associated with running a machine. Keep in mind that if you leave out a step, the operator will probably not catch the error, and the result may be serious. Also, you must alert readers to potentially dangerous operations by inserting the word WARNING in capital letters and by providing a short explanation of the danger. These warnings should always appear before the actual instruction. Sometimes generic warnings, which apply to any use of the mechanism, are placed in a special section at the front of the manual.

Other Sections

Manuals traditionally have several other sections, although not all of them appear in all manuals or in the exact arrangement shown here. These sections are the front matter, the body, and the concluding section. The front matter could include such elements as

▶ Title page
▶ Table of contents
▶ Safety warnings
▶ A general description of the mechanism
▶ General information, based on estimated knowledge level of the audience
▶ Installation instructions

The body could include this element:

▶ A theory of operation section

The concluding section could include such elements as

▶ Maintenance procedures
▶ Troubleshooting suggestions
▶ A parts list
▶ The machine's specifications

Test the Manual

Usability testing helps writers find the aspects of the manual that make it easier or harder to use, especially in terms of the speed and accuracy with which users perform tasks (Craig). You need to plan, conduct, and evaluate a usability test (Brooks). Figure 15.2 and Chapter 12 contain fully developed usability test reports. Whether you perform a test as detailed as those depends on a number of factors such as time, but you need to have someone unfamiliar with your object go through the manual to discover the places where your text and visuals are not clear.

Planning a Usability Test

Planning the test is selecting what aspects of the manual you want to evaluate, what method you will use, and who will be the test subjects.

Select the aspects of the manual that you want to study. The most important question is—Does this manual allow the readers to use the object easily and confidently? Consider using some or all of these questions (Bethke et al.; Queipo):

▶ *Time*
How long did it take to find information? To perform individual tasks? To perform groups of tasks?
▶ *Errors*
How many and what types of errors did the subject make?
▶ *Assistance*
How often did the subject need help?
At what points did the subject need help?
What type of help did the subject need?
▶ *Information*
▶ Was the information easy to find? Easy to understand? Sufficient to perform the task?
▶ *Format*
Is the format consistent?
Are the top-down areas (headings, introductions, highlighters) helpful?
Is the arrangement on the page helpful?
▶ *Audience Engagement*
Is the vocabulary understandable?
Is the text concrete enough?
Is the sequence "natural"? Does it seem to the learner that this is the "route to follow" to do this activity?

Select the method you will use to find the answers to your questions. Some questions need different methods. One method often is not sufficient to derive all the information you want to obtain. Test methods (Sullivan) include

▶ *Informal observation*—watching a person use the manual and recording all the places where a problem (with any of the topic areas you selected to watch for) arose.
▶ *User protocols*—the thoughts that the user speaks as he or she works with the manual and that an observer writes down, tapes, or video records.

> *Computer text analysis*—subjecting the text to evaluation features that a software program can perform, including word count, spelling, grammar, and readability scores (i.e., at what "grade level" is this material?)
> *Editorial review*—knowledgeable commentary from a person who is not one of the writers of the text.
> *Surveys and interviews*—a series of questions that you ask the user after he or she has worked with the manual.

Your goal is to match the test method with the kind of information you want. For instance, an editorial review would produce valuable information on consistency. User protocols or survey/interviews would help you determine if the vocabulary was at the appropriate level or if the page arrangement was helpful. Observation would tell you if the information given was sufficient to perform the task.

Select the test subjects. The test subjects are most often individuals who are probable members of the manual's target audience but who have not worked on developing the manual.

Conducting a Usability Test

Conducting the test is administering it. The key is to have a way to record all the data—as much feedback as possible as quickly as possible. You can use several methods:

> If you do an informal observation, you can use a tally sheet (Rubin) that has three columns—Observation, Expected Behavior or User Comment, and Design Implications. Fill out the observation column as you watch and the other two columns later.
> For user protocols, you can design a form like the one you use for informal observations, or you can audio or video record, although the difficulties with taping methods—setup procedure, use of the material after the session—require a clear decision on your part of whether you will really use the data you record.
> Computer and editorial analysis will tell you about the features of the text but will not tell the audience's reaction; these tests are relatively simple to set up, although telling an editor what to look for and setting a grammar checker to search for only certain kinds of problems are essential.
> For surveys and interviews, you can design a form (as outlined in Chapter 5) and administer it after the subject has finished the session. Sample questions include:

Were you able to find information on X quickly?
Did the comments in the left margin help you find information?
Did you read the introductions to the sequences?

Did the introduction to each sequence make it easier for you to grasp the point of sequence?

A typical way to record the answers is either yes/no/comments or some kind of recording scale (1 = highly agree, 5 = highly disagree).

Evaluating a Usability Test

Evaluating the test is determining how to use the results of the test (Sullivan). Your results could indicate a problem with

▶ The text (spelling, grammar, sufficiency of information)
▶ The text's design (consistency, usefulness of column arrangement, or highlighting techniques)
▶ The "learning style" of the audience (sequence of the text, basic way in which they approach the material)

If you have determined beforehand what is an acceptable answer (e.g., this procedure should take X minutes; this word is the only one that can be used to refer to that object), you will be able to make the necessary changes. For more help on this topic, consult John Craig (see Works Cited).

Worksheet for Preparing a Manual

☐ **Consider the audience.**

- How much do readers know about the general terms and concepts?
- Where and when will they use the manual?
- What should they be able to do after reading the manual?
- List all the terms a user must comprehend. Define each term.

☐ **Determine a schedule.**

- On what date is the last version of the manual due? By what date must each stage be completed?
- Who will review each stage?
- How long will each review cycle take?

☐ **List where you can obtain the knowledge you need to write the manual.**

- A person? Reading? Working with the mechanism?

☐ **Analyze the procedures a user must follow to operate the product.**

- What must be done to install it, to turn it on, to turn it off, and to do its various tasks?
- List the sequence for presenting the processes.

- Choose an organizational pattern for the sequence—chronological or most important to least important.
- Create a flow chart for each procedure the machine follows.
- Create a decision tree for each procedure the user follows.
- Name each part and its function.
- For a complicated product, you will discover far too many parts to discuss. Group them in manageable sections. Decide which ones your audience needs to know about.

☐ **Choose a visual aids strategy.**

- Will you use drawings or photographs?
- Will you use a visual aid for each instruction?
- Will you use a visual aid on each page?
- Will you use callouts?

☐ **Create a storyboard for your manual.**

☐ **Design pages by preparing a style sheet of up to four levels of heads, captions for visual aids, margins, page numbers, and fonts (typefaces).**

- Select rules, headers, and footers as needed to help make information easy to find on pages.

☐ **Write step-by-step instructions.**

- Clearly label any step that could endanger the person (WARNING!) or the machine (CAUTION!).

☐ **Field-test the manual.**

- Select the features of the manual you want to field-test.
- Select a method of testing those features. Be sure to create a clear method for recording answers. Determine what you think are acceptable results for each feature (e.g., How long should it take to perform the process?)
- Select subjects to use the manual.

Examples

The excerpts shown in Example 15.1 are several sections of an operator's manual for a baby stroller. The manual downloaded from the Web as a PDF. Presented here are the parts list, several pages of instructions for use including warnings about child safety, and the maintenance/troubleshooting section. These pages represent sections you will find in almost all manuals written for consumers. Example 15.2 is a usability report, written to explain the results of a usability test on a website.

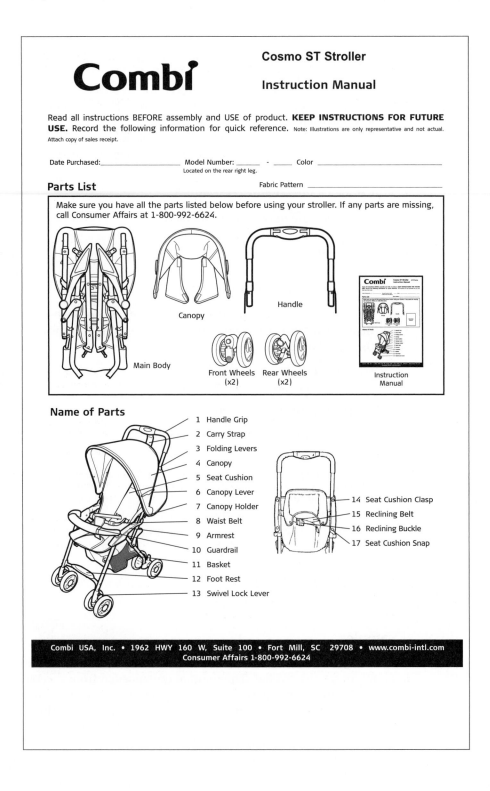

Cosmo ST Stroller

Instruction Manual

Combi

Read all instructions BEFORE assembly and USE of product. **KEEP INSTRUCTIONS FOR FUTURE USE.** Record the following information for quick reference. Note: Illustrations are only representative and not actual. Attach copy of sales receipt.

Date Purchased:_____ Model Number: _____ - _____ Color _____
Located on the rear right leg.

Parts List

Fabric Pattern _____

Make sure you have all the parts listed below before using your stroller. If any parts are missing, call Consumer Affairs at 1-800-992-6624.

Canopy

Handle

Main Body

Front Wheels
(x2)

Rear Wheels
(x2)

Instruction
Manual

Name of Parts

1 Handle Grip
2 Carry Strap
3 Folding Levers
4 Canopy
5 Seat Cushion
6 Canopy Lever
7 Canopy Holder
8 Waist Belt
9 Armrest
10 Guardrail
11 Basket
12 Foot Rest
13 Swivel Lock Lever

14 Seat Cushion Clasp
15 Reclining Belt
16 Reclining Buckle
17 Seat Cushion Snap

Combi USA, Inc. • 1962 HWY 160 W, Suite 100 • Fort Mill, SC 29708 • www.combi-intl.com
Consumer Affairs 1-800-992-6624

⚠ **WARNINGS:** FAILURE TO FOLLOW THESE INSTRUCTIONS AND WARNINGS COULD RESULT IN SERIOUS INJURY OR DEATH.

ALL OPERATING AND ASSEMBLY PROCEDURES SHOULD BE PERFORMED BY AN ADULT.

In the event of damage or problems, discontinue use and contact Consumer Affairs for instructions: 1-800-992-6624.

⚠ **AVOID SERIOUS INJURY FROM FALLING OUT OR SLIDING OUT. ALWAYS USE RESTRAINT SYSTEM.**

⚠ **DO NOT LIFT AND CARRY** the stroller while occupied. The stroller may fold.

⚠ Do not allow your child to stand in or on the stroller.

⚠ Never carry more than one child in stroller. Do not allow a child to ride or sit on areas of the stroller other than seat.

⚠ **NEVER LEAVE CHILD UNATTENDED.**

⚠ Do not use the stroller on steep slopes, stairs, escalators, beaches, rough roads, mud, etc.

⚠ Do not hang items on the handle or place items on canopy. This could cause the stroller to become unbalanced or tip over.

⚠ Always engage both brakes when parking the stroller.

⚠ Do not allow children to lean over sides of stroller. It may become unbalanced and tip over.

⚠ Excessive weight in basket (or items placed in basket which overflow the top or side of basket) may cause a hazardous condition. Do not load more than 10 lbs. (4.5kg) in the basket.

⚠ Completely remove all plastic bags and packing materials before using this product to prevent a possible choking hazard.

⚠ When Combi Centre infant seat is attached, **ALWAYS** ensure it is secure by pulling up on the handle.

⚠ **ALWAYS** secure your child with the infant carriers 5-point harness before using it with the stroller.

⚠ The guardrail **MUST** be attached to the stroller before attaching the infant carrier.

⚠ To ensure your child's safety, please refer to the Combi Centre instruction manual before using it with this stroller.

⚠ If unable to securely attach the Combi Centre to your stroller, discontinue use and contact Consumer Affairs.

⚠ **DO NOT** carry stroller by using the Combi Centre infant car seat handle.

⚠ **TIPPING HAZARD: DO NOT** place weight on rear of infant seat while attached to the stroller. This could cause the stroller to become unbalanced and tip over.

⚠ This stroller is designed to be pushed at a normal walking speed. This is not a jogging stroller.

Use only the Combi Centre infant carrier with this stroller.

Child must be 6 months old to use this stroller. This stroller does not have a full recline feature. Do not use this stroller with an infant.

Maximum weight of child must not exceed 45 lbs. (20 kg).

Total Maximum Weight of Canopy pocket MUST not exceed 1.0 lbs.(0.5 kg).

(Continued)

Assembling the Stroller

1. Remove stroller and rear wheel assemblies from the plastic bag.

2. Remove canopy by detaching rubber bands.

3. Remove front wheel assemblies, guardrail and stroller handle from inside the stroller.

4. Lift locking lever from stroller frame and open stroller.

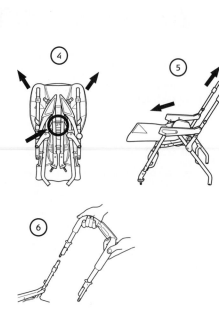

5. Grasp right and left steel tubes and snap the stroller away from you to open.

6. Insert stroller handle into frame by pressing chrome buttons into the back of the stroller frame. Push handle down until the chrome buttons lock into both sides of the handle.

7. While lifting up on gray tube (figure A), attach long gray tabs from handle to the round gray tabs on each side of stroller.

8. Secure the handle by lifting the folding levers on each side of stroller until the gray tabs lock into position.

(figure A)

Attaching & Removing the Front Swivel Wheels

1. Turn stroller over and remove packaging material from stroller frame.

2. Insert the front wheels onto the stroller leg. Make sure the swivel wheel lever is behind the stroller leg and released. You will hear a click when the wheel is properly attached.

Removing the Swivel Wheels

Press in the square button located at the bottom of the wheel holder while pulling the assembly off of the swivel shaft.

Release Button

Attaching the Rear Wheels

1. Stand behind the stroller.

2. Locate the small open-ing on each rear wheel assembly.

3. Locate the small chrome pin on each rear stroller leg.

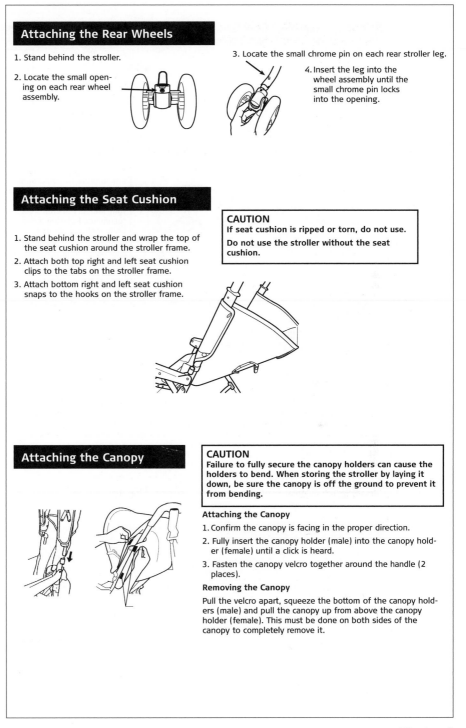

4. Insert the leg into the wheel assembly until the small chrome pin locks into the opening.

Attaching the Seat Cushion

1. Stand behind the stroller and wrap the top of the seat cushion around the stroller frame.

2. Attach both top right and left seat cushion clips to the tabs on the stroller frame.

3. Attach bottom right and left seat cushion snaps to the hooks on the stroller frame.

CAUTION
If seat cushion is ripped or torn, do not use.

Do not use the stroller without the seat cushion.

Attaching the Canopy

CAUTION
Failure to fully secure the canopy holders can cause the holders to bend. When storing the stroller by laying it down, be sure the canopy is off the ground to prevent it from bending.

Attaching the Canopy

1. Confirm the canopy is facing in the proper direction.

2. Fully insert the canopy holder (male) into the canopy hold-er (female) until a click is heard.

3. Fasten the canopy velcro together around the handle (2 places).

Removing the Canopy

Pull the velcro apart, squeeze the bottom of the canopy hold-ers (male) and pull the canopy up from above the canopy holder (female). This must be done on both sides of the canopy to completely remove it.

(Continued)

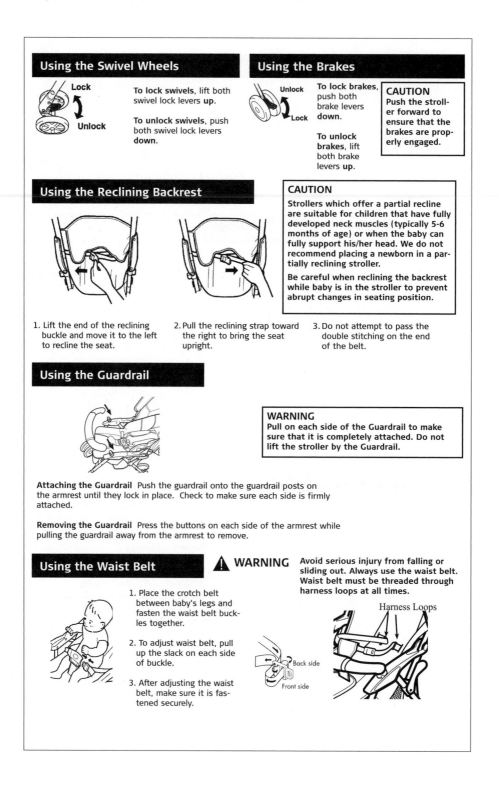

Using the Swivel Wheels

Lock

Unlock

To lock swivels, lift both swivel lock levers **up**.

To unlock swivels, push both swivel lock levers **down**.

Using the Brakes

Unlock

Lock

To lock brakes, push both brake levers **down**.

To unlock brakes, lift both brake levers **up**.

CAUTION
Push the stroller forward to ensure that the brakes are properly engaged.

Using the Reclining Backrest

CAUTION

Strollers which offer a partial recline are suitable for children that have fully developed neck muscles (typically 5-6 months of age) or when the baby can fully support his/her head. We do not recommend placing a newborn in a partially reclining stroller.

Be careful when reclining the backrest while baby is in the stroller to prevent abrupt changes in seating position.

1. Lift the end of the reclining buckle and move it to the left to recline the seat.

2. Pull the reclining strap toward the right to bring the seat upright.

3. Do not attempt to pass the double stitching on the end of the belt.

Using the Guardrail

WARNING
Pull on each side of the Guardrail to make sure that it is completely attached. Do not lift the stroller by the Guardrail.

Attaching the Guardrail Push the guardrail onto the guardrail posts on the armrest until they lock in place. Check to make sure each side is firmly attached.

Removing the Guardrail Press the buttons on each side of the armrest while pulling the guardrail away from the armrest to remove.

Using the Waist Belt

⚠ **WARNING**

Avoid serious injury from falling or sliding out. Always use the waist belt. Waist belt must be threaded through harness loops at all times.

1. Place the crotch belt between baby's legs and fasten the waist belt buckles together.

2. To adjust waist belt, pull up the slack on each side of buckle.

3. After adjusting the waist belt, make sure it is fastened securely.

Back side

Front side

Harness Loops

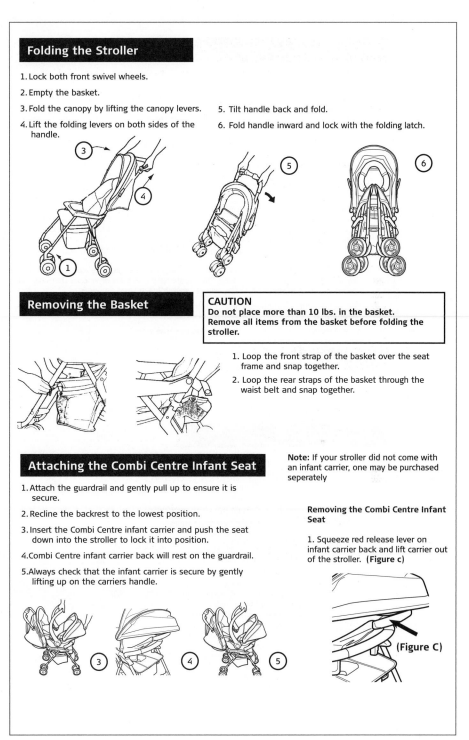

Folding the Stroller

1. Lock both front swivel wheels.

2. Empty the basket.

3. Fold the canopy by lifting the canopy levers.

4. Lift the folding levers on both sides of the handle.

5. Tilt handle back and fold.

6. Fold handle inward and lock with the folding latch.

Removing the Basket

CAUTION
Do not place more than 10 lbs. in the basket.
Remove all items from the basket before folding the stroller.

1. Loop the front strap of the basket over the seat frame and snap together.

2. Loop the rear straps of the basket through the waist belt and snap together.

Attaching the Combi Centre Infant Seat

1. Attach the guardrail and gently pull up to ensure it is secure.

2. Recline the backrest to the lowest position.

3. Insert the Combi Centre infant carrier and push the seat down into the stroller to lock it into position.

4. Combi Centre infant carrier back will rest on the guardrail.

5. Always check that the infant carrier is secure by gently lifting up on the carriers handle.

Note: If your stroller did not come with an infant carrier, one may be purchased seperately

Removing the Combi Centre Infant Seat

1. Squeeze red release lever on infant carrier back and lift carrier out of the stroller. **(Figure c)**

(Figure C)

(Continued)

Routine Maintenance

STORE IN A COOL AND DRY AREA, especially in the summer. Do not store the stroller in the trunk of a vehicle.

FABRIC CARE
Hand wash with brush. Wipe out excess water with a dry cloth and dry in a shaded area. For stubborn dirt, dilute a mild detergent and apply directly to the dirty area. Wipe out the detergent.

MAIN BODY
Briskly wipe off any mud and dirt. To clean the body, wipe with a damp cloth then wipe out any trace of moisture on the body with a dry cloth. The main body might become moldy and musty if not wiped dry.

Excessive exposure to the sun could cause premature color fading of stroller plastic and fabric.

WHEEL ASSEMBLIES
After prolonged normal use, the wheels of your stroller may not swivel or turn as easily as when the stroller was new. This is usually due to a buildup of dirt in the swivel mechanism. Before adding lubrication, remove the swivel wheels by pressing in the release button and wiping off mud and dirt. **Do not use soap and water to wash wheels.** Periodically check that all screws, bolts and nuts are tight.

Avoid lubricating the swivel wheels whenever possible. Excessive lubrication or adding lubricant to swivel wheels when new may cause them to develop a shimmy or wobble that may quicken their deterioration. Add silicone* to **the wheel assemblies** in small amounts.

Do not use lubricants because oil attracts dust and would hinder the movement rather than improve it.

*Silicone spray is available in varying size cans at most hardware stores.

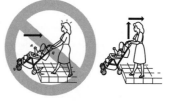

Front Swivel   Rear Wheel
Wheel

Trouble Shooting

STROLLER WILL NOT FOLD
- Check to see that the waist belt is not caught in the side.
- Make sure front swivels are locked.
- Remove all items from the basket.

DIFFICULT TO STEER/STROLLER PULLS TO ONE SIDE
- Remove any items hanging from the handle.
- Make sure weight in basket is distributed evenly.
- Check that the brakes are not in the locked position.
- Clean and add silicone to wheels, if necessary.

DIFFICULT TO GET UP CURBS
- Pulling stroller backward makes going up a curb difficult, and could pose a safety hazard. Pull the handle up to lift the rear wheels onto the curb, then pull the stroller backward until the front wheels clear.

LIMITED WARRANTY

1. Combi Juvenile Products, when purchased from an authorized retailer; are warranted to the original purchaser to be free of defect in materials and workmanship for a period of one (1) year from date of purchase. Warranty is valid for products purchased and used in the U.S. only, and used in compliance with the warrantor's use and maintenance instructions. Save your carton and receipt in case of return for warranty service is necessary.

2. During the warranty period your Combi product will either be repaired or replaced, at the warrantor's discretion, without charge if it is purchased from an authorized Combi dealer and is found to be defective in workmanship or materials. For service call 1-803-548-6633. The warrantor shall not be responsible for any costs incurred by the consumer in returning a defective product for repair.

   This limited warranty shall not apply to normal wear to the product, or if the product has been:

   a. Damaged due to accidents, misuse, abuse or neglect as determined by Combi.
   b. Tampered with or repaired by other than an authorized Combi repair center or with parts other than Combi approved parts.
   c. Leased or rented for commercial use.

THE WARRANTY PROVIDED HEREIN IS EXCLUSIVE AND IS IN LIEU OF ALL OTHER EXPRESS WARRANTIES. ANY IMPLIED WARRANTIES, INCLUDING ANY IMPLIED WARRANTY OF MERCHANTABILITY OR FITNESS FOR A PARTICULAR PURPOSE ARE EXPRESSLY EXCLUDED. WARRANTOR'S OBLIGATIONS AND THE CONSUMER'S OR OTHER BUYER'S REMEDIES UNDER THE LIMITED WARRANTY ABOVE ARE EXCLUSIVE AND ARE STRICTLY LIMITED TO REPAIR OR REPLACEMENT AS STATED. ANY INCIDENTAL OR CONSEQUENTIAL DAMAGES ON THE PART OF THE WARRANTOR AND ANY SUPPLIER OR DEALER ARE HEREBY ALSO EXPRESSLY EXCLUDED.

Some states limit or do not allow exclusions or limitations on how long an implied warranty lasts so the above limitation may not apply to you. Some states do not allow the exclusion or limitation of incidental or consequential damages, so the above limitation or exclusion may not apply to you. This warranty gives you specific legal rights, and you may also have other rights which vary from state to state.

(Continued)

Example 15.1

Excerpts From an Operator's Manual
Source: Combi USA, Inc., 1962 HWY 160 W, Suite 100, Fort Mill, SC 29708, www.combi-intl.com. Reprinted with permission.

Cosmo ST Stroller
Instruction Manual

If you would like to place an order for the replacement parts, please fill out the form below and return to the following address:

Combi
Attn: Consumer Affairs Department
1962 Highway 160 W, Suite 100
Fort Mill, SC 29708-8027

or   **Fax this completed form to**
803-548-3663

IMPORTANT: **We must have this information to process your order:**

Model Number: _____ - ____   Color & Pattern of Fabric:_____
Located on the rear right leg.

Parts List for the Cosmo ST (2400 Series) only.

	Price	Quantity	Total
1. Canopy	20.00		
2. Basket	10.00		
3. Standard Guardrail	15.00		
4. Guardrail Cover	5.00		
5. Rear Wheel Assemblies (left and right sides)	15.00/set		
6. Front Wheel Assemblies (left and right sides)	15.00/set		
7. Handle	25.00		
8. Carrying Strap	12.00		

If you need to order other parts for your stroller, please call us toll-free at 1-800-992-6624.

Shipping and Handling Fees: Add $8.00 for orders $34.99 and under. Add $13.00 for orders $35.00 and over. Note: Additional shipping charges will apply outside of the Continental U.S.

Combi USA reserves the right to change pricing at any time.

Amount of Order	
ADD Shipping & Handling	
Sales Tax (6%) (SC residents only)	
Total	

Ship To:

Customer Name

Street Address

Street Address

City, State, Zip

Telephone Number

Charge to my credit card:

☐ VISA   ☐ MASTERCARD
☐ AMERICAN EXPRESS   ☐ DISCOVER

Card Number

Expiration Date _____ - _____ - _____

Cardholder's Signature   071105

Combi USA, Inc. • 1962 HWY 160 W, Suite 100 • Fort Mill, SC 29708 • www.combi-intl.com
Consumer Affairs 1-800-992-6624

Example 14.2

Report on
NALC Branch
728 Website
Usability

1. EXECUTIVE SUMMARY

This report contains the results of a usability test for NALC Branch 728 website created by Hilary Peterson. I performed the test with three participants, all of whom indicated that they have at least a moderate familiarity with the Internet and with website use in general. All the test subjects are members of the target audience for the site, but had not seen the website before the test. I gave the test subjects a time limit of 15 minutes to complete three specific tasks; during that time, I observed the subjects and made careful notes of their actions and comments for later analysis. After the test, each subject completed a survey to express his overall response to the experience.

Analysis of the data I collected shows that the website is very effective overall. It is easy to navigate, and the page layout allows for easy scanning of the pages for important information. There is a need, however, to rethink the name of one of the lines, Get Smart!, and how it relates to the Letter Carrier Perfect Page to which it connects.

I make recommendations for the problem area.

2. INTRODUCTION

The goal of this report is to document the findings of a usability test of NALC Branch 728's website, created by Hilary Peterson. As the author of the report, I also conducted the usability test. The website is available at the following URL: http://fp1.centurytel.net/nalcbranch728. To serve and inform the members of Branch 728 and of other NALC branches in the area, there is a need for a useful, well-written website that includes an online version of the Branch's newsletter as well as a wealth of information for its users.

This report will interest the Web master, since it will indicate any improvements that should be made to the website. It will also interest my client, the editor of *The Sawdust Cities Satchel* and vice president of Union Branch 728. He will use this report to determine whether or not to request funding for the website.

These pages include a brief description of my methodology along with summaries of my test results, observations, and results of the post-test survey. I express my conclusions based on these data as well as my recommendations for improvement where I found problem areas during the test.

3. METHODOLOGY

3.1 Participant Selection

The primary audience for NALC Branch 728's website is members of Branch 728 who have at least a moderate familiarity with the Internet and with website use in general. The website was designed with this type of user in mind, so the pages are meant to be clean, simple, and easy to navigate.

(Continued)

My client chose three members of Branch 728 to participate in the usability test. The subjects possess varying levels of computer skills, but all of the subjects indicated that they were at least fairly familiar with the Internet. None of the subjects had ever seen or used the Branch 728 website before.

3.2 Task Selection

I gave each of the test subjects three test scenarios and three tasks to complete; these appear in Appendix A.

With question 1 of Task One, I aimed to test the readability of the homepage, where the answer to question 1 is found.

With question 2 of Task One, my goal was to test the usefulness of the bookmarking feature on the Constitution and By-Laws page.

Task Two tested the "scannability" of a very long page of text, the Get Smart! page. I also wanted to find out whether or not I could reasonably assume that most of my audience members were aware that "Get Smart!" is the new name for the Letter Carrier Perfect handbook.

Task Three tests the function and visibility of the Outside Links page.

3.3 Test Procedure

I scheduled 15 minutes for each participant to complete the assigned tasks. The subjects performed the test individually. Before the test, I gave an introduction to the test, emphasized the subject's role, and told each participant what to expect during the test. This introduction appears in Appendix A.

Next, I gave each subject scenarios and instructions for completing the tasks, encouraging them to talk aloud while they were performing the test. Hearing the participants say why they chose to click on a particular link helped to show the effectiveness of the links.

As the test's administrator, I did not interact with the test subjects or give hints when they were lost. Instead, I took careful notes on the subjects' behavior, demeanor, comments, and activities on screen.

3.4 Post-Test Survey

To obtain each participant's overall feelings about his experience, I administered a short post-test survey, shown in Appendix B.

4. RESULTS

4.1 Observations

All three of the test subjects completed the assigned tasks within the time limit of 15 minutes. The subjects completed the tasks in four minutes, six minutes, and nine minutes. The fastest test participant completed the tasks in the fewest possible number of steps. The slowest participant completed the tasks within the 15 minute time limit, but he did not use the most efficient path through the site.

All three of the subjects stayed on the homepage at the beginning of the test. The answer to the first question in Task One was there, and two of the three subjects found it within one minute. One subject did not find the answer right away, and he browsed unsuccessfully through a different page before returning to the homepage and finding the answer.

Question 2 of Task One was answered successfully by all three subjects within an appropriate amount of time. The subjects started by immediately clicking on the link that would take them to the correct page. However, two of the three subjects skipped over the bookmarking feature at the top of the Constitution and By-Laws page, which would have been the quickest way to complete the task. Of the two that did not immediately use the bookmarks, one scanned the long page of text and found the answer to the question. The other participant started to scan the page and then went back to the top and successfully used the bookmark to more quickly find the desired information.

4.2 Survey Results

The results of the post-test survey indicate that, overall, the website was easy to navigate, well designed, and easy to scan for important information. All of the test subjects indicated that they felt comfortable when using the site. All three test participants commented on the easy-to-use navigation bars. Two of the participants mentioned that "Get Smart!" and "Letter Carrier Perfect" are not yet synonymous for all of the people who will use this site, and that the links to the Get Smart! page should also indicate a connection to Letter Carrier Perfect. The post-test survey appears in Appendix B.

5. CONCLUSIONS

5.1 Positive Aspects

Two of the three test subjects found the answer to question 1 of Task One within a minute. They did this by scanning the homepage for the necessary information. I suspect that the subject who did not find the answer immediately felt pressured during the test and would have been more successful in a different situation. Question 2 of Task One went well for all of the test subjects. On a scale of 1–5 (1 = difficult; 5 = easy), two participants rated Task One a 5.

Task Three also went well for all of the participants. On a scale of 1–5 (1 = difficult; 5 = easy), completing Task Three was a 5 for all participants.

On the post-test survey, organization of the site, moving from page to page, ease of reading and scanning text, and page layout were rated at very easy or very effective by all subjects. All three participants indicated that they felt comfortable using the site and that it is a site that they would use often.

5.2 Problem Areas

The main problem the participants had during the test was completing Task Two. The task instructions ask the participant to find the online version of "Letter Carrier

(Continued)

Perfect," the name of which has recently been changed to "Get Smart!" I discovered during the test that this name change is not yet known to all of the website's audience members, and two of the subjects were not at all confident that clicking on the Get Smart! link would bring them to an online version of Letter Carrier Perfect.

6. RECOMMENDATIONS

Determine whether to continue using a link named "Get Smart!" to connect to a page that is an on-line version of "Letter Carrier Perfect."

Keeping the "Get Smart!" link is probably a good idea, because it will help to reinforce the name change of this handbook from "Letter Carrier Perfect" to "Get Smart!"

7. APPENDIX A

7.1 NALC Branch 728 Website Usability Test Introduction and Instructions

Thank you for participating in this project. Your input will help us to create a more useful website for Branch 728's members. You will be able to use this site in the near future to read *Satchel* articles and to take note of upcoming Branch meetings and activities, among other things.

The purpose of this test is to evaluate the usefulness and effectiveness of Branch 728's website.

To accomplish this, we will ask you to perform specific tasks. During the test, we will not make suggestions or be able to help in any way. Just use the site as you normally would to complete the tasks, and we will take notes for later analysis.

Please talk through your motions during the test. For example, let us know why you are choosing to click on a particular link. Then we'll have you fill out a short survey about your experience.

Please remember that the purpose of this test is to evaluate the usability of our website. We are not testing your personal abilities.

Task Scenarios and Instructions

Task One   As a member of NALC Branch 728, you are preparing for the next union meeting. You have decided to use the Branch's new website to gather some information. These are the questions you need to answer:

1. When is the next union meeting?
2. According to the Branch By-Laws, how many members constitute a quorum?

Task Two   As a Letter Carrier, you would like to find out more about your rights and responsibilities concerning safety, service, and mail security. The online version of Letter Carrier Perfect is arranged into three distinct sections. What are the three sections?

Task Three   You have taken some time to view the new Branch 728 website, but you need some more information from the Wisconsin State AFL-CIO. Use the Branch 728 website to connect to the Wisconsin State AFL-CIO website.

8. APPENDIX B

8.1 Post-Test Survey (highlighted numbers reflect the most common participant response)

1. What would have helped you complete these tasks more successfully?

2. What parts of the website did you find most helpful?

Please circle the number that best corresponds to your feelings about each question.

1. Completing Task One (next union meeting/quorum questions) was
    difficult   1   2   3   4   5   **easy**
2. Completing Task Two (Get Smart!/Letter Carrier Perfect question) was
    difficult   1   2   3   4   5   **easy**
3. Completing Task Three (connecting to WI State AFL-CIO site) was
    difficult   1   2   3   4   5   **easy**
4. Organization of the site was
    not effective 1   2   3   4   5   **very effective**
5. Moving from page to page was
    confusing   1   2   3   4   5   **easy**
6. The text on the website was generally
    difficult to read 1   2   3   4   5   **easy to read**
7. Scanning the pages for important information was
    confusing   1   2   3   4   5   **easy**
8. When I was looking for the pages I needed, I
    felt lost   1   2   3   4   5   **felt I knew where I was**
9. While working on the tasks, I
    felt confused   1   2   3   4   5   **felt comfortable**
10. I thought the layout of most of the pages was
    difficult to understand 1   2   3   4   5   **easy to understand**
11. NALC Branch 728's webpage seems like a site that I would
    never use   1   2   3   4   5   **use often**

Exercises

▶ You Analyze

1. Collect one or two professional (lawn mower, play station, power tool, automobile, appliance, digital camera, or smart phone) manuals and bring to class. Analyze them for page design, visual logic, text–visual interaction, sequence of parts, and assumptions made about the audience. Discuss these topics in groups of two to four, and then report to the class the strategies that you find most helpful and are most likely to use in Writing Assignment 1.

▶ Group

2. In groups of two to four, analyze the page layout of one of the manuals that appears in this chapter. Write a brief description of and reaction to this layout and share your reactions with the group. Your instructor will ask some groups to report their results.

▶ You Create

3. Using any machine or software program you know well, write a parts description.

4. Using any machine or software program you know well, create a flow chart for the sequences you want a reader to learn. Convert that flow chart into step-by-step instructions.

5. Review the types of decisions included in creating a template (pp. 454–455). Then create your own design for what you wrote in Exercise 3 or 4.

6. For the manual you are creating for Writing Assignment 1 or for the section you wrote for Exercise 4, create a storyboard.

7. Write the introduction to the parts description and sequences you created in Exercises 3 and 4.

8. For the manual you are creating for Writing Assignment 1, complete the following exercises. Your instructor will schedule these steps at the appropriate time in your project.

   a. Consider how consistently it handles all details of format.
   b. Consider how precisely it explains how to perform an action. Read closely to see whether everything you need to know is really present.
   c. Conduct a field test by asking a person who knows almost nothing about the product to follow your manual. Accompany the tester, but do not answer questions unless the action is dangerous or the tester is hopelessly lost (say, in a software program). Note all the problem areas, and then make those changes. Discuss changes that would help the user.

Writing Assignments

1. Write an operator's manual. Choose any product that you know well or one you would like to learn about. The possibilities are numerous—a bicycle, a sewing machine, part of a software program such as Microsoft Word® or Dreamweaver®, a computer system, any laboratory device, a welding machine. If you need to use high-quality photographs or drawings, you may need help from another student who has the necessary skills. Your manual must include at least an introduction, a table of contents, a description of the parts, and the instructions for procedures. You might also include a trouble-shooting section. Give warnings when appropriate. Complete this chapter's worksheet or the appropriate exercises.

2. Write a learning report for the writing assignment you just completed. See Chapter 5, Writing Assignment 8, page 137, for details of the assignment.

Web Exercise

Create a mini-manual to publish on a website. Use a simple machine, say, a flashlight. Include one section describing the parts and one section presenting the appropriate sequences for operating. Include several visual aids.

Works Cited

Bethke, F. J., W. M. Dean, P. H. Kaiser, E. Ort, and F. H. Pessin. "Improving the Usability of Programming Publications." *IBM Systems Journal* 20.3 (1981): 306–320. Print.

Brooks, Ted. "Career Development: Filling the Usability Gap." *Technical Communication* 38.2 (April 1991): 180–184. Print.

Cohen, Gerald, and Donald H. Cunningham. *Creating Technical Manuals: A Step-by-Step Approach to Writing User-Friendly Manuals.* New York: McGraw-Hill, 1984. Print.

Cooper, Marilyn. "The Postmodern Space of Operator's Manuals." *Technical Communication Quarterly* 5.4 (1996): 385–410. Print.

Craig, John S. "Approaches to Usability Testing and Design Strategies: An Annotated Bibliography." *Technical Communication* 38.2 (April 1991): 190–194. Print.

Queipo, Larry. "Taking the Mysticism Out of Usability Test Objectives." *Technical Communication* 38.2 (April 1991): 185–189, 190–194. Print.

Redish, Virginia. "Writing for People Who Are 'Reading to Learn to Do.'" *Creating Usable Manuals and Forms: A Document Design Symposium.* Technical Report 42. Pittsburgh, PA: Carnegie-Mellon Communications Design Center, 1988. Print.

Riney, Larry A. *Technical Writing for Industry: An Operations Manual for the Technical Writer.* Englewood Cliffs, NJ: Prentice-Hall, 1989. Print.

Rubens, Phillip M. "A Reader's View of Text and Graphics: Implications for Transactional Text." *Journal of Technical Writing and Communication* 16.1/2 (1986): 73–86. Print.

Rubin, Jeff. "Conceptual Design: Cornerstone of Usability." *Technical Communication* 43.2 (May 1996): 130–138.

Sullivan, Patricia. "Beyond a Narrow Conception of Usability Testing." *IEEE Transactions of Professional Communication* 32.4 (December 1989): 256–264. Print.

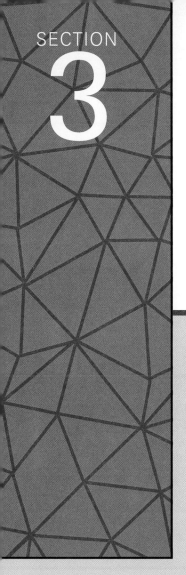

Professional Communication

Presentations

CHAPTER CONTENTS

CHAPTER 16 **IN A NUTSHELL**

Oral reports range from brief answers to questions at meetings, to hour-long speeches to large audiences.

Follow these guidelines:

▶ Analyze your audience.

▶ In PowerPoint presentations rely on images rather than screen after screen of lists.

▶ Speak in a normal voice. Help yourself speak normally by not memorizing—practice enough so you can speak from notes.

▶ Practice with any technology (laptops, blue tooth devices, projectors) before you give the speech.

▶ Be presentable. Dress appropriately; if you don't know what a professional should wear in this situation, ask someone who does. Avoid irritating mannerisms (smacking lips, shaking keys in pockets, saying "um" repeatedly).

Throughout your career, you will give oral reports to explain the results of investigations, propose solutions to problems, report on the progress of projects, make changes to policy, create business plans, justify requests for such items as more employees and equipment, or persuade clients to purchase your services and merchandise. Oral reports are almost always accompanied by a slide presentation created in PowerPoint or Keynote or some other presentation software.

As a speaker you need to master slide presentation. You must "train" it to follow your commands, and you must learn to interact with it so that you and the slides enhance the experience for the audience.

This chapter explains how to plan and deliver an oral presentation, assisted by PowerPoint.

Planning the Presentation

Planning includes decisions about audience, situation, organizational pattern, and presentation.

Plan for Your Audience

Because of the popularity of the slide presentation, many listeners have been subjected to the same kind of presentation many times. Indeed, the use of PowerPoint is so common that most oral reports are simply called "Power-Points" or "PowerPoint reports." Edward Tufte calculates that trillions of slides are produced yearly (3). Paradi ("Results") found that 38.7 percent of his respondents saw at least one presentation a day. For almost all of these presentations, the standard method as Tufte says ". . . is to talk about a list of points organized onto slides projected up on the wall" (3). The upshot of these numbers and this method is that many audiences anticipate that any presentation is likely to be another dreary time-waster. One phrase that encapsulates that dreariness is "Death by PowerPoint" (Corbin).

Ask Audience Analysis Questions

In presentations the audience analysis questions are the same as for reports or other documents. Ask these questions (based on Laskowski):

- Who are they?
- How many will be present?
- What is their knowledge level?
- What task will your presentation help them complete?
- What do they need?
- Why are they there?

Focus on Annoying PowerPoint Issues

Dave Paradi ("Results"), a frequent author about PowerPoint topics, creates annual surveys that investigate what annoys listeners about PowerPoint presentations. The most frequent complaint is "The speaker read the slides to us." Others in the top five annoyances are using full sentences rather than brief phrases, projecting text that is too small to read, choosing colors that make the slide hard to read, and using overly complex diagrams or charts. These complaints should give you a good framework for making decisions on how to create and present an effective presentation.

The key to preventing audience apathy or even a downright hostility is to focus on them, not on the technology of your presentation. The central goal for your presentation is, simply, that you be relevant. One speech expert says, "People will pay close attention to something they perceive as having relevance to their own lives and concerns" (Bacall). In order to remain aligned to your audience, ask the same audience analysis questions that you ask in report writing.

The answers to these questions will help you develop a presentation that keeps your audience attentive to your points, so that they walk away feeling that their time with you has been well spent and productive. For instance suppose you had to give a presentation on your company's new data analysis software to these two audiences: three experienced managers seated around a conference table, deciding whether to place an order with you; and 50 sales reps seated in a lecture hall, eager to familiarize themselves with the product prior to making sales calls. Obviously, these situations would require two very different presentations.

Plan for the Situation

Presentations are made in many venues, from small rooms in which people sit around a conference table to large auditoriums packed with conference attendees. Spend time investigating the physical layout of the room. Follow these guidelines:

▶ Determine the size of the room and where you will stand in relation to the audience, the screen, and the computer controls.

▶ Determine the location of the electrical outlets and the electrical cords on the floor.

▶ Learn how sound carries in the room. Will you have to use a microphone? If so, do you know how to adjust it so that your voice carries well without ringing or buzzing?

▶ Determine whether you will have to bring a disk to use in a computer already present in the room, or whether you need to bring your laptop.

▶ If you have to bring your own laptop, determine how to hook it up to the overhead projector system located in the room.

Plan Your Organizational Pattern

The organizational pattern that you choose depends on the needs of your audience and the actual content of the material to be presented. In effect you need to write a script for yourself and your audience. Common "scripts" or organizational patterns are

- ▶ story
- ▶ goal-methods-results-discussion

The story approach (Shaw, Brown, and Bromiley; Markowitz; Reynolds, "Organization"; "Presentation") has three stages: set the stage, introduce the dramatic conflict, resolve the conflict. To *set the stage*, the speaker defines the current situation by analyzing relevant factors—for example, market forces, company objectives, and/or technological changes. To *introduce dramatic* conflict, the speaker explains the challenges that face the company in the current situation. What are the obstacles (the bad guys) in this situation? Poorly functioning technology? New competitor? Market share losses? To *resolve the conflict*, the speaker must show how the audience can overcome the obstacles to win—by reversing the issues with the technology, competitor, market share. Of course the topics could be anything, not just business-related.

The story structure is effective because it "defines relationships, a sequence of events, cause and effect, and a priority among items—and those items are likely to be remembered as a complex whole" (Tufte 4). Creating the story will cause you to think clearly about the complexity of the ideas. The upshot will be listeners who grasp the significance of the main point because they have been engaged by the story that expounds it.

The goal-method-results-conclusion works in nearly any situation. In effect you are presenting a verbal IMRD (see Chapter 10). This type of presentation is most effective when you decide which of the sections the listeners need most. If all they care about are the results, you can focus on results and minimize the other sections.

Plan Your Presentation

To plan the presentation, determine your relationship to the slides, create a storyboard, and finally create the series of slides.

Determine Your Relationship to the Slides

A presentation consists of you and the information on the slides. You need to determine your relationship. Who is in charge? You or the slides? The slide is not the main source of information in the speech—you are. You place the content of the slide into a relevant context for the audience.

Consider these two aspects of your relationship:

> ▶ Is the information on the slide in the foreground or background?
> ▶ What items should be on the slide?

Foreground and Background. It is helpful to see the slide in either "the foreground" or "the background." If it is in the foreground, it provides information, either visual aid or textual, that you explain. Project a visual aid in order to explain it. For instance, if you have details of sales made during a quarter, you might show a bar graph with its columns of numbers, and then discuss the reasons for the differences in the heights of the bars. Project text in order to emphasize it. If, for instance, you want the audience to remember a certain point ("42 percent of defective parts came from machine 3"), you can project just that point on the screen and then discuss it.

With foreground material, then, you speak to the slide. The idea is that you will explain in some detail what the audience sees on screen. The combination of the slide content and your explanation is the point for the audience to get. For instance, Figure 16.1 is a pie chart that breaks down costs in a program. The speaker projected the slide, and then spoke for several minutes explaining those costs. The speaker could address each slice of the pie without having to move on to another slide.

When a slide is in the background, you do not speak to it. Instead it is present in order to summarize key points, either what is about to be covered or what has just been covered (Miller, "Presentation"). At its best, this method helps the audience stay on track and follow and remember the points that have been made. If you have four subpoints to make about a topic, projecting them on a slide, then discussing each in turn will help the listeners follow along.

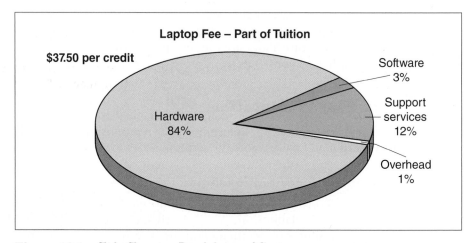

Figure 16.1 Slide Showing Breakdown of Costs

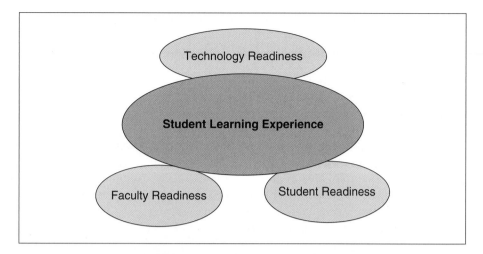

Figure 16.2   Slide Showing an Outline

Figure 16.2 shows a slide that functions as an outline. The slide does not use the conventional list but shows the four topics in a way that previews their relationship. The speaker is easily able to create a sense of the upcoming sections of the presentation while at the same time indicating the relationship of the points she will make.

What Should Be on the Slide?

Later in this chapter a section (pp. 498–503) details how to make a slide look good, or easily accessible to the listeners. In this section the focus is the contents of the slide. This section is really the most important part of this chapter. Sections cover content presented by images and content presented by text.

Content Presented Predominantly by Image.   Consider using the "Predominantly Image" approach, displaying exclusively or primarily images ("Creating"; "Effective Presentations"; Alley). In other words put images into the foreground, and speak to a series of foregrounded images. This strategy takes more time in the creation stage, but pays dividends in audience retention and interest. It is a key way around "Death by PowerPoint" because it makes you interact with the audience. Point out to the audience what they should see in the image and why it is relevant to them. For instance a photograph of a dilapidated desk, could helpfully convince an audience that this desk, and others like it, need to be replaced. An oral explanation of the relationship among the percentages that affect a pay increase is hard to follow, but a graph makes it clear.

To use the Predominantly Image approach, you have to answer the question—What strategy will best engage the audience with the image? Here are seven suggestions:

▶ Assertion-Evidence Structure. (See Figure 16.3.) In this strategy the title is a one-sentence assertion that states what you want the audience to know as a result of the slide. The body of the slide consists of visual evidence for that assumption ("Visual"; Alley and Neeley; "Rethinking"). The visual evidence can be a photograph, a table, a graph, a screen capture, or an equation.

▶ Context. (See Figures 16.4 and 16.5 [p. 491].) Give the viewer a big picture so that the small details make sense. For instance, displaying a map of the way streets intersect could help the city council better understand the proposed site of a park or problems with a watershed. If the council members see the big picture, they can see why your recommendation makes sense.

▶ Example. (See Figure 16.6, p. 492.) The example is an illustration of what you are talking about. Use this strategy so that the audience does not have

Figure 16.3 Assert-Evidence Slide

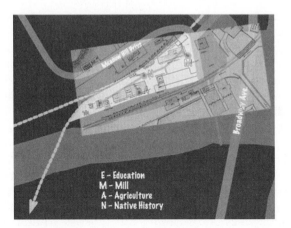

Figure 16.4 Slide That Provides Context

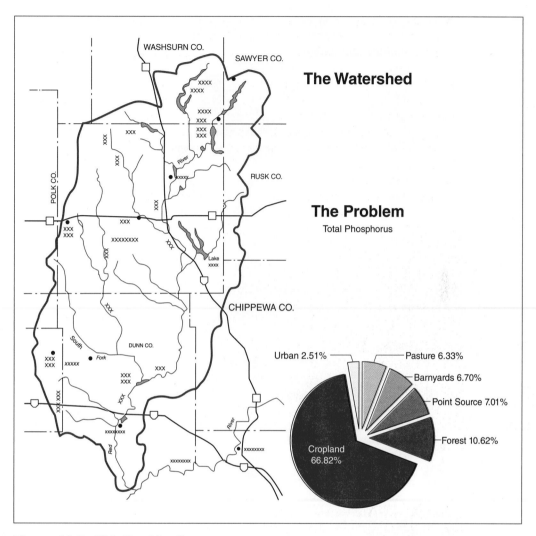

Figure 16.5 Slide Provides Context

to read your mind or "feel like they had to be there." If the kitchen you were asked to analyze had a defective sink arrangement for the required cooking tasks, show a picture of it so listeners know what you mean. If you are trying to explain the concept of immune compatible replacement organs, you might use a slide showing the elements of the concept.

▶ Reduced Learning Time. (See Figures 16.7 and 16.8, pp. 492–493). An image can help a viewer access the complexities of a process. For instance an exploded view of an assembly, perhaps an unfamiliar object like an antique spinning wheel can help viewers reassemble it without any parts left over.

Applications of Nuclear Transfer

**I. Create immune-compatible
   replacement organs**

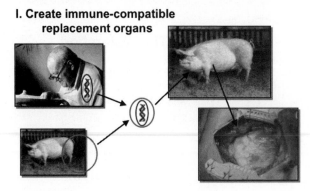

Figure 16.6   Slide Provides Example

Flyer Wheel/Bobbin Assembly of 18th Century Spinning Wheel

Figure 16.7   Slide Reduces Learning Time

▶ Comparison Contrast. (See Figure 16.9, p. 493.) Images can show the difference between any two objects. Before and after is a common use of compare and contrast. For instance a bar graph could use bars to show change in student critical thinking abilities. Two photographs could illustrate erosion on a hill side.

▶ Analogy. (See Figure 16.10, p. 494.) An analogy helps a person connect to something new, especially a new concept. For instance planning a report could be compared to using a GPS screen. To navigate from one position to another you have to have a good idea of where you are and where you hope to go. Displaying a map could help audiences retain and use that concept as they work on their writing.

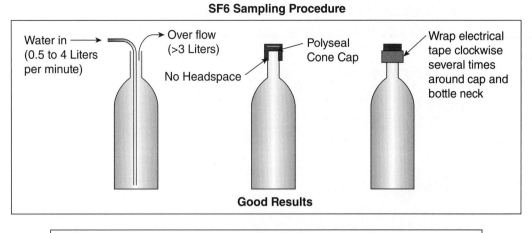

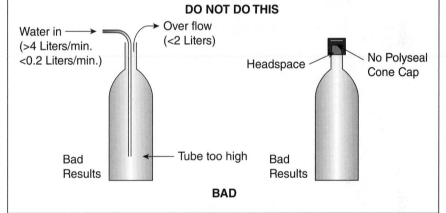

Figure 16.8   Slide Reduces Learning Time

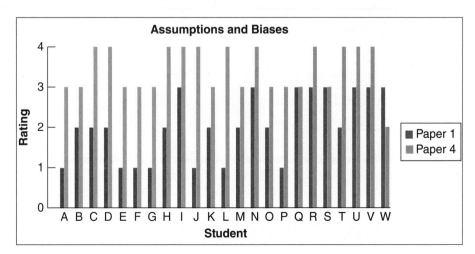

Figure 16.9   Comparison/Contrast Slide

You can get there a number of ways.
Before you start to write, plan your route.

Figure 16.10   Slide Uses Analogy

Source: Google

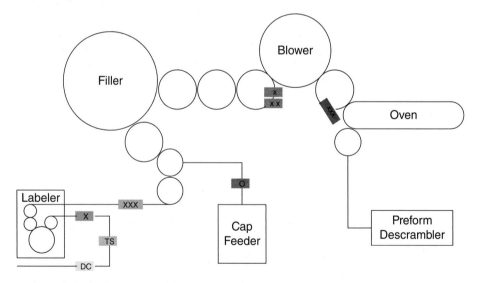

Figure 16.11   Demonstrate a Process

Source: PressCo

> Demonstrate a Process. (See Figure 16.11). A process has a sequence of steps. A series of slides could illustrate what happens at each step; for instance a series of slides could explain the actions of each stage of moving a bottle through a bottle-filler machine. Or one slide could show the overall layout of a plant while you point out to viewers the importance of each section, or show where a bottle neck occurs.

Content Presented by Text

Text-only slides are outlines, lists of words arranged in a hierarchy (level 1, level 2, etc.). While slides of this type show listeners the sections and subsections of the presentation, they project very little depth or indication of relationship of ideas. Compared to the image method outlined earlier, this method has limitations especially in the audience's retention of ideas and in their level of interest (Alley and Neeley). Place the outline text in the speaker notes section of the slide so that you have them available as you speak about the content of the slide.

If you do use text slides, remember your relationship to them. When you display text, what purpose should the text serve in relation to your speaking? Remember the annoyances focus earlier in this chapter—don't read the slides to the audience. Follow these suggestions:

▶ Use the text as an introductory overview. Set the transition so that all the text appears at the same time. The viewers will immediately discern the points you will cover in the next few minutes. As you speak, the words or phrases on screen will serve as an outline and also as a reminder of what you have already mentioned.

▶ Use the text as topic introductions. Set the transition so that only one point appears at a time. The appearance of the text announces the topic for the next point in the speech. No need for you to read the text to the audience—they can see it. You can launch right into your discussion.

Our Model
- Laptops are part of tuition—per credit cost
- More than a box—
    Software, network, wireless, servers storage, service and support
- Two platforms—Apple and PC
- "Refreshed" every 2 years
- UW-Stout leases laptops from vendors

Figure 16.12   Sample Text Slide

▶ Give audiences time to read the slide to themselves. If you have a quotation that makes a number of points that you want to expand, consider showing it on one slide. Stop talking. Give the audience time to read it. After they have had time to digest it, point out the phrases or the patterns in the thinking that you wish to develop. Since a quotation on a slide almost always violates all the rules about simplicity, use this strategy sparingly. If the text is too small to read, consider deleting large portions of the quotation or hand out a paper version.

Use a paper handout. Paper handouts can more effectively show complex text, numbers, and data graphics (Tufte 22). A handout can replace or supplement projected visual aids. Pass out copies of a key image, perhaps a table. Listeners can make notes on it as you speak. One speech expert suggests this: "If what you say when you expand the bullet points is useful for the audience to take away, put it in the handout" (Stratten).

Create a Storyboard

A storyboard is a text and graphics outline of your presentation ("How"; Johnson; "Working"). One expert says, "You should know what you intend to say and then figure out how to visualize it" (Wax). A storyboard will help you do that. A storyboard can be as simple as a two-column table that lists topics on the left and visuals on the right (Figure 16.13). A storyboard can also be a three-column sequence of lines, boxes, and comments. Place the slide's text in the line; sketch the look of the slide in the box, and add explanatory material in the comments space (Figure 16.14, p. 497). Variations of storyboards also exist in presentation programs. PowerPoint, for instance, has both an outline and a slide sorter function that allow you to see the entire sequence of your presentation (Figure 16.15, p. 497).

Topic	Visual Aid
Introduction	
Source of assignment	
Recommendation	List of recommendations
Preview	Outline of main topics
	List of main methods for each type (use of a two-column slide)
Section 1	
Method of researching	Cross-sectional view
Section 2	
Three types of laminates	
Advantages of each	List of advantages
Section 3	
Cost	Table of costs
Conclusion	
Summary	

© Cengage Learning

Figure 16.13   Sample Text Storyboard

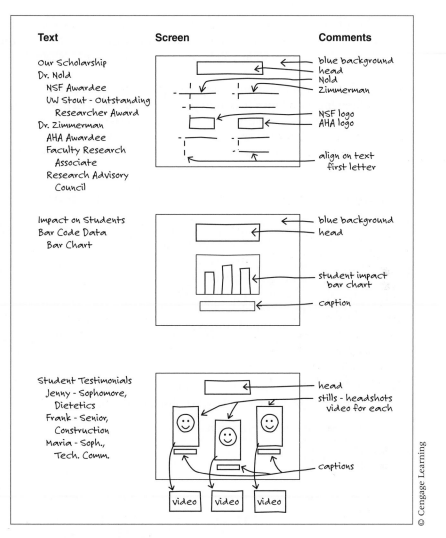

Figure 16.14   Sample Three-Column Storyboard

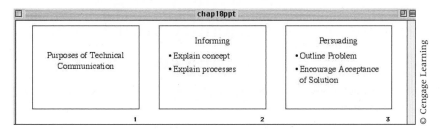

Figure 16.15   PowerPoint Outline View and Slide Sorter View

Create the Slides

After you have the storyboard outline of your presentation, create the slides. A well-designed presentation will help convey your point pleasingly to an attentive audience. But, as Edward Tufte points out, "If your words or images are not on point, making them dance in color won't make them relevant" (22). The following guidelines (based in part on Welsh) will help you create pleasing slides and sequences that help to convey information rather than to distract attention from the main points. As you create slides, understand the parts of a slide, determine what should be on the slide, and how the slide should look.

Understand the Parts of a Slide. The parts of the slide are the title, the text or graphics, the footer, and the background (see Figure 16.16).

▶ The title appears at the top, usually in the largest type size. Use it to explicitly identify the contents of the slide.
▶ The text makes the points you want to highlight. Use phrases that convey specific content rather than generic topics.
▶ The graphic consists of a table, chart, or drawing.
▶ The footer, the area at the bottom of the slide, contains dates, slide numbers and other pertinent information.
▶ The background is the color or design that appears behind the text or graphics.

Design the Slide to help the Reader. The major "rules" of slide design (Paradisi, "Ten"; Zahoursky; "PowerPoint"; Chen) are

Make the Content Big Enough
Keep It Simple
Be Consistent

Figure 16.16 Parts of a Visual

Big Enough means the text is easy to read from a distance and details in visuals are apparent from a distance. *Simple* means that the slide has no extraneous "fluff" like odd colors, lines, or images that detract listeners from content. Viewers absorb more if you show them less. *Consistent* means each item of the same type is presented in the same way. People notice inconsistencies. If the inconsistency is present to make a point (e.g. highlighting an important word to make it stand out), you focus the audience's attention. If the inconsistency has no apparent reason, you confuse the audience. You can help yourself follow those rules by following these guidelines:

▶ Learn to use the Slide Master to Create a Visual Logic. Experienced PowerPoint users become adept at using the slide master. You guarantee that all the slides will be consistent. If you select a template and then use the slide master, you guarantee consistency for font, font size, font color, background, and bullet type. You can add "footer" text such as slide numbers and dates that will appear on each slide in the same place. In addition, if you find a design that you find effective, save it as a template. The next time you create a slide presentation, you will have all the formatting decisions finished.

▶ Evaluate templates before using them. Although many templates are colorful and cute or catchy, choose one with your audience in mind. Look at it from their standpoint—will this combination of colors and fonts help them understand your point? If you're not certain, don't use it. See Figures 16.17 and 16.18 (p. 500) which illustrate the use of a template and the effect of deviating from the template.

▶ Title each slide so your audience will have a quick reference to the topic at hand.

Fonts.   Keep your text font simple. Fonts can portray a wide range of emotions, from casual to authoritative, from serious to comic. Selecting and using a font that will elicit the desired response from your audience is important. Follow these guidelines when selecting fonts:

▶ Use what is known to work well. Black text on white background is the classic combination. Yellow backgrounds with black lettering work well in most situations (think about school buses). Other good combinations are deep-blue backgrounds with yellow letters, or gray backgrounds with black letters.

▶ Use only one font, preferably a sans serif font like Helvetica or Arial.

▶ Make sparing use of different sizes, boldfacing, or italics. For instance, a long italicized quote is difficult to read. Too much boldfacing or too many different sizes gives a cluttered, "ransom note" effect that is very distracting.

▶ Use larger font sizes and/or different colors for your titles. When using larger fonts, be sure to use those sizes that are easy to read (24–36 points), but not so large that they become distracting.

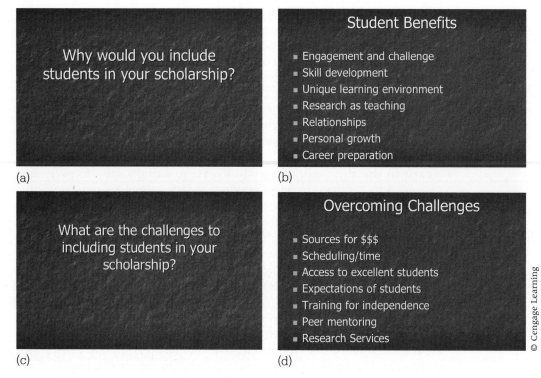

(a)   (b)

(c)   (d)

Figure 16.17   Slide Template

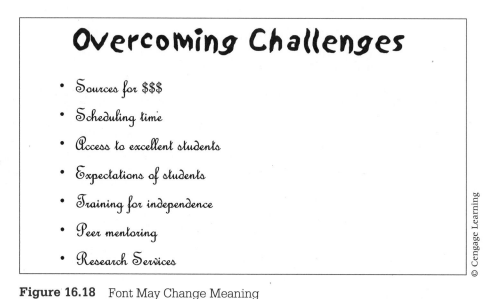

Figure 16.18   Font May Change Meaning

TIP ‖‖‖

Line Length in Visuals

Many texts (see George Miller) encourage using no more than seven lines and limiting lines to seven words. It is more helpful to think about whether you want the audience to read a long quote or to present a group of short lines, summarizing what you are saying. Shorter text lines keep the words in the background as you speak, focusing attention on you.

- How large should the font be? Your viewers have to read the text or see the details of the visual. So the font must be big enough. But how big is big? PowerPoint defaults to 44 points for titles and 32 points for "first level" (the items closest to the left margin of the body) text. Usually those sizes can be seen easily from all seats in any room. Sometimes those sizes take up too much room on the slide for your purposes; in that case, reduce the size. An easy way to test the size of the fonts is to use the "eight-foot rule:" "Print out your slide. Tape it to the wall. Step back 8 feet (2.5 m). If you can read it easily, it's the right size" ("Creating").

Colors. Use color to enhance your presentation. Color combinations should help viewers focus on key points, not on the combination itself. To use color effectively, consider these guidelines:

- ▶ Use color intelligently to establish visual logic. Use one of the software's templates and it creates the logic for you. Give each item (title, text, border) in the template its own color. Use only one color for emphasizing key words. (See "Focus on Color," pp. 171–178).
- ▶ Consider the setting. If you will speak in a well-lit room, use a light background and dark text. If you will speak in a dark room, use a dark background and a light text.
- ▶ Use green and red sparingly. Ten percent of the population is color blind, and can't distinguish between red and green. It is best to assume that at least one person in every ten of your audience will be limited in his or her ability to translate color.
- ▶ Avoid hard-to-read color combinations, such as yellow on white and black on blue. Violet can also be very hard to read.
- ▶ Select combinations with an awareness of technology. Colors that look well together in a sign or in print, as in a magazine or newspaper, will probably not work the same way projected onto a screen. For example, ambient light, which is what is produced by an LCD projector, will affect contrast greatly; it will turn a dark color, such as burgundy or a deep green, into a pastel.

Animation. Animation makes items move. Text can appear and depart from the slide by many routes—move to the left or right, up or down, or disintegrate

Use Builds to Reveal Information as You Need It

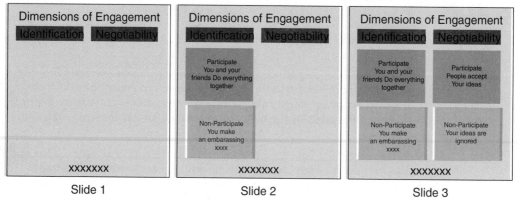

Figure 16.19   Series of Slides That "Build"

or blossom out from the center. Your goal, however, is to assist the audience grasp content, not provide gimmicky entertainment. For the most part, just use the "appear" animation. If you do use animations, follow these guidelines:

▶ Use only one type of text animation. For example, a simple "wipe-right" text animation—the text appears to move off the screen left to right—will keep the reader's eye going in the normal reading direction.
▶ Treat previous lines carefully. Fading or subduing the previous bullets when the new information appears will help to keep the audience focused.
▶ Use builds. You can make a build, a sequence of slides that change the visual slightly, each time adding a bit more information. If you have a complicated concept to convey, you can illustrate the relationship of new information to previous information. In the three slides that follow, the next slide adds a new concept to the ones explained in the previous slide. By using the build, you keep the basic plan in front of the audience but you also illustrate the relationship and the way in which the new concept builds on and is related to the previously explained concepts.

Sound Effects.   Basically, don't use sound effects. They add little or nothing to the impact the content makes on the audience. If you do use them

▶ be consistent (use the same one for the same operation such as introducing an new topic)
▶ be subtle. Use a quiet sound. Gimmicky sounds like typewriters typing or bullets ricocheting can get a reaction the first time but then quickly become distracting.

▶ If you have a special point that you want to emphasize, or if you want to use a sound effect for some comic relief in a deadly serious situation, go ahead, but use these strategies sparingly (see "PowerPoint Do's").

Slide Transitions.　Slide transitions, like scene cuts in a movie, move an old slide off and a new slide onto the screen. The goal of using this device is to create a sequence logic—each instance of the event means that another type of data is about to appear. If your audience concentrates on the transition rather than on the message, you've lost them. Follow these guidelines:

▶ Be consistent. Use the same transition for the same kind of event. Make each slide move off to the left or right or whatever you've chosen.
▶ Use only one or two simple transitions.
▶ Select transitions in order to aid the viewer. For example a simple fade-to-black between sections of a presentation signals that a new topic is being considered.

Making an Effective Presentation

To make an effective presentation, learn to "dance" with your slides, develop your introduction, navigate the body, develop your conclusion, rehearse your presentation, and deliver your presentation.

Learn to "Dance" with Your Slides

You and the screen are interacting with the viewers. What should you do? One helpful, and common, model appears regularly on TV news shows—the weather report. Watch how the weather person relates to the screen, using images and voice to tell you what happened today and what will probably happen tomorrow. The weather person presents a weather map as a still image, tells the viewer what to notice, then often creates a moving image to show the pattern and uses his or her voice to inject the complexity (that is, what it means for your plans) into the situation. Because the weather person "dances' with the screen, the words and the moving images together tell the story.

　You might also want to watch the effective presentation from Baruch College (http://media.baruch.cuny.edu/faculty/jbelland/powerpoint/video_begin.htm). In it a teacher works with a number of slides, one of them a map. In one version he "dances," carefully pointing out the meaning of the color coding on the map, thus creating a clear vision of which countries were on opposing sides in the debate that he is explaining. In the other version he skips over the color coding. In the version where he "dances," the map and his words bring the point home to the viewer. In the other version he merely adds confusion to the point. To return to the analogy used at the beginning of this chapter, he and the screen are the unruly pet at the party. You remember the antics, not the point.

Develop the Introduction

The introduction establishes your relation to the audience and names the point of your presentation. You can be funny or serious; you can treat the audience as experts or beginners. Your goal is to create a rapport with them so that they feel you understand why the topic is important to them and that you have treated it in the correct manner. Be explicit about your purpose and the sections of the presentation. In other words, follow the old advice of "Tell 'em what you are going to tell 'em" (Bacall; Tracy). Keep introductions short but use the time to develop the audience's trust.

Follow these guidelines (you may not need to include all these points):

- Keep the introduction short.
- Explain your credibility. What is your background with the topic? What did you do to find out about the topic?
- Indicate your special knowledge of or concern with the subject.
- Explain why your report is important to your audience.
- Identify the situation that required you to prepare the report (or the person who requested it).
- Present your conclusions or recommendations right away. Then the audience will have a viewpoint from which to interpret the data you present.
- Preview the main points so your listeners can understand the order in which you will present your ideas.

Navigate the Body

Many studies have shown that listeners simply do not hear everything the speaker says. Therefore, you should give several minutes to each main idea—long enough to get each main point across, but not long enough to belabor it.

Use Transitions Liberally

Clear transitions are very helpful to an audience of listeners. Your transitions remind them of the report's structure, which you established in the introduction. Indicate how the next main idea fits into the overall report and why it is important to know about it. For instance, a proposal may seem very costly until the shortness of the payback period is emphasized.

Emphasize Important Details

Presumably, if you have created a storyboard, you know the details that you want to emphasize for the audience, and you have placed them on slides. Choose details that are especially meaningful to the audience. For instance, explain any anticipated changes in equipment, staff, or policy, and show how these changes will be beneficial.

Impose a Time Limit

Find out how long the audience expects the presentation to last and fit your speech into that time frame. If they expect 15 minutes and you talk for 15 minutes, they will feel very good. Generally, speak for less time than is required. It is much better to present one or two main ideas carefully than to attempt to communicate more information than your listeners can comfortably grasp.

Develop a Conclusion

Keep the conclusion short. Perhaps announce "In conclusion . . . ." Summarize if necessary. Perhaps place on the last screen the "takeaway" points, those one or two items you want the audience to remember. For instance, "This type of bottle filler will increase production by 20 percent." It is common practice to also ask for questions at the end of a presentation.

Rehearse Your Presentation

Practice your presentation (Markowitz). Do not just "wing it." Practice by going straight through the presentation. Use note cards or the "Notes" function of your software. If it is a formal presentation, when you practice, wear the same clothes you will wear in the actual presentation.

Practice Developing a Conversational Quality

When you make your presentation, sound like a person speaking to people, and use both voice and gestures to emphasize important points. Even the best information will fall on deaf ears if it is delivered like a robotic time-and-weather announcement. Rehearse until you feel secure with your presentation, but always stop short of memorizing it. If you memorize, you will tend to grope for memorized words rather than concentrating on the listeners and letting the words flow.

Practice Handling Your Technology and Visual Aids

Understand how to open and navigate your presentation. If necessary, have the presentation on several media. Often, speakers have the same file both on a flash drive, on the Web, and on their laptops. If you are unfamiliar with the technology, practice opening the files from all the locations of your file. Because technological arrangements in new places can be difficult to navigate, have a backup plan in case your technology does not work. Practice giving the presentation so that you know how to open the software program and advance and reverse the slides. If you use a Bluetooth device, practice using it beforehand. Practice talking to the audience and looking at the screen only for those slides that you will use as foreground slides.

TIP ||||

Use the B Key

At points during your presentation, and especially during a question/answer session, you can 'turn off' the screen, by pushing the b key. The screen turns black. Push the key again and the contents of the screen reappear (Reynolds).

If you have paper handouts, decide whether to distribute them before or during the presentation. Distributing them before the presentation eliminates the need to interrupt your flow of thought later, but because the listeners will flip through the handouts, they may be distracted as you start. Distributing them during the presentation causes an interruption, but listeners will focus immediately on the visual.

Deliver Your Presentation

You will increase your effectiveness if you use notes and adopt a comfortable extemporaneous style.

Use Notes

Experienced speakers have found that outlines prepared on a few large note cards (5 by 8 inches, one side only) are easier to handle than outlines on many small note cards. Some speakers even prefer outlines on one or two sheets of standard paper, mounted on light cardboard for easier handling.

The outline should contain clear main headings and subheadings. Make sure the outline has plenty of white space so you can keep your place.

Adopt a Comfortable Style

The extemporaneous method results in natural, conversational delivery and helps you concentrate on the audience. Using this method, you can direct your attention to the listeners, referring to the outline only to jog your memory and to ensure that ideas are presented in the proper order. Smile. Take time to look at individual audience members and to collect your thoughts. Instead of rushing to your next main point, check whether members of the audience understood your last point. Your word choice may occasionally suffer when you speak extemporaneously, but reports delivered in this way still communicate what you want to say better than those memorized or read.

The following suggestions will help as you face your listeners and deliver the presentation:

1. Look directly at each listener at least once during the report. With experience, you will be able to tell from your listeners' faces whether you are communicating well. If they seem puzzled or inattentive, repeat the main idea, give additional examples for clarity, or solicit questions. Don't proceed in lockstep through your notes. Adapt.

2. Make sure you can be heard, but also try to speak conversationally. You should feel a sense of your voice as a round, full tone, projecting with conviction. You should also feel that your voice fills the space of the room, with the sound of your voice bouncing back slightly to your own ears. The listeners should get the impression that you are talking to them rather than just presenting a report. Inexperienced speakers often talk too rapidly.

3. Try to become aware of—and to eliminate—your distracting mannerisms. No one wants to see speakers brush their hair, scratch their arms, rock back and forth on the balls of their feet, smack their lips. If the mannerism is pronounced enough, it may be all the audience remembers of your presentation. Stand firmly on both feet without slumping or swaying.

4. To point out some aspect of a visual projected by an overhead projector, lay a pencil or an arrow made of paper on the appropriate spot of the transparency.

5. When answering questions, make sure everyone hears and understands each question before you begin to answer it. If you cannot answer a question during the question-and-answer session, say so, and assure the questioner that you will find the answer and provide it at a later time.

Globalization and Oral Presentations

Giving a presentation to a foreign or non-English-speaking audience is easier if you give some thought to relating to an audience whose culture is not your own. A key idea for your planning is that although English is commonly studied as a second language, "English proficiency within a given audience can vary widely, so the best approach is to simplify and clarify content at every turn" (Zielinski). In order to simplify and clarify content, follow these tips: Use simple sentences, make clear transitions, avoid digressions, reduce use of potentially confusing pronouns, restate key points, pause periodically, use subject–verb–object word order, repeat phrases using the exact wording. If you call it a "plan" the first time, continue to use that word; don't switch to "proposal" or "map" or "vision" (Zielinski).

Also, be aware that the international audience's reaction to you may differ greatly from what you are used to. For example, in Japan, it is not unusual for audience members to close their eyes in order to convey concentration and

(Continued)

attentiveness, while in the United States closed eyes are a sign that you are lulling the audience to sleep. Applause is a generally universal sign of approval, but whistling in Europe is a negative reaction to your presentation. Finally, know that other cultures have a different sense of acceptable personal space than Americans have. Middle Easterners and Latin Americans tend to stand much closer than Americans find comfortable, while many Asian cultures stand quite far away from each other. Keep this in mind if you have others onstage with you or if you will be going into the audience for your presentation (McKinney, "International").

Be aware of body language conventions. Hand gestures that are accepted in the United States, such as the A-OK symbol (the circle formed with your index finger and thumb), or the thumbs-up gesture, are considered obscene in some countries. Pointing with a finger can be impolite; use a fully extended hand. In some countries, emphatic gestures are poorly received. Body language that is unwittingly offensive can cause an audience to focus on what is inappropriate and lose the content of your presentation (Merritt).

Plan for differences in technology. Bring pictures of the equipment that you will need during your presentation. Bring a voltage converter. Remember that many countries have differently sized standard paper and may use a two-hole instead of a three-hole punch. Most importantly, have a backup plan and keep a sense of humor (McKinney, "Public," "Professional").

For further reference, check out these websites:

Executive Planet at <www.executiveplanet.com/> is a guide to all aspects of conducting international business: etiquette, customs, and culture. If you'll be presenting in a foreign country, you can find out the details of greetings, business attire, and meeting formalities as well as general information for many different countries.

For insight into the cultural dynamics of other countries and regions and how this may impact your business dealings abroad, refer to "International Business Etiquette and Manners for Global Travelers" at <www.cyborlink.com/>. Your presentations will be more effective if you know of the appropriate way you should act.

Although addressed to potential conference speakers, this article offers an invaluable overview of preparing to give a presentation to an international audience: Gagnon, Michael, and Raymond Wallace. "Making a Presentation in English at a European Conference." Federation of European Chemical Societies. Division of Chemical Education. July 2001. Web. 24 Mar. 2012 <http://www.chemia.uj.edu.pl/dydaktyka_a/EuCheMS.pdf>.

Worksheet for Preparing an Oral Presentation

☐ **Identify your audience.**

- What is your listeners' level of knowledge about the topic?
- What is their level of interest in the entire speech?
- Why are they attending?
- What do they need?

☐ **Create an outline showing the main point and subpoints.**

☐ **Which strategy will best help the audience? Problem-solution? Narrative? IMRD?**

☐ **Assign a time limit to each point.**

☐ **Create a storyboard.**

☐ **What visual aid will illustrate each point most effectively?**

☐ **Decide whether you need any kind of projection or display equipment.**
Laptop? LCD projector? Flip chart?

☐ **Review the speaking location. Do you know how to make your technology (laptop, disk, projector) interact with the technology resident at the site?**

☐ **Determine your relationship to the slides. Will they be foreground or background for you?**

☐ **Prepare clearly written note cards—with just a few points on each.**

☐ **Rehearse the speech several times, including how you will actually handle the technology.**

Worksheet for Evaluating an Oral Presentation

☐ **Answer these questions:**

1. Clarity
   Did the speaker tell you the point early in the speech? Could you tell when the speaker moved to a new subpoint?
2. Tone
   Did the speaker sound conversational? Did the speaker go too fast? Go too slow? Speak in a monotone?
3. Use of technology
   Did the speaker interact effectively with the slides?
   Did the slides help you understand the content, or were they distracting

Exercises

▶ You Create

1. Create a PowerPoint presentation of two or three slides that illustrates a problem in one of your current projects. Give a brief speech (two to three minutes) explaining the problem. Alternate 1: Prepare two or three PowerPoint slides that illustrate the solution or its effects, and present the entire problem and solution to the class in a four- to five-minute speech. Alternate 2: Prepare the PowerPoint presentation in groups of two to four. Select a speaker for the group. Give the speech.

2. Report on a situation with which you are involved. Your work on an assignment for this class is probably most pertinent, but your instructor will provide his or her own requirements. Depending on the available time, draw a visual on the board, make a transparency, create a handout, or prepare one or two PowerPoint slides. In two minutes, explain the point of the visual aid. Class members will complete and/or discuss the evaluation questions.

3. For Exercises 1 and 2, each member of the audience should prepare a question to ask the speaker. Conduct a question-and-answer session. When the session is finished, discuss the value and relevance of the questions that were asked. What constitutes a good question? Also evaluate the answers. What constitutes a good answer?

4. Make a storyboard for the speech you will give for the following Speaking Assignment. Divide a page into these three columns and fill them in, following this example:

Point	Visual Aid	Time
Method of extrusion	Cross section of laminate	2 minutes

5. Use the generic graph (below) to make a two-minute speech. Title the graph and explain its source, its topic, and the significance of the pattern. Choose any topic that would change over time.

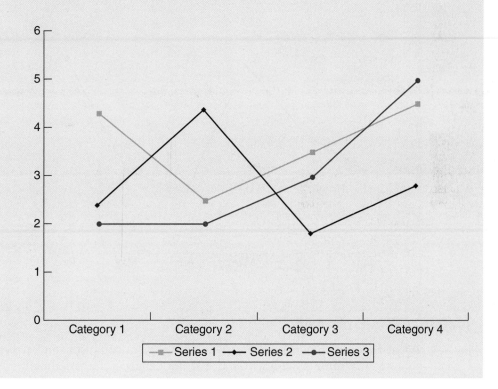

6. Give a brief speech in which you freely use technical terms. The class will ask questions that will elicit the definitions. If there is time, redeliver the speech at a less technical level.

Speaking Assignment

Your instructor may require an oral presentation of a project you have written during the term. The speech should be extemporaneous and should conform to an agreed-upon length. Outline the speech, construct a storyboard, make your visuals, and rehearse. Follow your presentation with a question-and-answer session.

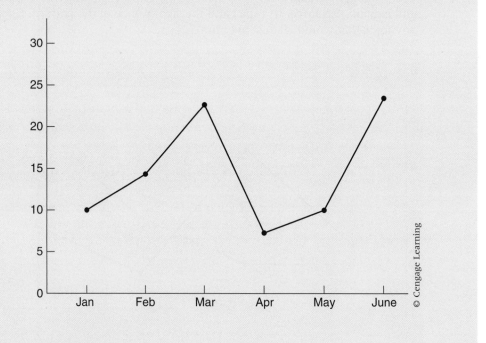

Writing Assignment

Write a learning report for the speaking assignment you just completed. See Chapter 5, Writing Assignment 8. p. 137, for details of the assignment.

Web Exercises

1. Create a PowerPoint version of a document you have previously written for this class. Upload the PowerPoint to the Web. Give a four- to five-minute speech using the online slides. Alternate: Upload the PowerPoint presentation created above. Have classmates, as assigned by your teacher, read and evaluate the report. Use the evaluation sheet for the appropriate type of report in the appropriate chapter.

2. Using screens that you download from the Web, create a PowerPoint presentation to your classmates in which you do either of the following:
   a. Explain the effective elements of a well-designed screen.
   b. Explain how to use the screen to perform an activity (order plane tickets, contact a sales representative, perform an advanced search).

Works Cited

Alley, Michael and Neeley, Kathryn A. "Rethinking the Design of Presentation Slides: A Case for Sentence Headlines and Visual Evidence." Technical Communication 52.4 (Nov. 2005): 417–426.

Alley, Michael. Rethinking the Design of PowerPoint. Leonhard Center for the Enhancement of Engineering Education at Penn State. Web. 3 March 2012. <http://www.writing.engr.psu.edu/slides.html>.

Bacall, Robert. "How Attention Works for Audiences." CKS PowerPoint.com. Web. 14 May 2012. <www.onppt.com/ppt/article1035.html>.

Ball, Corbin. "Avoiding 'Death by PowerPoint.'" Corbin Ball Associates. 2002. Web. 14 May 2012.

Chen, Victor. "Designing Effective PowerPoint Presentations." Slideshare. Web. 24 Mar. 2012. <http://www.slideshare.net/vinhha25580/designing-effective-power-point-presentations>.

"Creating Visual Aids That Really Work: Designing Effective Slides Using PowerPoint." ECG. Web. 28 May 2012. <http://ecglink.com/library/ps/powerpoint.html>.

"Effective Presentations in Engineering and Science: Guidelines and Video Examples—Visual Aids." Leonhard Center for the Enhancement of Engineering Education at Penn State. Web. 3 March 2012. <http://www.engr.psu.edu/speaking/Visual-Aids.html>.

"Effective Use of PowerPoint." Online Tutorial. Baruch College. City College of New York. Web. 28 May 2012. <http://media.baruch.cuny.edu/faculty/jbelland/powerpoint/outline.htm>.

"How to Storyboard a PowerPoint Presentation." *Resources for Management Consultants & Commercial Professionals.* Web. 14 May 2012. <http://www.strategyexpert.com/articles/Storyboard>.

Lane, Robert and Andre Vicek. "Speaking Visually: Eight Roles Pictures Play in Presentation. Microsoft Office. Web. 22 March 2012. <http://office.microsoft.com/en-us/powerpoint-help/speaking-visually-eight-roles-pictures-play-in-presentation-HA010361422.aspx>.

Laskowski, Lenny. "A.U.D.I.E.N.C.E. Analysis: It's Your Key to Success." LJLSeminars.com. 1996. Web. 14 May 2012. <www.ljlseminars.com/audiences.htm>.

Markowitz, Ed. "How to Create a Great PowerPoint Presentation." Inc.com. 2011. Web 14 May 2012. <http://www.inc.com/guides/201102/how-to-create-a-great-powerpoint-presentation.html>.

McKinney, C. "Public Speaking: Bilingual Help." 2003. Advanced Public Speaking Institute. Web. 24 Mar. 2012. <www.public-speaking.org/public-speaking-bilingual-article.htm>.

_____. "Public Speaking: International Perspective on Humor." 2003. Advanced Public Speaking Institute. Web. 24 Mar. 2012. <www.public-speaking.org/public-speaking-international-article.htm>.

_____. "Public Speaking: Equipment Photographs." Advanced Public Speaking Institute. 2003. Web. 24 Mar. 2012. <www.public-speaking.org/public-speaking-equipmentphotos-article.htm>.

Merritt, Anne. "10 Common Gestures Easily Misunderstood Abroad." Matador Abroad. 2010. Web. 29 May 2012. <http://matadornetwork.com/abroad/10-common-gestures-easily-misunderstood-abroad/>.

Miller, George. "The Magical Number Seven, Plus or Minus Two: Some Limits on Our Capacity for Processing Information." *Psychological Review* 63 (1956): 81–97.

Miller, Glenn. "Presentation Disasters: Conference Style." 08 Mar. 2004. PowerPointAnswers.com. Web. 1 May 2012. <www.onppt.com/ppt/article1036.html>.

Paradi, Dave. Results of the 2011 Annoying PowerPoint survey. ThinkOutsideTheSlide.com. 2011. Web. 3 March 2012. <http://www.think-outsidetheslide.com/articles/annoying_powerpoint_survey_2011.htm>.

PowerPoint Do's and Don'ts. Heart Rhythm Society. 24 Mar. 2012. <http://www.hrsonline.org/Education/WomensLeadership/ProfGrowth/EffectivePres/PPT/DoDont/>.

"Presentation Storyboard—Tell Them a Story!" Excellence in Presentations. 2007. Web.14 May 2012. <http://nobullets.wordpress.com/2007/12/04/presentation-storyboard-tell-them-a-story/>.

Reynolds, Garr. Top Ten Delivery Tricks. Garr Reynolds. Com. 2005. Web. 3 Mar. 2012. <http://www.garrreynolds.com/Presentation/delivery.html>.

_____. "Organization & Preparation Tips." Garr Reynolds.com. 2005. Web. 15 May 2012. <http://www.garrreynolds.com/Presentation/prep.html>.

Shaw, Gordon, Robert Brown, and Philip Bromiley. "Strategic Stories": How 3M Is Rewriting Business Planning." *Harvard Business Review* 76 (May–June 1998): 42–44. Reprint 98310.

Johnson, Naomi. "Storyboarding." Sierra College CIS 100-PowerPoint. Web. 14 May 2012. <http://sites.google.com/site/cis100powerpoint/Home/lectures/understanding-storyboarding>.

"Storyboard" WWWNetSchool. Web. 14 May 2012. <www.thirteen.org/edonline/software/earthinflux/orgc.html>.

Stratten, Scott. "Business Tip: Giving Effective PowerPoint Presentations." About.com Small Business:Canada. Web. 14 May 2012. <http://sbinfocanada.about.com/cs/management/qt/powerptpres.htm>.

Tracy, Larry. "Preparing a Presentation." Presentation-Pointers.com. 2000. Web. 11 Oct. 2003. < http://www.presentation-pointers.com/showarticle/articleid/216/>.

Tufte, Edward. *The Cognitive Style of PowerPoint.* Cheshire, CT: Graphics Press, 2003.

Wax, Dustin. "10 Tips for More Effective PowerPoint Presentations." StepcaseLifehack. 2011. Web. 15 May 2012. <http://www.lifehack.org/articles/technology/10-tips-for-more-effective-powerpoint-presentations.html>.

Welsh, Theresa. "Presentation Visuals: The Ten Most Common Mistakes." *intercom* 43.6 (1996): 22–43.

"Working Smarter with Presentation Storyboards." Smart Draw. 2008. Web. 22 May 2012. <http://www.smartdraw.com/>.

Zahorsky, Darrell. Business Presentations—8 Secrets to a Knockout Business Presentation. About.com Small Business Information. Web. 3 March 2012. <http://sbinformation.about.com/od/sales/a/presentationtip.htm>.

Job Application Materials

CHAPTER CONTENTS

CHAPTER 17 IN A NUTSHELL

The goal of the *letter of application* and the résumé is to convince someone to offer you a job *interview*.

Basic letter strategies. Know the elements of block letter format. Relate to the potential employer's needs. Show how you can fill those needs. If, in the job announcement, an employer lists several requirements, your letter should include comment on each. In those comments, present a convincing and memorable detail: "At Iconglow I managed the group that developed the online Help screens. Under my direction, we analyzed what topics were needed and which screen design would be most effective."

Write in small chunks, putting the employer's keywords at the beginning of each chunk. Pay close attention to spelling and grammar—mistakes could cost you an interview.

Basic résumé strategies. Design your *résumé* so that key topics jump out. Include sections on

▶ Your objective (one brief line).

▶ How to contact you.

▶ Your education (college only).

▶ Your work history (most relevant jobs at the top; list job title, employer, relevant duties, and responsibilities).

Most résumés place the major heads at the left margin and indent the appropriate text about one inch.

Basic interview strategies. At the *interview*, you talk to people who have the power to offer you the job. Impress them by knowing about their company. Tell the truth—if you don't know the answer to a question, say so.

Electronic application. Many corporations now ask for your résumé online, either e-mailed to them or entered into their Web-based application site. While many sites accept—and prefer—that the résumé be sent as a Microsoft Word document, others require use of a plain text document. For that type of document do not use bold, italic, tabs, bullets, or multiple columns. Set all sections at the left margin. So that the lines do not "break" oddly, set the margins of the document to 1 inch on the left and 2.5 inches on the right.

This chapter explains the process of producing an effective résumé and letter. You must understand the basic format of a letter, analyze the situation, plan the contents of the résumé and letter, present each in an appropriate form, and perform effectively at an interview.

The Basic Format of a Letter

The two basic formats are the block format and the modified block format (*Merriam Webster's*).

Block Format and Modified Block Format

Block format is the basic letter format. Place all the letter's elements flush against the left margin. Single space the paragraphs and double space between them. Do not indent the first word of each paragraph (see Fig. 17.1). The *modified block format* is the same as the full block format with two exceptions: The date line and closing signature are placed on the right side of the page, five spaces to the right of the center line.

Elements of a Letter

Internal Elements

This section describes the basic elements of a letter from the top to the bottom of a page.

Heading

The heading is your address.

> 4217 East Eleventh Avenue
> Post Office Box 2701
> Austin, TX 78701

- Spell out words such as *Avenue, Street, East, North,* and *Apartment* (but use *Apt.* if the line would otherwise be too long).
- Put an apartment number to the right of the street address. If, however, the street address is too long, put the apartment number on the next line.
- Spell out numbered street names up to *Twelfth.*
- To avoid confusion, put a hyphen between the house and street number (1021-14th Street).
- Either spell out the full name of the state or use the U.S. Postal Service Zip Code abbreviation. If you use the Zip Code abbreviation, note that the

Heading	4217 East Eleventh Avenue Post Office Box 2701 Austin, TX 78701
	(skip 2 lines)
Date	February 24, 2005
	(skip 2 lines)
Inside address	Ms. Susan Wardell Director of Planning Acme Bolt and Fastener Co. 23201 Johnson Avenue Arlington, AZ 85322
	(double-space)
Salutation, mixed punctuation	Dear Ms. Wardell:
	(double-space)
Subject line	SUBJECT: ABC CONTRACT (optional)
	(double-space)
Body paragraphs flush left	_____
	(double-space between paragraphs)
	_____
	(double-space)
Closing, mixed punctuation	Sincerely yours,
Signature	*John K. Palmer* *(skip 3 lines)*
Typed name	John K. Palmer
Position in company	Treasurer
	(double-space)
Typist's initials	abv
Enclosure line	enc: (2)
	(skip 1 or 2 lines; depends on letter length)
Copy line	c: Ms. Louise Black

Figure 17.1

Block Format

state abbreviation has two capital letters and no periods and that the Zip Code number follows one space after the state (NY 10036).

▶ Note on letterhead: place the date two lines below the last line of the letterhead, in the position required by the format (e.g., flush left for block).

Date

Dates can have one of two forms: February 24, 2015, or 24 February 2015.

▶ Spell out the month.
▶ Do not use ordinal indicators, such as 1st or 24th.

Inside Address

The inside address is the same as the address that appears on the envelope.

Ms. Susan Wardell
Director of Planning
Acme Bolt and Fastener Co.
23201 Johnson Avenue
Arlington, AZ 85322

▶ Use the correct personal title (Mr., Ms., Dr., Professor) and business title (Director, Manager, Treasurer).
▶ Write the firm's name exactly, adhering to its practice of abbreviating or spelling out such words as *Company* and *Corporation.*
▶ Place the reader's business title after his or her name or on a line by itself, whichever best balances the inside address.
▶ Use the title *Ms.* for a woman unless you know that she prefers to be addressed in another way.

Salutation

The salutation always agrees with the first line of the inside address.

▶ If the first line names an individual (Ms. Susan Wardell), say, "Dear Ms. Wardell:" If the name is "gender neutral" (Robin Jones), say "Dear Robin Jones:"
▶ If the first line names an office (Director of Planning), address the office.

Dear Director of Planning:

▶ If you know only the first initial of the recipient, write "Dear S. Wardell,"
▶ If you know only a Post Office box (say, from a job ad), use a subject line.

Box 4721 ML
The Daily Planet
Gillette, WY 82716
Subject: APPLICATION FOR OIL RIG MANAGER

Body

Single-space the body, and try to balance it on the page. It should cover the page's imaginary middle line (located 5.5 inches from the top and bottom of the page). Use several short paragraphs rather than one long one. Use 1-inch margins at the right and left.

Complimentary Closing and Signature

Close business letters with "Sincerely" or "Sincerely yours." Add the company name if policy requires it.

Sincerely yours,

[handwritten signature]

John K. Palmer

▶ Capitalize only the first word of the closing.
▶ Allow three lines for the handwritten signature.

Optional Lines

▶ Add an enclosure line if the envelope contains additional material. Use "Enclosure:" or "enc:". Place the name of the enclosure (résumé) after the colon.

Enclosure: résumé

Succeeding Pages

For succeeding pages of a letter, place the name of the addressee, the page number, and the date in a heading.

Susan Wardell	-2-	February 24, 2015

Analyzing the Situation

To write an effective résumé and letter of application, you must understand your goals, your audience, the field in which you are applying for work, your own strengths, and the needs of your employers. Note: Be sure to review the Focus box (pp. 555–557) at the end of this chapter for detail on creating and sending résumés via e-mail.

Understand Your Goals

Your goals are to get an interview and to provide topics for discussion at that interview. If you present your strengths and experiences convincingly in the letter and résumé, prospective employers will ask to interview you. To be convincing, you must explain what you can do for the reader, showing how your strengths fill the reader's needs.

The letter and résumé also provide topics for discussion at an interview. It is not uncommon for an interviewer to say something like "You say in your résumé that you worked with material requirements planning. Would you explain to us what you did?"

Understand Your Audience

Your audience could be any of several people in an organization—from the human resources manager to a division manager, one person or a committee. Whoever they are, they will have only a limited amount of time to read your letter and résumé and so will want to see immediately your qualifications stated in a professional manner.

The Reader's Time

Employers read letters and résumés quickly. A manager might have 100 résumés and letters to review. On the initial reading, the manager spends only 30 seconds to 3 minutes on each application, quickly sorting them into "yes" and "no" piles.

Skill Expectations

Managers want to know how the applicant will satisfy the company's needs. They look for evidence of special aptitudes, skills, contributions to jobs, and achievements at the workplace (Harcourt and Krizar). Suppose, for instance, that the manager placed an ad specifying that applicants need "experience in materials resource planning." Applications that show evidence of that experience probably will go into the "yes" pile, but those without evidence will go into the "no" pile.

Professional Expectations

Managers read to see if you write clearly, handle details, and act professionally. Clean, neat documents written in clear, correct English and formatted on high-quality paper demonstrate all three of these skills.

Assess Your Field

Find out what workers and professionals actually do in your field, so that you can assess your strengths and decide how you may fill an employer's needs. Answer the following questions:

1. What are the basic activities in this field?

2. What skills do I need to perform them?

3. What are the basic working conditions, salary ranges, and long-term out-looks for the areas in which I am interested?

Talk to professionals, visit your college placement office, and use your library. To meet professionals, set up interviews with them, attend career conferences, or join a student chapter or become a student member of a professional orga-nization. Your college's placement service probably has a great deal of career and employer information available.

Two helpful books, among many that describe career areas, are the *Dictionary of Occupational Titles (DOT,* available at http://www.occupationalinfo. org/)* and the *Occupational Outlook Handbook (OOH,* available at http://www. bls.gov/ooh/)*, both issued by the U.S. Department of Labor. The *DOT* pres-ents brief but comprehensive discussions of positions in industry, listing the job skills that are necessary for these positions. You can use this infor-mation to judge the relevance of your own experience and course work when considering a specific job. Here, for instance, is the entry for manu-facturing engineer:

012.167-042 MANUFACTURING ENGINEER (profess. & kin.) Plans, directs, and coordinates manufacturing processes in industrial plant: Develops, evaluates, and improves manufacturing methods, utilizing knowledge of product design, materials and parts, fabrication processes, tooling and production equipment ca-pabilities, assembly methods, and quality control standards. Analyzes and plans work force utilization, space requirements, and workflow, and designs layout of equipment and workspace for maximum efficiency [INDUSTRIAL ENGINEER (profess. & kin.) 012.167-030]. Confers with planning and design staff concern-ing product design and tooling to ensure efficient production methods. Confers with vendors to determine product specifications and arrange for purchase of equipment, materials, or parts, and evaluates products according to specifica-tions and quality standards. Estimates production times, staffing requirements, and related costs to provide information for management decisions. Confers with management, engineering, and other staff regarding manufacturing capabilities, production schedules, and other considerations to facilitate production processes. Applies statistical methods to estimate future manufacturing requirements and potential.
GOE: 05.01.06 STRENGTH: L GED: R5 M5 L5 SVP: 8 DLU: 89

The *OOH* presents essays on career areas. Besides summarizing necessary job skills, these essays contain information on salary ranges, working conditions, and employment outlook. This type of essay can help you in an interview. For instance, you may be asked, "What is your salary range?" If you know the appropriate figures, you can confidently name a range that is in line with industry standards.

Assess Your Strengths

To analyze your strengths, review all your work experience (summer, part-time, internship, full-time), your college courses, and your extracurricular activities to determine what activities have provided specific background in your field.

Prepare this analysis carefully. Talk to other people about yourself. List every skill and strength you can think of; don't exclude any experiences because they seem trivial. Seek qualifications that distinguish you from your competitors. Here are some questions (based in part on Harcourt and Krizar) to help you analyze yourself.

1. What work experience have you had that is related to your field? What were your job responsibilities? In what projects were you involved? With what machinery or evaluation procedures did you work? What have your achievements been?

2. What special aptitudes and skills do you have? Do you know advanced testing methods? What are your computer abilities?

3. What special projects have you completed in your major field? List processes, machines, and systems with which you have dealt.

4. What honors and awards have you received? Do you have any special college achievements?

5. What is your grade point average?

6. How have you paid for your college expenses?

7. What was your minor? What sequence of useful courses have you completed? A sequence of three or more courses in, for example, management, writing, psychology, or communication might have given you knowledge or skills that your competitors do not possess.

8. Are you willing to relocate?

9. Are you a member of a professional organization? Are you an officer? What projects have you participated in as a member?

10. Can you communicate in a second language? Many of today's firms are multinational.

11. Do you have military experience? While in the military, did you attend a school that applies to your major field? If so, identify the school.

Assess the Needs of Employers

To promote your strengths, study the needs of your potential employers. Review their websites, and, if they have them, their Facebook and Twitter sites. You can easily discover the names of persons to contact for employment information and details describing the company, as well as its location(s) and the career opportunities, training and development programs, and benefits it offers.

Planning the Résumé

Your résumé is a one-page (sometimes two-page) document that summarizes your skills, experiences, and qualifications for a position in your field. Plan it carefully, selecting the most pertinent information and choosing a readable format.

Information to Include in a Résumé

The information to include in a résumé is that which fills the employer's needs. Most employers expect the following information to appear on applicants' résumés (Harcourt and Krizar; Hutchinson and Brefka):

- Personal information: name, address, phone number
- Educational information: degree, name of college, major, date of graduation
- Work history: titles of jobs held, employing companies, dates of employment, duties, a career objective
- Achievements: grade point average, awards and honors, special aptitudes and skills, achievements at work (such as contributions and accomplishments)

Résumé Organization

Traditionally, the information required on a résumé has usually been arranged in chronological order, emphasizing job duties. Because employers are accustomed to this order, they know exactly where to find information they need and can focus easily on your positions and accomplishments (Treweek; "Results").

The chronological résumé has the following sections:

- Personal data
- Career objective
- Summary (optional)
- Education
- Work experience

Personal Data

The personal data consist of name, address, telephone number (always found at the top of the page), place to contact for credentials, willingness to relocate, and honors and activities (usually found at the bottom of the page). If appropriate add an e-mail address and personal website URL.

List your current address and phone number. Tell employers how to acquire credentials and letters of reference. If you have letters in a placement file at your college career services office, give the appropriate address and phone numbers. If you do not have a file, indicate that you can provide names on request.

Federal regulations specify that you do not need to mention your birth date, height, weight, health, or marital status. You may give information on hobbies and interests. They reveal something about you as a person, and they provide topics of conversation at a surprising number of interviews.

Career Objective

The career objective states the type of position you are seeking or what you can bring to the company. A well-written objective reads like this: "Management Consulting Position in Information Systems" or "Position in Research and Development in Microchip Electronics" or "To use my programming, testing, and analysis skills in an information systems position." Many résumé readers prefer a statement that includes the name of their company and the title of the job you are applying for, for example "Manufacturing Engineer for BignStrong Box Corporation."

Summary

The summary, an optional section, emphasizes essential points for your reader (Parker). In effect, it is a mini-résumé. List key items of professional experience, credentials, one or two accomplishments, and one or two skills. If you don't have room for the summary in the résumé, consider putting it into your accompanying letter.

SUMMARY OF QUALIFICATIONS

Strong operations and client relationship management background with proven expertise in leading an operations team for multimillion-dollar retail organization. Well-developed customer relations skills that build lasting client loyalty. Proven new business development due to excellent prospecting and client rapport building skills. Able to develop processes that increase productivity, profitability, and employee longevity.

Education

The education section includes pertinent information about your degree. List your college or university, the years you attended it, and your major, minor, concentration, and grade point average (if good). If you attended more than one school, present them in reverse chronological order, the most recent at the top. You can also list relevant courses (many employers like to see technical writing in the list), honors and awards, extracurricular activities, and descriptions of practicums, co-ops, and internships. You do not need to include your high school.

EDUCATION

Bachelor of Science, University of Wisconsin–Stout, May 2014.
Major: Food Systems and Technology; Emphasis: Food Science
Minor: Chemistry
Associate of Applied Science Degree, Georgia Military College, Brunswick,
Georgia 2007.

ACADEMIC ACCOMPLISHMENTS

Phi Theta Kappa—International Honor Society of the Two Year College Academic
National Honor Society

Work Experience

The work experience section includes the positions you have held that are relevant to your field of interest. List your jobs in reverse chronological order—the
most recent first. In some cases, you might alter the arrangement to reflect the
importance of the experience. For example, if you first held a relevant eight-
month internship and then took a job as a dishwasher when you returned to
school, list the internship first. List all full-time jobs and relevant part-time
jobs—as far back as the summer after your senior year in high school. You do
not need to include every part-time job, just the important ones (but be prepared to give complete names and dates).

Each work experience entry should have four items: job title, job description, name of company, and dates of employment. These four items can be
arranged in several ways, as the following examples show. However, *the job
description is the most important part* of the entry. Describe your duties, the
projects you worked on, and the machines and processes you used. Choose
the details according to your sense of what the reader needs.

Write the job description in the past tense, using "action" words such as
managed and *developed*. Try to create pictures in the reader's mind (Parker).
Give specifics that he or she can relate to. Arrange the items in the description
in order of importance. Put the important skills first. The following example
illustrates a common arrangement of the four items in the entry:

PROFESSIONAL EXPERIENCE

Job title
Company
details

Sales/Marketing Director, Information Services Group, LLC, Milwaukee, WI
2010–2012

Duties and
accomplishments

Established client database that led to strong relationships with key accounts.

Extensive coordination over all advertising including: writing, proofreading, detail
organization, layouts, designs, and productions.

Personally responsible for several major accounts, doubling sales revenues for
fiscal year 2011.

Developed new accounts due to excellent prospecting and follow-through abilities.

Extensive telemarketing, cold calls, and sales presentations.

Organized, developed, and implemented new employee handbook; responsible for material and design.

Order of Entries on the Page

In the chronological résumé, the top of any section is the most visible position, so you should put the most important information there. Place your name, address, and career objective at the top of the page. In general, the education section comes next, followed by the work section. However, if you have had a relevant internship or full-time experience, put the work section first. Figure 17.2 shows a chronological résumé.

Writing the Résumé

Drafting your résumé includes generating, revising, and finishing it. Experiment with content and format choices. Ask a knowledgeable person to review your drafts for wording and emphasis. Pay close attention to the finishing stage, in which you check consistency of presentation and spelling.

The résumé must be easy to read. Employers are looking for essential information, and they must be able to find it on the first reading. To make that information accessible, use highlight strategies explained in Chapter 6: heads, boldface, bulleting, margins, and white space. Follow these guidelines (and compare them to the sample résumés in this chapter):

- Usually limit the résumé to one page.
- Indicate the main divisions at the far left margins. Usually, boldface heads announce the major sections of the résumé.
- Boldface important words such as job titles or names of majors; use underlining sparingly.
- Use bulleted lists, which emphasize individual lines effectively.
- Single-space entries and double-space above and below. The resulting white space makes the page easier to read.
- Control the margins and type size. Make the left margin 1 inch wide.
- Use 10- or 12-point type.
- Treat items in each section consistently. All the job titles, for example, should be in the same relative space and in the same typeface and size.
- Print résumés on good-quality paper; use black ink on light paper (white or off-white). Note: many employers want you to submit a digital copy, not a hard copy of your résumé. Follow the instructions in their ad. To find out more about sending résumés via e-mail, review Focus on Electronic Résumés on pp. 555–557.

Use LinkedIn

As of this writing (2012) LinkedIn was quickly developing as a likely source for being hired. Kathy Bernard suggests these strategies for improving your chances of being contacted for a job as a result of your LinkedIn presence.

Add a color photograph of yourself.
Seek recommendations from others to post on LinkedIn.
Join LinkedIn groups related to your field and your university.
Add your résumé to your profile.
Use your status bar to let people know that you are job seeking.
Add relevant keywords throughout the Experience section of your profile.

Michelle L. Stewart

2837 Main Street   (715) 421-8765
Eau Claire, WI 54701   michstew27@yahoo.com

CAREER OBJECTIVE

To obtain a position in the food industry as a Consumer Scientist.

SUMMARY OF QUALIFICATIONS

Strong operations and client relationship management background with proven expertise in leading an operations team for multimillion-dollar retail organization. Well-developed customer relations skills that build lasting client loyalty. Proven new business development due to excellent prospecting and client rapport building skills. Able to develop processes that increase productivity, profitability, and employee longevity.

EDUCATION

Bachelor of Science Degree, University of Wisconsin–Stout, May 2013
Major: Food Systems and Technology; Emphasis: Food Science
Minor: Chemistry
Associate of Applied Science Degree, Georgia Military College, Brunswick, Georgia 2006.

ACADEMIC ACCOMPLISHMENTS

Phi Theta Kappa—International Honor Society of the Two Year College
Academic National Honor Society

PROFESSIONAL EXPERIENCE

Sales/Marketing Director, Information Services Group, LLC, Milwaukee, WI 2010–2011.

Established client database that led to strong relationships with key accounts.

Extensive coordination over all advertising including: writing, proofreading, detail organization, layouts, designs, and productions.

Personally responsible for several major accounts, doubling sales revenues for fiscal year 2010.

Developed new accounts due to excellent prospecting and follow-through abilities.

Extensive telemarketing, cold calls, and sales presentations.

Organized, developed, and implemented new employee handbook; responsible for material and design.

Sales Professional, IKON Technology Services, Milwaukee, WI 2009–2011

Responsible for maintaining client relationships with several key business accounts.

Coordinated delivery of products and services to ensure on-time completion of projects.

Operations/Merchandise Manager, Best Buy Co., Inc, Madison, WI 2007–2009

Managed the daily processes of operations team including: development for three-person staff, flow delegation, implementation of new systems, processing of all transactions, and effective cost management.

Managed major departmental merchandising reorganizations according to company standards.

Created new performance standards and assessment tools that ranked employee performance and provided training and employee coaching to help employees meet new goals.

Primarily responsible for scheduling staff of 105+ and maintaining monthly labor budgets based on sales volume/store performance/job functions resulting in improved productivity.

Trained staff on company procedures to increase their ability to provide excellent customer service resulting in an increase in team morale, customer loyalty, and profitability.

Coordinated human resource activities including: hiring, training, performance evaluations, and team development.

(Continued)

Administered employee benefits, compensation, and payroll; mediated employee legal conflict/disputes; and dealt with employee terminations.

Developed new processes to reduce controllable expenses, resulting in an increase in net profit.

RELATED EXPERIENCE

Front End Operations Manager, Sam's Club, Madison, WI 2006–2007

Pharmacy Department Manager, Wal-Mart Corporation, St. Mary's, GA 2002–2006

© Cengage Learning

Figure 17.2   Chronological Résumé

Planning a Letter of Application

The goal of sending a letter of application (also called a *cover letter*) is to be invited to an interview. To write an effective letter of application, understand the employer's needs, which are expressed in an ad or a job description. Planning a specific letter requires you to analyze the ad or description and match the stated requirements with your skills.

Analyze the Employer's Needs

To discover an employer's needs, analyze the ad or analyze typical needs for this kind of position. To analyze an ad, read it for key terms. For instance, a typical ad could read, "Candidates need 1+ years of C++. Communication skills are required. Must have systems analysis skills." The key requirements here are 1+ years of C++, communication skills, and systems analysis.

If you do not have an ad, analyze typical needs for this type of job. A candidate for a manufacturing engineer position could select pertinent items from the list of responsibilities printed in the *Dictionary of Occupational Titles* (see p. 522).

Match Your Capabilities to the Employer's Needs

The whole point of the letter is to show employers that you will satisfy their needs. If they say they need 1+ years of C++, tell them you have it.

As you match needs with capabilities, you will develop a list of items to place in your letter. You need not include them all; discuss the most important or interesting ones.

Ethics and Résumés

In writing a résumé, you want to engender confidence in your abilities, and avoid either underselling or overselling your experience. Recruiters are often looking through scores of résumés in search of measurable accomplishments that sound relevant for the position being filled. Using creative words to enhance the sound of an otherwise mundane job may result in your résumé being passed over, as the manager spends extra seconds trying to decipher the hidden meanings. What you say about your experience should be defensible and logical, and can be creative, but not outlandish. The résumé that honestly and straightforwardly presents the candidate's experience with a positive spin has the best chance of being read and landing you an interview (Truesdell). For example, stretching dates of employment to cover a jobless period is lying, and can cost you the offer or get you fired later on. Saying that you "specialized in retail sales and assisted in a 10 percent increase in sales at the store level," when you worked at the local video store and the store saw a 10 percent increase in sales is embellishing, but is not a lie. Claiming that you were store manager when you were not is lying (Trunk).

Résumé padding, telling lies on résumés, is becoming more and more common. Lying about experience or accomplishments on a résumé is risky. Résumé padding may get you a job for the short term if the employer hiring you does not check your résumé carefully, but résumé padding can come back to haunt you later (Callahan). *Inventing experiences, educational degrees, and accomplishments shouldn't be done due to the damage it can cause the writer down the road, even if the writer does not have a moral problem with lying.* False information on your résumé sits like a land mine waiting to explode (Callahan). In the recent past, lies on résumés ruined prestigious and lengthy careers. Companies have fired successful CEOs upon discovering they had misrepresented university credentials (Callahan). When share prices plummet as a result, investors may even sue the CEO for fraud. An otherwise stellar career that may never have needed a made-up degree can be brought to an unceremonious and humiliating end. Being a candidate without a master's degree or with a gap in employment is not out of the ordinary. Being a candidate who got caught stretching dates of employment to cover gaps or inventing a degree that never existed is inexcusable, ruinous, and unethical (Callahan; Truesdell).

Writing a Letter of Application

A letter of application has three parts: the introductory application, the explanatory body, and the request conclusion. The usual strategy is to organize your letter's body by skills, matching your experience to the employer's key word. This section first reviews the parts of a letter of application and then presents the entire letter in block format (see Figure 17.3, pp. 535–536).

Apply in the Introduction

The introductory application should be short. Inform the reader that you are applying for a specific position. If it was advertised, mention where you saw the ad. If someone recommended that you write to the company, mention that person's name (if it is someone the reader knows personally or by name). You may present a brief preview that summarizes your qualifications.

Apply

Tell source

Qualification
preview

I am interested in applying for the patient services manager position recently advertised in your Web homepage. I will complete a bachelor's degree in Dietetics in May 2014 from the University of Wisconsin–Stout. The skills I have developed from my academic background support my strong interest in working with your leading food and facility management services. I feel that my career goals and strong beliefs in assisting others to achieve a higher quality of life make me an excellent candidate for this position.

Convince in the Body

The *explanatory body* is the heart of the letter. Explain, in terms that relate to the reader, why you are qualified for the job. This section should be one to three paragraphs long. Its goal is to show convincingly that your strengths and skills will meet the reader's needs. Write one paragraph or section for each main requirement.

Base the content of the body on your analysis of the employer's needs and on your ability to satisfy those needs. Usually the requirements are listed in the ad. Show how your skills meet those requirements. If the ad mentioned "experience in software development," list details that illustrate your experience. If you are not responding to an ad, choose details that show that you have the qualifications normally expected of an entry-level candidate.

The key to choosing details is "memorable impact." The details should immediately convince readers that your skill matches their need. Use this guideline: In what terms will they talk about me? Your details, for instance, should show that you are the "development person." If you affect your reader that way, you will be in a positive position.

Globalization and Job Applications

Applying for a job overseas may open the way for an exciting professional and personal adventure. Understanding how to reformat your résumé to fit the needs of your potential employers will make the process faster and easier.

The term *curriculum vitae* (CV) is often used in other countries. A CV is generally the same as a résumé—a document detailing your education and work experience. However, different countries, employers, and cultures call for different areas of information and different levels of detail. Most countries have specific formats that they find acceptable. If you are unsure about which format you should use, ask; or use the standard, reverse-chronological format—put the most recent job highest in the list and work backward. Many countries expect information on résumés and CVs considered unprofessional or illegal in the United States: date of birth, marital status, and nationality ("Résumés").

When applying to a Japanese company, you may submit either a two-page, standard format résumé in Japanese, called a *rirekisho,* or a cover letter and résumé in English. Center your name and contact information at the top of the page and begin your résumé with a summary of your major qualifications. Your next section will be Employment Experience, followed by Education. Finally, you will end your résumé with personal information—date of birth, marital status, and nationality (Thompson).

If you are interested in working and living in the United Kingdom, you will format your résumé a bit differently. Include a cover letter addressed to a specific person. Your résumé will start with your personal information: name, contact information, date of birth, marital status, and nationality. Three major sections follow. The Profile section describes your professional designation, your immediate ambitions, and, in bulleted list form, your relevant skills and work-related achievements. Your Employment History section begins with your current position and then provides the name, location, and focus of the companies you have worked for. In the Education section, list your schools in reverse chronological order and include degrees awarded, additional courses and training, and any special skills (Thompson).

If you do write your résumé in the language of the country to which you are applying, find a native speaker of the language to review it. If it is written in a language other than English, one of your goals is to show that you are familiar with culturally appropriate language. Furthermore, if you e-mail your résumé, keep in mind that the American standard paper size is 8.5 by 11 inches, whereas the European A4 standard is 210–297 mm. Be sure to reformat your document in your word processing program for those parameters to ensure that the receiver doesn't lose any of your information.

For further reference, check out these websites:

Goinglobal at <www.goinglobal.com> contains lots of information on applying, interviewing, and working abroad. You can find advice on how you can expect to fit into the culture whether you want to work in Australia, Belgium, or China.

Jobweb at <www.jobweb.com> is the online complement to the *Job Choices* magazine series. It has good, general information on all aspects of your job search, including an international search. It includes sample résumés and CVs and has articles written by professionals.

Monster Workabroad at <http://workabroad.monster.com> includes practical advice on getting a job overseas. You'll find specific job information as well as general information on your country of choice. You can also receive newsletters and chat with other international job hunters.

Skills Section

Source of skill

In my courses, I have learned many cost control methods. For instance Institutional Food Purchasing and Food Beverage Cost Controls required me to create budgets, track daily expenditures, and manage time. I understand how to monitor costs in relation to patient food service. I understand the qualities that a good manager possesses and believe that it is essential to strive to achieve and maintain a high level of patient satisfaction.

I have also had experience in food safety. Through the educational aspects of Microbiology and Food Science, I've developed a strong background in microorganisms and deciphering what control methods to use that will ensure the quality of the organization and its services. I have acquired a solid background in relation to Hazard Analysis Critical Control Point (HACCP) practices and methods.

Convince

I am knowledgeable about menu delivery systems and the proper procedures that must be followed to assure compliance with regulations. After completing my JACHO certification, I am aware that this organization sets the standards by which health care quality is measured. I will thus work to continuously improve the safety and quality of care provided to the public through the health care accreditation standards of JACHO.

Use of skill

Skill activities

Request an Interview

In the final section, ask for an interview and explain how you can be reached. The best method is to ask, "Could I meet with you to discuss this position?" Also explain when you are available. If you need two days' notice, say so. If you can't possibly get free on a Monday, mention that. Most employers will try to work around such restrictions. If no one is at your house or dorm in the morning to answer the phone, tell the reader to call in the afternoon. A busy employer would rather know that than waste time listening to a phone ring. Thank your reader for his or her time and consideration. Readers appreciate the gesture; it is courteous and it indicates that you understand that the reader has to make an effort to fulfill your request.

Request

How to contact writer

Thank you

I would welcome the opportunity to meet with you and discuss my qualifications for the position. I have enclosed a copy of my résumé. If you have any questions or would like to talk with me, I can be reached by phone at (715) 555-1224 or e-mail at michshan@wisuniv.edu. Thank you for considering me for this position. I look forward to hearing from you soon.

Select a Format

To make a professional impression, follow these guidelines:

- ◗ Type the letter on 8.5-by-11-inch paper.
- ◗ Use white, 20-pound, 100 percent cotton-rag paper.
- ◗ Use black ink.
- ◗ Use block or modified block format explained above.
- ◗ Sign your name in black or blue ink.
- ◗ Proofread the letter carefully. Grammar and spelling mistakes are irritating at best; at worst, they are cause for instant rejection.

Examples 17.1–17.2 (pp. 540–541) show two application letters organized by skills. Examples 17.3 and 17.5 (pp. 542 and following pages) show two résumé styles.

1427 Crestview Street
Menomonie, WI 54751

November 5, 2013

Johnson-United Hospital
2715 Jamestown Avenue
Gaithersburg, Maryland 20878
Subject: Patient Services Manager Position

Clear application

Use of key words at the beginning of the paragraph.

I am interested in applying for the patient services manager position recently advertised in your Web homepage. I will complete a bachelor's degree in Dietetics in May 2014 from the University of Wisconsin–Stout. The skills I have developed from my academic background support my strong interest in working with your leading food and facility management services. I feel that my career goals and strong beliefs in assisting others to achieve a higher quality of life make me an excellent candidate for this position.

Details of work experience and classes used to create interest.

As a student, I have learned many cost control methods. The courses Institutional Food Purchasing and Food Beverage Cost Controls have allowed me to become familiar with proper operating techniques, accounting for the monitoring of costs in relation to patient food service. I understand the qualities that a good manager possesses and believe that it is essential to strive to achieve and maintain a high level of patient satisfaction.

Use of key words at the beginning of the paragraph.

I have also had experience in food safety. Through the educational aspects of Microbiology and Food Science, I've developed a strong background in microorganisms and deciphering what control methods to use that will ensure the quality of the organization and its services. I have acquired a solid background in relation to Hazard Analysis Critical Control Point (HACCP) practices and methods.

I am knowledgeable about menu delivery systems and the proper procedures that must be followed to assure compliance with regulations. After completing my JACHO certification, I am aware that this organization sets the standards by which health care quality is measured. I will thus work to continuously improve the safety and quality of care provided to the public through the health care accreditation standards of JACHO.

Asks for interview.

I would welcome the opportunity to meet with you and discuss my qualifications for the position. I have enclosed a copy of my résumé. If you have any questions or would like to talk with me, I can be reached by phone at (715) 555-1224 or

(Continued)

e-mail at michshan@uw.edu. Thank you for considering me for this position. I look forward to hearing from you soon.

Sincerely,

Shannon M. Michaelis

Shannon M. Michaelis, R.D.

Enclosure: résumé

© Cengage Learning

Figure 17.3 Complete Letter of Application in Block Format

Interviewing

The employment interview is the method employers use to decide whether to offer a candidate a position. Usually the candidate talks to one or more people (either singly or in groups) who have the authority to offer a position. To interview successfully, you need to prepare well, use social tact, perform well, ask questions, and understand the job offer (Stewart and Cash).

Prepare Well

To prepare well, investigate the company and analyze how you can contribute to it (Spinks and Wells). To investigate the company, consult their website, read their Facebook page, perhaps sign up to follow them on Twitter. For a more in-depth look, read annual reports, descriptions in *Moody's,* items from *Facts on File, F&S Index, Wall Street Journal Index,* or *Corporate Report Fact Bank.* After you have analyzed the company, assess what you have to offer. Answer these questions:

- What contributions can you make to the company?
- How do your specific skills and strengths fit into its activities or philosophy?
- How can you further your career goals with this company?

Use Social Tact

To use social tact means to behave professionally and in an appropriate manner. Acting too lightly or too intensely are both incorrect. First impressions are extremely important; many interviewers make up their minds early in the interview. Follow a few common sense guidelines:

▶ Shake hands firmly.

▶ Dress professionally, as you would on the job.

▶ Arrive on time.

▶ Use proper grammar and enunciation.

▶ Watch your body language. For instance, sit appropriately; don't lounge or slouch in your chair.

▶ Find out and use the interviewers' names.

Perform Well

Performing well in the interview means to answer the questions directly and clearly. Interviewers want to know about your skills. Be willing to talk about yourself and your achievements; if you respond honestly to questions, your answers will not seem like bragging. For a successful interview, follow these guidelines:

▶ Be yourself. Getting a job based on a false impression usually ends badly.

▶ Answer the question asked.

▶ Be honest. If you don't know the answer, say so.

▶ If you don't understand a question, ask the interviewer to repeat or clarify it.

▶ In your answers, include facts about your experience to show how you will fit into the company.

Ask Questions

You have the right to ask questions at an interview. Make sure you have addressed all pertinent issues (Spinks and Wells). If no one has explained the following items to you, ask about them:

▶ Methods of on-the-job training

▶ Your job responsibilities

▶ Types of support available—from secretarial to facilities to pursuit of more education

▶ Possibility and probability of promotion

▶ Policies about relocating, including whether you get a promotion when you relocate and whether refusing to relocate will hurt your chances for promotion

▶ Salary and fringe benefits—at least a salary range, whether you receive medical benefits, and who pays for them

Understand the Offer

Usually a company will offer the position—with a salary and starting date—either at the end of the interview or within a few days. You have the right to request a reasonable amount of time to consider the offer. If you get another offer from a second company at a higher salary, you have the right to inform the first company and to ask whether they can meet that salary. Usually you accept the offer verbally and sign a contract within a few days. This is a pleasant moment.

Writing Follow-Up Letters

After an interview with a particularly appealing firm, you can take one more step to distinguish yourself from the competition. Write a follow-up letter. It takes only a few minutes to thank the interviewer and express your continued interest in the job.

> Thank you for the interview yesterday. Our discussion of Ernst and Young's growing MIS Division was very informative, and I am eager to contribute to your team.
> I look forward to hearing from you.

Worksheet for Preparing a Résumé

☐ **Write out your career objective; use a job title.**

☐ **List all the postsecondary schools you have attended.**

☐ **List your major and any minors or submajors.**

☐ **List your GPA if it is strong.**

☐ **Complete this form. Select only relevant courses or experiences.**

College Courses	Skills Learned	Projects Completed

☐ **List extracurricular activities, including offices held and duties.**

☐ **Complete this form for all co-ops, internships, and relevant employment.**

Job Title	Company	Dates	Duties	Achievements

☐ **List your name, phone number, current address, and permanent address if it is different. If appropriate, add e-mail address and personal website URL.**

☐ **Review standard résumé format. See pages 528–530.**

☐ **Choose a layout design.**

Worksheet for **Writing** a **Letter** of **Application**

☐ **State the job for which you are applying.**

☐ **State where you found out about the job.**

☐ **Complete this form:**

Employer Need (such as "program in C++")	Proof That You Fill the Need (show yourself in action: "developed two C++ programs to test widget quality")

☐ **Select a format; the block format is suitable.**

☐ **Write compelling paragraphs:**

- An introduction to announce that you are an applicant
- A body paragraph for each need, matching your capabilities to the need
- Select details that cause "measurable impact"—ones that cause readers to remember you because you can fill their needs
- A conclusion that asks for an interview

☐ **Purchase good-quality paper and envelopes, and get a new cartridge for your printer if you produce the letter yourself.**

Worksheet for **Evaluating** a **Letter** of **Application**

Answer these questions about your letter or a peer's:

a. Are the inside address and date handled correctly? Are all words spelled out?

b. Do the salutation and inside address name the same person?

c. Is there a colon after the salutation?

d. Does the writer clearly apply for a position in paragraph 1?

e. Does each paragraph deal with an employer need and contain an "impact detail"?

f. Does the closing paragraph ask for an interview? In an appropriate tone?

g. Would you ask this person for an interview? Why or why not?

Examples

The Examples 17.1 and 17.2 illustrate ways in which applicants can show how their skills meet the employers' needs. Examples 17.3 and 17.5 illustrate advance student and professional resumes. Example 17.4 shows how Example 71.3 looks as an ASCII text.

Example 17.1

Letter
Organized by
Skills

Slightly
modified
block format
uses indented
paragraphs to
save space.

Specific
application.

Each body
paragraph has
in first line a
key word or
phrase from
the ad.

Brief narrative
demonstrates
skills.

Two examples
illustrate skills.

Narrative
illustrates
skills.

Request for
contact.

1503 West Second Street
Menomonie, WI 54751

ABC Global Services
1014 Michigan Avenue
Chicago, IL 60605

November 3, 2013

SUBJECT: I/T Specialist–Programmer Position

I would like to apply for the IT Specialist–Programmer Position for ABC Global Services in Chicago. I learned of this position through the University of Wisconsin–Stout Placement and Co-op Office. Because of my past co-op experiences and educational background, I feel I am an ideal candidate for this position.

I am very familiar with all the steps of the application life cycle and processes involved in each of these steps. In my Software Engineering course, I and four other students developed a Math Bowl program that is used at the annual Applied Math Conference. We analyzed the problem through an analysis document, created a design document, coded the Math Bowl program from the design document, and maintained the software. In my past co-op with IBM, I was also involved in requirements planning, design reviews, coding, testing, and maintenance for my team's projects.

I have experience with low-level languages such as C++ and Assembler Language through my work experience and course study. In my Computer Organization class, I developed a CPU simulator in C++, which manipulated the Assembler Language. During my Unisys co-op in the Compiler Products department, I developed test programs in C, which manipulated the low-level compiler code.

I have strong customer relations and communication skills. Every day I deal with students and faculty at my job as a Lab Assistant at the Campus Computer Lab. It is my job to help them learn the available software and troubleshoot user's problems.

Enclosed is my résumé for your consideration. I am interested in talking with you in person about this position and a possible interview. Please call me at (715) 242-3368 at your convenience. Thank you for your consideration.

Sincerely,

Heather Miller

Heather Miller

Enclosure: résumé

Example 17.2

Letter
Organized by
Skills

Dana Runge
461-19th Avenue West Apt. 1
Menomonie, WI 54751

Standard
block format

October 29, 2012

Rachel Rizzuto
Campus Relations Specialist
Target Headquarters
1000 Nicollet Mall
Minneapolis, MN 55403

Dear Ms. Rizzuto:

Specific
application

I would like to apply for the Business Analyst position registered at the University of Wisconsin–Stout Placement and Co-op Office. I believe that my past retail experience and education make me an ideal candidate for the position.

Each body
paragraph has
in first line a
key word or
phrase from
the ad.

I have excellent analytical skills that were developed further from my education and work experience. In my Managerial Accounting course, my group researched the Pepsi Corporation's performance through their 2011 Annual Report. We applied that research by converting numbers and statistics into financial ratios and analysis. We then compared that information to industry norms and made conclusions. Our results were reported to the class as a PowerPoint presentation.

Narrative
demonstrates
skills.

I have great experience with planning and organizing. As the Events Coordinator for the Stout Retail Association, I researched, planned, and implemented a three-day trip to Chicago. I organized everything, from the hotel and transportation to the tours and activities. I had to make several contacts and reservations, as well as create a detailed time and activity agenda. I was responsible for the schedule of sixty members.

Narrative
demonstrates
skills.

I demonstrated clear and effective communication for a group country report in my Environmental Science class. I led the group in outlining what each member was responsible for doing, as well as the time line for completion dates. We set up meetings and exchanged communication channels, such as phone numbers and e-mail addresses. Our group had a clear understanding of what each needed to do as an individual, as well as in a group. We then compiled our information into a class presentation.

Request for
contact.

A copy of my résumé is enclosed for your review. I am available for an interview at your convenience. If you have any questions, please feel free to call

(Continued)

(715-237-1421) or e-mail me (rungdan@uws.edu). Thank you for your time and consideration. I look forward to your reply.

Sincerely,

Dana Runge

Dana Runge

Enclosure: résumé

Example 17.3

Chronological
Resume

LAURA E. KREGER

2027 SE Taylor St
503.210.5381
Portland, OR 97214
laura.kreger@gmail.com

OBJECTIVE   To earn a Graduate Research Assistantship or grading position in the School of Community Health

EDUCATION   **Portland State University**, Portland, OR
Master of Public Health—Health Promotion, June 2013 (anticipated)
Cumulative GPA: 4.00

UC Berkeley Extension, online
Certificate Program: Professional Sequence in Editing, April 2011

University of Wisconsin-Stout, Menomonie, WI
BS in Technical Communication, May 2005
Cumulative GPA: 3.98, summa cum laude
Applied Field: Health Sciences
Studied Abroad: Southern Cross University, Lismore, Australia, *Feb.–June 2003*
University Honors Program

Significant Courses
• Foundations of Public Health
• Substantive Editing
• Writing for the Media
• Environmental Health
• Intermediate Copyediting
• Critical Writing

PROFESSIONAL EXPERIENCE

Marketing Copywriter, *Feb. 2007–April 2011*
Content Editor, *Sept. 2005–June 2006; Sept. 2006–Feb. 2007*
Fred Meyer, Inc., Portland, OR

- Wrote food and pharmacy print and Web advertisements, e-mails, signage, etc.
- Edited the work of other writers at all stages of the proofing process.
- Collaborated with account managers, buyers, designers, and other writers.
- Interviewed corporate office and store department managers for story content.
- Researched food trends, nutrition principles, and natural and organic standards.

Freelance Writer & Copyeditor, *2002–present*

- Published in *Intentionally Urban* magazine and Llewelyn's *Green Living Guide*
- Copyedited *Welcome to the Table* by Tony Kriz

Office Assistant, *2003–2005*
Office of International Education, UW-Stout, Menomonie, WI

- Organized and led predeparture orientation for students studying abroad
- Coordinated student teaching program in Australia
- Communicated with students, parents, faculty, embassies, and universities abroad

Grader, *2002–2005*
Dr. Nancy Schofield, UW-Stout, Menomonie, WI

- Graded computer-aided drafting and design (CADD) assignments, tests, and finals
- Supervised open lab hours

Public Health Communicator, *June–Sept. 2004*
Practicum Experience with the UW-Stout Obesity Research Team

- Wrote and designed a flyer, press release, and grant proposal
- Researched, wrote, and designed health education handouts and activity sheets

(Continued)

VOLUNTEER EXPERIENCE	**Volunteer**, Red Cross Emergency Warming Center, *November 2010–present*
	Writers Group Leader, Imago Dei Community, *Sept. 2007–present*
	Lawn & Garden Volunteer, Holden Village, Chelan, WA, *June–Aug. 2006*
AWARDS	**2011 Urban & Public Affairs Graduate Memorial Scholarship**
	UW-Stout Chancellor's Award for Academic Excellence, Each semester 2000–2005

LAURA E. KREGER
4833 North Astoria St.   503.444.4444
Portland, OR 97227   lkreger@charternet.com

OBJECTIVE   To earn a Graduate Research Assistantship or grading
                 position in the School of Community Health

EDUCATION   Portland State University, Portland, OR
                Master of Public Health—Health Promotion, June 2013
                (anticipated) Cumulative GPA: 4.00

                UC Berkeley Extension, online
                Certificate Program: Professional Sequence in Editing, April
                2011

                University of Wisconsin-Stout, Menomonie, WI
                       BS in Technical Communication, May 2005
                Cumulative GPA: 3.98, summa cum laude
                       Applied Field: Health Sciences
                       Studied Abroad: Southern Cross University, Lis-
                       more, Australia, Feb.–June 2003
                University Honors Program

                       Significant Courses
Foundations of Public Health Substantive Editing   Writing for the Media
Environmental Health   Intermediate Copyediting Critical Writing

PROFESSIONAL   Marketing Copywriter, Feb. 2007–April 2011
EXPERIENCE   Content Editor, Sept. 2005–June 2006; Sept. 2006–Feb. 2007
                 Fred Meyer, Inc., Portland, OR

- Wrote food and pharmacy print and web advertisements, emails, signage, etc.
- Edited the work of other writers at all stages of the proofing process
- Collaborated with account managers, buyers, designers, and other writers
- Interviewed corporate office and store department managers for story content
- Researched food trends, nutrition principles, and natural and organic standards

Freelance Writer & Copyeditor, 2002–present
- Published in Intentionally Urban magazine and Llewelyn's Green Living Guide
- Copyedited Welcome to the Table by Tony Kriz

Office Assistant, 2003–2005
Office of International Education, UW-Stout, Menomonie, WI
- Organized and led pre-departure orientation for students studying abroad
- Coordinated student teaching program in Australia
- Communicated with students, parents, faculty, embassies, and universities abroad

Grader, 2002–2005
Dr. Nancy Schofield, UW-Stout, Menomonie, WI
- Graded computer-aided drafting and design (CADD) assignments, tests, and finals
- Supervised open lab hours

Public Health Communicator, June–Sept. 2004
Practicum Experience with the UW-Stout Obesity Research Team
- Wrote and designed a flyer, press release, and grant proposal
- Researched, wrote, and designed health education handouts and activity sheets

VOLUNTEER EXPERIENCE	Volunteer, Red Cross Emergency Warming Center, November 2010–present
	Writers Group Leader, Imago Dei Community, Sept. 2007–present
	Lawn & Garden Volunteer, Holden Village, Chelan, WA, June–Aug. 2006
AWARDS	2011 Urban & Public Affairs Graduate Memorial Scholarship
	UW-Stout Chancellor's Award for Academic Excellence, Each semester 2000–2005

Example 17.4

LAUREL VERHAGEN
3747 Flintville Rd,
Green Bay, WI 54313 | 920-246-4036 |
verhagen.laurel@mcrf.mfldclin.edu

Technical communicators act as a filter between experts and their intended audience. We are trained to be genre-sensitive, audience-focused, well-organized, critical thinkers. We are trained in the art of communication.

EXPERIENCE

Marshfield Clinic Research Foundation, Biomedical Informatics Research Center, Marshfield, WI

Applications Analyst–Senior, *Infrastructure and Central Resource Team*   **04/2012–Present**

— Mentored programmer and application analyst team members focusing on productivity, efficiency, and effectiveness. (Currently focused on the Grant Budgeting Development effort for Sponsored Programs group.)

— Presented two posters at AMIA Joint Summit on Clinical Research Informatics.

— Contributed as co-author on a manuscript submitted to JAMIA in April 2012.

Applications Analyst, *Infrastructure and Central Resource Team*   **02/2010–03/2012**

— Worked with Legal Services from Marshfield Clinic, Security Health Plan and Ministry Health Care to develop strategic plan for infrastructure project.

— Managed fiscal accounts, tracked personnel time, and adjusted scope to accommodate project budget and resource availability.

— Fostered relationships with internal and external collaborators, including the Informatics Advisory Committee and scientists at University of Wisconsin–Madison.

— Drafted and maintained protocols and application submissions to the Institutional Review Board for human subjects' research.

— Coordinated multiple concurrent development efforts with a focus on daily meetings and rapid prototyping. Prioritized critical path and mandatory deliverables.

— Created and distributed Request for Proposals, evaluated vendors, and coordinated a complete security evaluation of Trusted Broker and the WGI Portal.

— Provided process improvement, user support, assistance on administrative projects for the ICR team.

Marshfield Clinic, Information Systems, Marshfield, WI

Software Product Analyst, *Financial*
Development Team   **01/2006–05/2009**

— Supported a software development team composed of seven programmers.

— Analyzed user work flow and needs; worked closely with staff and directors from billing/reimbursement, variance, data entry, and patient assistance center.

— Wrote functional specifications for new projects.

— Designed minor to moderately complex systems.

— Wrote software documentation in the form of online help and user guides.

— Managed software releases: tested applications for usability and function, followed change management procedure, communicated change to users, provided follow-up user support.

— Mentored software product analysts, interns, and application analysts on the financial systems.

— Member: IT Personnel Committee, Software Product Analyst Training Special Interest Group, Icon Standardization Special Interest Group, Color Standardization Special Interest Group, IS Recruitment webpage SIG (chairperson), Online Documentation Special Interest Group (chairperson), Online Documentation Tool Selection Workgroup, GUI Presentation Committee.

Software Product Analyst Intern, *Special Projects*
Development Team   **05/2005–08/2005**

— Wrote scriptable regression test plans for software applications that track medications and preventative service reminders.

— Tested applications for consistency, accuracy, and bug-reduction.

— Created full-featured help systems using RoboHelp and Techsmith SnagIt.

— Edited and reorganized group documentation.

University of Wisconsin-Stout, Menomonie, WI

Advanced Programmer, *University Publications*   **05/2004–12/2005**

— Designed clean, usable webpages using cascading style sheets and (x)html.

— Trained faculty and staff to use software and implement Web standards.

Project Manager, *website Usability Testing Center*   **01/2004–12/2005**

— Coordinated center projects including budget, timeline, and resource. Maintained and trained student employees on center equipment (Windows, Macintosh, video, etc.)

(Continued)

— Tested and assessed usability based on ease of navigation, use, and understanding.

Lab Assistant, *Technical Communication*
Resource Center   **02/2004–12/2005**

— Provided support to fellow students for software and hardware issues.

EDUCATION

University of Wisconsin-Stout, Menomonie, WI

B.S. in Technical Communication   **12/2005**

Applied Field: Biomedical Engineering

AWARDS

Distinguished Student Service Award, *Society for Technical Communication*
2005

Sigma Tau Chi, Honor Fraternity of the *Society for Technical Communication*
2005

RELATED EXPERIENCE

University of Wisconsin-Stout, Menomonie, WI

Industry Advisory Board Member, *Technical Communication Program*
2005–Present

Society for Technical Communication

Member   **2003–2010**

President, *University of Wisconsin-Stout*
Student Chapter   05/2005–12/2005

Treasurer, *University of Wisconsin-Stout Student Chapter*

Exercises

▶ Group

1. Your instructor will arrange you in groups of two to four by major. Each person
    should photocopy relevant material from one source in the library. Include at
    least the *Dictionary of Occupational Titles* and the *Occupational Outlook Handbook.*

In class, make a composite list of basic requirements in your type of career. Use that list as a basis for completing the Writing Assignments that follow.

▶ You Analyze

2. Analyze one of the letters in the Sample Documents section of the Instructor's Resource Manual (Examples xxx–xxx). Comment on format and effectiveness of tone, detail, and organization.

3. Analyze this letter. It responded to an ad for a consumer scientist. The general requirements included abilities to evaluate and analyze new products; to write reports; and to assist with training, test kitchen organization, cooking demonstrations, and developing recipes and guides.

> 2837 Main Street
> Eau Claire, WI 54701
>
> April 9, 2013
> Wolf Appliance Company, LLC
> Attention: Human Resources
> P.O. Box 44848
> Fitchburg, Wisconsin 53719
>
> Subject: Consumer Scientist Position
>
> I am extremely interested in becoming a part of Wolf Appliance Company as a Consumer Scientist. In May 2013, I will be graduating from the University of Wisconsin—Stout with a Bachelor of Science degree in Food Systems and Technology, with a concentration in Food Science. I am also knowledgeable in the development and design of marketing tools for new products.
>
> My background includes extensive experience with analyzing data, evaluating results, and developing reports. Through course work, I have had to collect, analyze, and develop scientific reports in various formats utilizing MS Office, adapting the report to the specific situation. As an operations manager, I also had the responsibility of gathering data, reporting the information, and explaining variances to the Corporate Office for profit and loss statements.
>
> I developed excellent organizational, training, and evaluating techniques as an operations manager for a $40-million-a-year business. With an average staff of 105 employees, I was responsible for ensuring the training and evaluations were done promptly and accurately.
>
> As a merchandise manager, I was responsible for managing several remerchandising projects simultaneously. The projects I was responsible for were successful because of my organization and leadership abilities. The enclosed résumé outlines my credentials and accomplishments in further detail.

If you are seeking a Consumer Scientist who is highly self-motivated and is a definite team player wanting to be a part of a successful team, then please consider what I have to offer. I would be happy to have a preliminary discussion with you or member of your committee. You can reach me at (725) 421-8765. I look forward to talking with you and exploring this opportunity further.

Sincerely,

Michelle Stewart

4. Analyze an ad and yourself by filling out the third point in the Worksheet for Writing a Letter of Application (p. 539).

5. Analyze yourself by using the fifth and seventh points in the Worksheet for Preparing a Résumé (p. 538).

6. After completing Writing Assignment 2, read another person's letter. Ask these questions: What do you like about this letter? What do you dislike about this letter? How would you change what you dislike?

7. After completing Writing Assignment 1, read the ad and résumé of a classmate. Read the ad closely to determine the employer's needs. Read the résumé swiftly—in a minute or less. Tell the author whether he or she has the required qualifications. Then switch résumés and repeat. This exercise should either convince you that your résumé is good or highlight areas that you need to revise.

▶ You Revise

8. Revise this rough draft.

1221 Lake Avenue
Menomonie, WI 54751

March 21, 2014

Human Resources
Polaris Industries Inc.
301–5th Avenue SW
Roseau, MN 54826

Subject: Inquiring about a summer co-op in the mechanical design area.

My name is Josh Buhr and I am a junior at UW–Stout majoring in Mechanical Design. I would like this co-op to learn and grow in my field.

Working with your company as a design engineer, I would be able to fulfill many of the requirements that you have listed in your ad. I am able to design and lay out

[handwritten annotations: p-532; I'm interested in applying for the co-op in the mechanical design area for Polaris Industries Inc. in Roseau.; background, the skill I have gained from my academic encourage me to apply for this job to be part of this prestigious mechanical design company]

[handwritten: I have strong communication skills / problem solver]

drawings on AutoCAD in 3-D so I can easily verify that there is freedom of movement between the parts. With AutoCAD 14 I can even animate the object to see even clearer the movement between parts.

If I find a problem with one of the designs I will be able to talk to the people necessary and bring up the topic in a productive manor. The course at Stout "discussion" has taught me how to work in groups and even how to seat people to get the most out of everyone's mind, in a positive atmosphere. *[handwritten: through the education of course at stout]*

I also need to mention that the AutoCAD experience at Stout has given me many needed skills for design. In the second semester of AutoCAD, I learned and used geometric tolerancing on a complete assembly and layout of a two cycle engine. This engine was fully dimensioned to clearance fits and I even made the piston and a few other parts animated to see that everything was in a freedom of movement. *[handwritten: for instance]*

If you like the qualifications that I have listed above please feel free to call me at (705) 236-9037 to set up an interview. I would be more that happy to meet with you. I have wanted to work with your company for several years after I bought my 95 500; and was never passed since.

Sincerely,

[handwritten: I'm looking forward to meet with you and discuss my qualifications for the position / I can be reached by phone or ... / Thank you for considering me for this position]

Josh Buhr

9. **Revise this rough draft.**

871 17th Avenue East
Menomonie, WI 54751
February 12, 2013

Parker Hannifin Corporation
Personnel Department
2445 South 25th Avenue
Broadway, IL 60513

Dear Personnel Department:

I wrote to apply for the entry level Plant Engineering position that you advertised in the Minneapolis Star Tribune January 27, 2013.

I will graduate from the University of Wisconsin–Stout in December with a Bachelor of Science degree in Industrial Technology with a concentration in Plant Engineering.

I have been in charge of projects dealing with capital equipment justification and plant layout while on my co-op with Kolbe and Kolbe Millwork, Company. Some specific work that I completed include: installation of a portable blower system and light design of jigs, conveyor beds and machines. I worked in the machine shop my first month there and have a working knowledge of equipment and procedures in that environment.

I am a self-motivated individual who prides in excelling in everything I take on. This quality is reflected in my résumé through my increased job responsibilities topping with my co-op "experience" and my 3.5 cumulative grade point average.

I would appreciate an opportunity to interview with you. Please contact me at the above address or call me at (725) 624-1629 after 2:00 p.m. on weekdays, or I can return your call. Thank you for your time and consideration. I look forward to hearing from you.

Sincerely,

Keith Munson

▶ You Create

10. Create a work experience résumé entry (p. 543) for one job you have held. Select an arrangement for the four elements. Select details based on the kind of position you want to apply for. Alternate: Create a second version of the entry but focus one version on applying for a technical position and one version on applying for a managerial position.

11. Write a paragraph that explains a career skill you possess.

Writing Assignments

1. Using the worksheet on page 538 as a guide, write your résumé following one of the two formats described in this chapter.

2. Find an ad for a position in your field of interest. Use newspaper Help Wanted ads or a listing from your school's placement service. On the basis of the ad, decide which of your skills and experiences you should discuss to convince the firm that you are the person for the job. Then, using the worksheet on page 539 as a guide, write a letter to apply for the job.

3. Write a learning report for the writing assignment you just completed. See Chapter 5, Assignment 8, p. 137, for details of the assignment.

Web Exercise

Visit at least two websites at which career positions are advertised relative to your expertise. Do one or more of the following, depending on your instructor's directions:

    a. Apply for a position. Print a copy of your application before you send it. Write a brief report explaining the ease of using the site. Include comments about any response that you receive.

    b. Analyze the types of positions offered. Is it worth your time and energy to use a site like this? Present your conclusions in a report or an oral report, as your instructor designates. Print copies of relevant screens to use as visual aids to support your conclusions.

    c. As part of a group of three or four, combine your research in part b into a large report in which you explain to your class or to a professional meeting the wide range of opportunities available to the job seeker.

Works Cited

Bernard, Kathy. "12 quick tips to improve your LinkedIn profile and presence right now, plus job postings." GetaJobTips.com. 2012. Web. 16 May 2012. <http://getajob-tips-for-getting-hired.blogspot.com/2010/10/12-quick-tips-to-improve-your-linkedin.html>.

Callahan, David. "Résumé Padding." *The Cheating Culture*. Web. 26 May 2012. <www.cheatingculture.com/resume-padding.htm>.

Curtis, Rose and Warren Simmons. *The resume.com Guide to Writing Unbeatable Resumes*. New York: McGraw-Hill: 2004. Print.

Dictionary of Occupational Titles. 4th ed. Rev. Washington, DC: U.S. Dept. of Labor, 2011. Web. 29 May 2012. <http://www.occupationalinfo.org/>.

Gonyea, James C., and Wayne M. Gonyea, *Electronic Résumés: A Complete Guide to Putting Your Résumé On-Line*. New York: McGraw-Hill, 1996. Print.

Harcourt, Jules, and A. C. "Buddy" Krizar. "A Comparison of Résumé Content Preferences of Fortune 500 Personnel Administrators and Business Communication Instructors." *Journal of Business Communications* 26.2 (1989): 177–190. Print.

Hilden, Eric. "The 2010 Orange County Resume Survey." Saddleback College. <www.saddleback.edu/.../2010OrangeCountyResumeSurveyResults.pdf>. Also available at Pittsley, Megan. "Results from 2010 Employer Resume Preference Survey." examiner.com. 2010. Web. May 16, 2012. http 16 May 2012. <http://www.examiner.com/article/results-from-2010-employer-resume-preference-survey>.

Hutchinson, Kevin L., and Diane S. Brefka. "Personnel Administrators' Preferences for Résumé Content: Ten Years After." *Business Communication Quarterly* 60.2 (1997): 67–75. Print.

Ireland, Susan. "How to Email Your Resume." Susan Ireland's Resume Site. Web. 16 May 2012. <http://susanireland.com/resume/online/email/>.

Kendall, Pat. "ASCII/Plain Text Resumes." Advanced Resume Concepts. 2012. Web. 14 May 2012. <http://www.reslady.com/ASCII-ABCs.html>.

Parker, Yana. *The Résumé Catalog: 200 Damn Good Examples.* Berkeley, CA: Ten Speed Press, 1996. Print.

"Prepare Your Resume for Email and On-line Posting." The Riley Guide. Web. 14 May 2012. <http://www.rileyguide.com/eresume.html>.

"Résumés/CVs." Goinglobal.com. 7 Feb. 2004 Web. 16 May 2012. <www.goinglobal.com/topic/resumes.asp>.

Skarzenski, Emily. "Tips for Creating ASCII and HTML Résumés." *intercom* 43.6 (1996): 17–18. Print.

Spinks, Nelda, and Barron Wells. "Employment Interviews: Trends in the Fortune 500 Companies—1980–1988." *The Bulletin of the Association for Business Communications* 51.4 (1988): 15–21. Print.

Stewart, Charles J., and William B. Cash, Jr. *Interviewing Principles and Practices* 8th ed. Dubuque, IA: Brown, 1997. Print.

Thompson, Mary Anne. "Writing Your International Résumé." Jobweb. Web. 31 Jan. 2004. <www.jobweb.com/Resources/Library/International/Writing_Your_185_01.htm>.

Treweek, David John. "Designing the Technical Communication Résumé." *Technical Communications* 38.2 (1991): 257–260. Print.

Truesdell, Jason. "Honesty." *Tech.Job.Search.* 26 Feb. 2004. Web. 20 May 2012. <www.jagaimo.com/jobguide/resume/g-honest.htm>.

Trunk, Penelope. "Resume Writing: Lies v. Honesty." 2 June 2003. *The Brazen Careerist.* 29 May 2012. <http://www.bankrate.com/brm/news/career/20030602a1.asp>.

Whitcomb, Susan Britton. *Resume Magic: Trade Secrets of a Professional Writer.* Indianapolis: JIST Publishing, 2007. Print.

Focus on
Electronic Résumés

Electronic résumés have changed the job search. In many businesses before a human reads the résumé, technology intervenes to do much of the initial sorting. As a result, "keyword strategies" are very important. This section explains briefly how the sorting works, keyword strategies, and online, scannable, and ASCII résumés.

How the Sorting Works

Using one of the methods explained later, the candidate submits a résumé that, because it is electronic, is put into a searchable database. When an employer wants to find candidates to interview, he or she searches the database with a software program that seeks those keywords that the employer says are important. For example, the employer might want someone who can design websites and who knows Dreamweaver and CSS coding. Every time the search program finds a résumé with those words in it, it pulls the résumé into an electronic "yes" pile, which the human can then read.

Gonyea and Gonyea explain the process this way: "If the computer finds the same word or words [that describe the candidate the company is attempting to find] anywhere in your résumé, it considers your résumé to be a match, and will then present your résumé, along with others that are also considered to be a match, to the person doing the searching" (62).

Thus, your use of effective keywords is the key to filling out such a form. As Gonyea and Gonyea say, "To ensure that your résumé will be found, it is imperative that you include as many of the appropriate search words as are likely to be used by employers and recruiters who are looking for someone with your qualifications" (62).

Keyword Strategies

The technological sorters search for keywords. Your odds of being one of the "hits" in the search are increased by including a lot of keywords in your résumé. In addition to using keywords as you describe yourself in the education and work history sections of your résumé, you can also include a keyword section right in your résumé.

Put the keyword section either first or last in your résumé, or in the box supplied by the résumé service. Use words that explain skills or list aspects of a job. The list should include mostly nouns of the terms that an employer would use to determine if you could fill his or her need—job titles, specific job duties, specific machines or software programs, degrees, major, and subjective skills, such as communication abilities. You can include synonyms; for instance, in the list next, Web design and DreamWeaver are fairly close in meaning, because you use one to do the other, but including both increase your chances of the software's choosing your résumé.

A short list might look like this:

> C++, software engineering, HTML, programmer, needs analysis, client interview, team, Web design, Photoshop, AuthorIT, design requirements.

Remember, the more "hits" the reading software makes in this list, the more likely that your résumé will be sent on to the appropriate department.

Online Résumés/Job Searches

Job-posting sites allow employers to post employment opportunities; résumé-posting sites allow candidates to post their information. The exact way in which sites work varies, but all of them work in one of two ways. On job-posting sites, like America's Job Bank, employers post ads, listing job duties and candidate qualifications, and candidates respond to those ads. On résumé-posting sites, like Monster.com, candidates post résumés in Web-based databases, and employers search them for viable applicants.

(Continued)

You have two options. You can begin to read the "Web want ads," and you can post your résumé.

"Web Want Ads" are posted at the job-posting site. For instance, America's Job Bank <www.ajb.dni.us> lists job notices posted by state employment offices, and Careerbuilder. com <http://www.careerbuilder.com/> posts ads for companies in all lines of work. You simply access the site and begin to read. In addition, many of the résumé-posting services have a want ad site. For instance, Monster.com <www.monster.com> has an extensive listing of jobs in all categories. In all of these sites, job seekers can search by city, by job type, by level of authority (entry level, manager, executive). Candidates can search free of charge, but companies pay the sites to post the ads.

Post Your Résumé

Many sites provide this service; usually, it is free. Each site has you create your résumé at the site. You open an account, then fill in the form that the site presents to you. For instance, you will be asked for personal information (e.g., name and address) and also such typical items as job objective, work experience, desired job, desired salary, and special skills. Usually, filling out the form takes about 30 minutes.

Scannable Résumés

Many companies use optical character recognition (OCR) software (McNair; Quible) to scan résumés. First, paper résumés are scanned, turning them into ASCII (ASCII is the same as .txt) files, which are entered into a database. Second, when an opening arises, the human resources department searches the database for keywords.

Those résumés that contain the most keywords are forwarded to the people who will decide whom to interview. This development means that job seekers must now be able to write résumés that are scannable and that effectively use keywords.

Scannable résumés are less sophisticated looking than traditional ones, because scanners simply cannot render traditional résumés correctly. These documents contain all the same sections as traditional résumés but present them differently.

ASCII/Plain Text/.txt Résumés

Often companies ask you to send your résumé by e-mail. The best way to do so is to send the résumé as an ASCII (or plain text or .txt, the terms are synonymous) file, one that contains only letters, numbers, and a few punctuation marks but does not contain formatting such devices as boldfacing and italics (Skarzenski; Whitcomb; a note—one survey ["2010"] showed that two-thirds of managers preferred to just get a .doc file). ASCII résumés maintain all the traditional sections; you just present them so that they will interact smoothly with whatever software program is receiving them.

An ASCII file is easy to create. After you create your résumé, presumably in MS Word or Apple Pages, save a copy as Plain Text (.txt). You may have to change the placement of some of the elements of the original document. Example 17.4 (pp. 544–545) illustrates part of example 17.3 converted to plain text. Note that by using the most recent version of Microsoft word much of the formatting was retained.

The key items to be aware of are (Kendall; Whitcomb):

Use one column. Start all heads and text at the same left-hand margin. Place heads ABOVE not next to the explanatory text.

Keep the line length to fewer than 65 characters. Some software systems have difficulty with longer lines. The easiest way to achieve this line length is to set your left margin to 1 inch and the right margin to 2.5 inches.

Use a 12-point fonts. Use the same font throughout the document.

Do not use italics, underlining, bold face, tabs, bullets. Substitute hyphens for bullets.

E-mail yourself and a friend your résumé. The résumé you receive should be an exact duplicate of what you saw on your computer screen before you e-mailed it.

To e-mail the résumé, you have three choices:

Send it as an attachment ("Hilden").

Copy it into the body of the e-mail ("Prepare").

Copy it into the body of the e-mail and also include it as an attachment (Ireland).

Opinion varies on which choice to make. If it is in the body of the e-mail, the reviewer immediately sees it, and does not have to open another document. If it is an attachment it is easy to send off to someone else. If it is both, you have all your bases covered. The best advice is read the job announcement. It probably states which way the company wants to receive the résumé. Follow their preference.

Send the file to the receiver in two ways. You can send a file as part of an e-mail message or as an attachment to an e-mail message. Some programs can read the message both ways, some only one way. If the recipient's program does not have the capabilities, it will not be able to read your message.

Brief Handbook for Technical Writers

APPENDIX CONTENTS

Handbooks like this appendix are easy to use if you already know the grammar concepts discussed in them, but what if you don't? Sorry, but there is no easy answer. If you have a problem with one of these "Red Ink" strategies (RIS), you probably habitually do it. The easiest way to alter your habit is to have a friend, teacher, or tutor help you. Ask him or her to read over some of your work and tell you whether you regularly employ one of these RISs. Some of the issues discussed later have "flags": for instance use of *this* or use of *–ing* words or pronouns. Your helper can point out the flags so that you learn to question yourself whether you have handled the situation correctly.

But the sentence errors don't have flags. Other than that there is a comma in a comma splice, there is no easy way to identify these problems. You can help yourself by setting the grammar checker in your word processor, or you can work on sentences till you learn to recognize the problems, or you can use the grammar way—learn to identify a clause (see "How Do You Know It Is a Comma Splice or Run-on" pp. 561–562 and "Focus on Pronouns" p. 566).

This appendix presents the basic rules of grammar and punctuation. It contains sections on problems with sentence construction, agreement of subjects and verbs, agreement of pronouns with their antecedents, punctuation, abbreviations, capitalization, and numbers.

Problems with Sentence Construction

The following section introduces many common problems in writing sentences. Each subsection gives examples of a problem and explains how to convert the problem into a clearer sentence. No writer shows all of these errors in his or her writing, but almost everyone makes several of them. Many writers have definite habits: They often write in fragments, or they use poor pronoun reference, or they repeat a word or phrase excessively. Learn to identify your problem habits and correct them.

Identify and Eliminate Comma Splices

A *comma splice* occurs when two independent clauses are connected, or spliced, with only a comma. You can correct comma splices in four ways:

1. Replace the comma with a period to separate the two sentences.

Splice

The difference is that the NC machine relies on a computer to control its movements, a manual machine depends on an operator to control its movements.

Correction

The difference is that the NC machine relies on a computer to control its movements. A manual machine depends on an operator to control its movements.

2. Replace the comma with a semicolon only if the sentences are very closely related. In the following example, note that the word *furthermore* is a conjunctive adverb. When you use a conjunctive adverb to connect two sentences, always precede it with a semicolon and follow it with a comma. Other conjunctive adverbs are *however, also, besides, consequently, nevertheless,* and *therefore*.

Splice

The Micro 2001 has a two-year warranty, furthermore the magnetron is covered for seven years.

Correction

The Micro 2001 has a two-year warranty; furthermore, the magnetron is covered for seven years.

3. Insert a coordinating conjunction (*and, but, or, nor, for, yet,* or *so*) after the comma, making a compound sentence.

Splice

The engines of both cranes meet OSHA standards, the new M80A has an additional safety feature.

Correction

The engines of both cranes meet OSHA standards, but the new M80A has an additional safety feature.

4. Subordinate one of the independent clauses by beginning it with a subordinating conjunction or a relative pronoun. Frequently used subordinating conjunctions are *where, when, while, because, since, as, until, unless, although, if,* and *after*. The relative pronouns are *which, that, who,* and *what*.

Splice

Worker efficiency will increase because of lower work heights, lower work heights maximize employee comfort.

Correction

Worker efficiency will increase because of lower work heights that maximize employee comfort.

Exercises

Correct the following comma splices:

1. Different models of computers, software programs, and text formats are incompatible, data processing and information retrieval is slow and inefficient.

2. Web content development is becoming more necessary as businesses rely on the Internet to provide product information, advertising, and purchasing options for consumers, good writers can convey all this information accurately, stylishly, and in a concise manner.

3. For example, a millimeter is one-thousandth of a meter, therefore, a nanometer is one million times smaller.

4. Positive displacement pumps produce a pulsating flow, their design provides a positive internal seal against leakage.

5. In the printing business there are two main ways of printing, the first is by using offset and the second is by using flexography.

6. Two methods currently exist to enhance fiber performance, one method is to orient the fibers.

Identify and Eliminate Run-On Sentences

Run-on, or fused, sentences are similar to comma splices but lack the comma. The two independent clauses are run together with no punctuation between them. To eliminate run-on sentences, use one of the four methods explained in the preceding section and summarized here.

- ▶ Place a period between the two clauses.
- ▶ Place a semicolon between them.
- ▶ Place a comma and a coordinating conjunction between them.
- ▶ Place a relative pronoun or subordinating conjunction between them.

Exercises

Correct the following run-on sentences:

1. Biology is not the only field of science that nanotechnology will permeate this revolution can quite possibly influence all sciences and most likely create more.

2. Nonpositive displacement pumps produce a continuous flow because of this design, there is no positive internal seal against leakage.

3. Offset is also known as offset lithography or litho printing the offset process uses a flat metal plate with a smooth printing surface that is not raised or engraved.

4. The countries that import OCC at the highest rates often produce corrugated board of 100 percent recycled fibers due to the lessened performance of bogus board, a dilemma may be created.

5. Images are engraved around the cylinder the plate is not stretched and distorted like traditional plates.

How Do You Know It Is A Comma Splice or Run-on?

It is a comma splice or run-on if it has two or more *independent clauses* improperly connected. In order to identify the problem you have to know what an independent clause is and how to recognize one. After that the problem is pretty easy to fix. If you don't know those terms, spend time with your writing lab, or a knowledgeable friend, or try a site like The Purdue Online Writing Lab http://owl.english.purdue.edu/.

You can practice by creating and changing some sentences. Follow the Methods explained next.

The basic method is that when you place clauses one after the other in a paragraph, you have to *join them appropriately*, or the Grammar Etiquette Team gets you for acting inappropriately.

How do you join them appropriately?

The answer is any of next three methods, BUT NOT with JUST a comma or with nothing at all:

Method 1. Just write two sentences.
  You will earn points with this card. You can redeem them for travel rewards.

Method 2. Place a semicolon between them.
  You will earn points with this card; you can redeem them for travel rewards.

Method 3. Use a coordinating conjunction between them. (This is probably the easiest way to fix these.)
  You will earn points with this card, and you can redeem them for travel rewards.

Method 4. Here they are, joined *inappropriately* with only a comma.
  You will earn points with this card, you can redeem them for travel rewards.

Method 5. And here they are joined *inappropriately* with NOTHING!
  You will earn points with this card you can redeem them for travel rewards.

(Continued)

Methods 1–3 are OK. The other two are not. The only way to fix those splices and run-ons yourself is to know two things: the definition of an independent clause and the appropriate way to join clauses. The grammar checker in your word processing program may have a setting, probably for "comma use" which could be a great help as you wrestle with this problem (the grammar checker in MS Word 2011 caught the errors in Methods 4 and 5).

Identify and Eliminate Sentence Fragments

Sentence fragments are incomplete thoughts that the writer has mistakenly punctuated as complete sentences. Subordinate clauses, prepositional phrases, and verbal phrases often appear as fragments. As the following examples show, fragments must be connected to the preceding or the following sentence.

1. Connect subordinate clauses to independent clauses.

   a. The fragment below is a subordinate clause beginning with the subordinating conjunction *because*. Other subordinating conjunctions are *where, when, while, since, as, until, unless, if,* and *after*.

Fragment

We should accept the proposal. Because the payback period is significantly less than our company standard.

Correction

We should accept the proposal because the payback period is significantly less than our company standard.

   b. The following fragment is a subordinate clause beginning with the relative pronoun *which*. Other relative pronouns are *who, that,* and *what*.

Fragment

The total cost is $425,000. Which will have to come from the contingency fund.

Correction

The total cost of $425,000 will have to come from the contingency fund.

2. Connect prepositional phrases to independent clauses. The fragment below is a prepositional phrase. The fragment can be converted to a subordinate clause, as in the first example below, or made into an *appositive*—a word or phrase that means the same thing as what precedes it.

Fragment

The manager found the problem. At the conveyor belt.

Correction 1

The manager discovered that the problem was the conveyor belt.

Correction 2

The manager found the problem—the conveyor belt.

3. Connect verbal phrases to independent clauses.

   a. Verbal phrases often begin with *-ing* words. Such phrases must be linked to independent clauses.

Fragment	The crew will work all day tomorrow. Installing the new gyroscope.
Correction	Tomorrow the crew will work all day installing the new gyroscope.

b. Infinitive phrases begin with *to* plus a verb. They must be linked to independent clauses.

Fragment	I contacted three vendors. To determine a probable price.
Correction	I contacted three vendors to determine a probable price.

Exercises

Correct the following sentence fragments:

1. The National Nanotechnology Initiative 247 million dollars in federal funding in 1999.
2. The fourth aspect of the quality control function making adjustments to the process, in order to bring specifications into line.
3. Thickness availability of Celotex from 1/20 to 21/40.
4. Through orientation and the alkali process secondary fiber performance enhanced to levels above current performance.
5. While cost savings initiatives and quality have become critical for businesses to remain competitive today.
6. The virgin fiber sought by foreign corrugated producers to supply their customers with a higher quality product.

Place Modifiers in the Correct Position

Sentences become confusing when modifiers do not point directly to the words they modify. Misplaced modifiers often produce absurd sentences; worse yet, they occasionally result in sentences that make sense but cause the reader to misinterpret your meaning. Modifiers must be placed in a position that clarifies their relationship to the rest of the sentence.

1. In the next sentence, *that is made of a thin, oxide-coated plastic* appears to refer to *the information.*

Misplaced modifier	The magnetic disk is the part that contains the information that is made of a thin, oxide-coated plastic.
Correction	The magnetic disk, which is made of a thin, oxide-coated plastic, is the part that contains the information.

2. In the next sentence, the modifier says that the horizontal position must be tested, but the meaning clearly is something different.

Misplaced modifier   Lower the memory module to the horizontal position that requires testing.

Correction   Lower the memory module that requires testing to the horizontal position.

Exercises

Correct the misplaced modifiers in the following sentences:

1. ADA noted that VCLDs reduced energy endurance without carbohydrate supplementation.

2. Three topic areas related to printing were formulated to determine Web feasibility.

3. Although EPA legislation contains strong support, it cannot make up for the lack of information in the two other areas for oil field service engineers.

4. Technology must continue to improve in order to fully benefit economically in the recovery and preparation processes.

Use Words Ending in *-ing* Properly

A word ending in *-ing* is either a present participle or a gerund. Both types, which are often introductory material in a sentence, express some kind of action. They are correct when the subject can perform the action that the *-ing* word expresses. For instance, in the next sentence, the *XYZ computer table* cannot *compare* cost and durability.

Unclear   Comparing cost and durability, the XYZ computer table is the better choice.

Clear   By comparing cost and durability, you can see that the XYZ computer table is the better choice.

Exercises

In the following sentences, the participle (the -ing word) is used incorrectly; revise them.

1. While walking across the parking lot, the red convertible had its hood raised.

2. When filling out an online document, personal data always appear first.

3. When using a laptop computer, the mouse can be difficult to manage.

4. After comparing the weight and the features of the two laptops, both laptops seem to be of equal value.

5. Reviewing the internship, visual displays floor plans, and merchandising were my main duties.

6. By eliminating unnecessary wording, this would decrease cost by $10.00 per book.

Make the Subject and Verb Agree

The subject and the verb of a sentence must both be singular or both be plural. Almost all problems with agreement are caused by failure to identify the subject correctly.

1. When the subject and verb are separated by a prepositional phrase, be sure you do not inadvertently make the verb agree with the object of the preposition rather than with the subject. In the following sentence, the subject *bar* is singular; *feet* is the object of the preposition *of.* The verb *picks* must be singular to agree with the subject.

Faulty   A bar containing a row of suction feet pick up the paper.

Correction   A bar containing a row of suction feet picks up the paper.

2. When a *collective noun* refers to a group or a unit, the verb must be singular. Collective nouns include such words as *committee, management, audience, union,* and *team.*

Faulty   The committee are writing the policy.

Correction   The committee is writing the policy.

3. Indefinite pronouns, such as *each, everyone, either, neither, anyone,* and *everybody,* take a singular verb.

Faulty   Each of the costs are below the limit.

Correction   Each of the costs is below the limit.

4. When compound subjects are connected by *or* or *nor,* the verb must agree with the nearer noun.

Faulty   The manager or the assistants evaluates the proposal.

Correction   The manager or the assistants evaluate the proposal.

Exercises

Correct the subject–verb errors in the following sentences:

1. Everybody in both classes were late in arriving to the classroom.
2. All the books in the bookstore is available for purchase.
3. When writing a technical manual, analysis of various audiences are very important.
4. The reasons the flat screen monitor should be used is well-documented in research.
5. Hypertext (HTML) is a method in which punctuation markings, spaces, and coding is used to create pictures on a webpage.

Use Pronouns Correctly

A pronoun must refer directly to the noun it stands for, its *antecedent*.

As in subject–verb agreement, a pronoun and its antecedent must both be singular or both be plural. Collective nouns generally take the singular pronoun *it* rather than the plural *they*. Problems result when pronouns such as *they, this,* and *it* are used carelessly, forcing the reader to figure out their antecedents. Overuse of the indefinite *it* (as in "*It* is obvious that") leads to confusion.

Controlling Pronouns

Focus on Pronouns

Pronouns create clarity problems because we use them so loosely in speech. Two rules govern pronouns: (1) They refer to an antecedent, and (2) they should remain consistent throughout a report or article.

Antecedents

The pronoun problem occurs when writers mention a role (e.g., a teacher) and then refer to that role with a pronoun. In speech many people simply use 'they' because the role is filled by many individuals.

So we get

The *teacher* must come to class prepared. *They* should have a lecture prepared or else *they* should have some in-class activities planned.

The issue in grammatical terms is that *teacher* is singular and *they* is plural. Thus, the pronoun and its antecedent do not *agree in number* (new term—*number* means one [singular] or more than one [plural]). In speech most people simply ignore the grammar error. In formal situations, like reports, the rule still holds.

So, you have to rewrite the sentences. One way to do so is to use *he or she*, which has become common. Another way is to just change the role word (*teacher*) to a plural (*teachers*):

> *Teachers* must come to class prepared. *They* should have a lecture prepared or else *they* should have some in-class activities planned.

Wandering Pronouns

To understand *wandering pronoun* consider this somewhat overdone example, but you will notice the usage. The pronouns wander back and forth from one to another. Don't write like this. Find a way to use just one pronoun.

> *The painter dips the brush into the paint in the can. They tap the sides of the brush against the rim. You get the excess paint off that way. Then we hold the brush roughly perpendicular to the surface we are painting. You spread a nice even coat. We now understand the basic hand moves that a painter makes.*

In speech often people mix up pronouns as in the paragraph above, but in writing the pronouns have to be consistent. You can pick any of the pronouns—*you, they, we*—but you have to pick one. Probably in this case *you* would be best.

> *Dip the brush into the paint in the can, and then tap the sides of the brush against the rim. You get the excess paint off that way. Hold the brush roughly perpendicular to the surface you are painting. You will spread a nice even coat. You now understand the basic hand moves that a painter makes.*

The only way to become comfortable with this issue is to watch your pronouns. When they start to wander, you need to revise by selecting just one of them.

Problems with Number

1. In the following sentence, the pronoun *It* is wrong because it does not agree in number with its antecedent, *inspections*. To correct the mistake, use *they*.

Vague   The inspections occur before the converter is ready to produce the part. It is completed by four engineers.

Clear   The inspections occur before the converter is ready to produce the part. They are completed by four engineers.

2. In current practice, it is now acceptable to deliberately misuse collective pronouns in an effort to avoid sexist writing.

Technically correct   Everyone must bring his or her card.

Correct for informal situations   Everyone must bring their card.

Problems with Antecedents

If a sentence has several nouns, the antecedent may not be clear.

1. In the following case, *It* could stand for either *pointer* or *collector*. The two sentences can be combined to eliminate the pronoun.

Vague   The base and dust *collector* is the first and largest part of the lead *pointer. It* is usually round and a couple of inches in diameter.

Clear   The base and dust collector, which is the largest part of the lead pointer, is usually round and several inches in diameter.

2. In the following case, *It* could refer to *compiler* or *software*.

Vague   The new *compiler* requires new *software. It* must be compatible with our hardware.

Clear   The new compiler, which requires new software, must be compatible with our hardware.

Problems with *This*

Many inexact writers start sentences with *This* followed immediately by a verb ("*This* is," "*This* causes"), even though the antecedent of *this* is unclear. Often the writer intends to refer to a whole concept or even to a verb, but because *this* is a pronoun or an adjective, it must refer to a noun. The writer can usually fix the problem by inserting a noun after *this*—and so turn it into an adjective—or by combining the two sentences into one. In the following sentence, *this* probably refers either to the whole first sentence or to *virtually impossible,* which is not a noun.

Vague   Ring networks must be connected at both ends—a matter that could make wiring virtually impossible in some cases. This would not be the case in the Jones building.

Clear   Ring networks must be connected at both ends—a matter that could make wiring virtually impossible in some cases. We can easily fill this requirement in the Jones building.

Exercises

Revise the following sentences, making the pronoun references clear:

1. Probably the computer will begin to have problems in three or four months. This is only an estimate but it would not be surprising, due to my experience with other computers. This would increase the amount of money spent on the computer.

2. The computers are in the warehouse and the boxes enter on a conveyor belt and then they load them into rail cars.

3. As more and more nodes are installed, it affects the time it takes for information to travel around the ring.

4. The pick-and-place robot places the part in the carton. It is sealed by an electronic heat seal device. [*Carton* is the intended antecedent.]

5. The computer malfunctions are generally minor problems that take one or two hours to fix. This costs the company about $1000.

6. As you can see, the IT department has a substantially higher absentee rate than any of the other departments. This shows a definite problem in this department. This is the only department that requires such extreme on-call requirements with little or no extra compensation.

Punctuation

Writers must know the generally accepted standards for using the marks of punctuation. The following guidelines are based on *The Chicago Manual of Style* and the U.S. Government's *A Manual of Style*.

Apostrophes

Use the apostrophe to indicate possession, contractions, and some plurals.

Possession

The following are basic rules for showing possession:

1. Add an *'s* to show possession by singular nouns.

   a machine's parts   a package's contents

2. Add an *'s* to show possession by plural nouns that do not end in *s*.

   the women's caucus   the sheep's brains

3. Add only an apostrophe to plural nouns ending in *s*.

   three machines' parts   the companies' managers

4. For proper names that end in *s,* use the same rules. For singular add *'s;* for plural add only an apostrophe.

   Ted Jones's job   the Joneses' security holdings

5. Do not add an apostrophe to personal pronouns.

   Theirs   ours   its

Contractions

Use the apostrophe to indicate that two or more words have been condensed into one. As a general rule, do not use contractions in formal reports and business letters.

I'll = I will   should've = should have   it's = it is   they're = they are

Plurals

When you indicate the plurals of letters, abbreviations, and numbers, use apostrophes only to avoid confusion. This recommendations follows *Chicago* (353–354).

1. Do not use apostrophes to form the plurals of letters.

Xs   Ys   Zs

2. Do not use apostrophes to form the plurals of abbreviations and numbers.

JPEGs   2010s

3. Use apostrophes to form the possessive of abbreviations.

OSHA's decision

Brackets

Brackets indicate that the writer has changed or added words or letters inside a quoted passage.

According to the report, "The detection distance [5 cm] fulfills the criterion."

Colons

Use colons:

1. To separate an independent clause from a list of supporting statements or examples.

The jointer has three important parts: the infeed table, the cutterhead, and the outfeed table.

2. To separate two independent clauses when the second clause explains or amplifies the first.

The original problem was the efficiency policy: We were producing as many parts as possible, but we could not use all of them.

Commas

Use commas:

1. To separate two main clauses connected by a coordinating conjunction (*and, but, or, nor, for, yet,* or *so*). Omit the comma if the clauses are very short.

Two main clauses The Atlas carousel has a higher base price, but this price includes installation and tooling costs.

2. To separate introductory subordinate clauses or phrases from the main clause.

Clause If the background is too dark, change the setting.

Phrases As shown in the table, the new system will save us over a million dollars.

3. To separate words or clauses in a series.

Words Peripheral components include scanners, external hard drives, and external fax/modems.

Phrases With this program you can send the fax at 5 p.m., at 11 p.m., or at a time you choose.

Clauses Select equipment that has durability, that requires little maintenance, and that the company can afford.

4. To set off nonrestrictive appositives, phrases, and clauses.

Appositive AltaVista, a Web search engine, has excellent advanced search features.

Phrase The bottleneck, first found in a routine inspection, will take a week to fix.

Clause The air flow system, which was installed in 1979, does not produce enough flow at its southern end.

Dashes and parentheses also serve this function. Dashes emphasize the abruptness of the interjected words; parentheses deemphasize the words.

5. To separate coordinate but not cumulative adjectives.

Coordinate He rejected the distorted, useless recordings.

Coordinate adjectives modify the noun independently. They could be reversed with no change in meaning: *useless, distorted recordings.*

Cumulative An acceptable frequency–response curve was achieved.

Cumulative adjectives cannot be reversed without distorting the meaning: *frequency–response acceptable curve.*

6. To set off conjunctive adverbs and transitional phrases.

Conjunctive
adverbs

The vice-president, however, reversed the recommendation.

The crane was very expensive; however, it paid for itself in 18 months.

Therefore, a larger system will solve the problem.

Transitional
phrases

On the other hand, the new receiving station is twice as large.

Performance on Mondays and Fridays, for example, is far below average.

Dashes

You can use dashes before and after interrupting material and asides. Dashes give a less formal, more dramatic tone to the material they set off than commas or parentheses do. The dash has four common uses:

1. To set off material that interrupts a sentence with a different idea

The fourth step—the most crucial one from management's point of view—is to ring up the folio and collect the money.

2. To emphasize a word or phrase at the end of a sentence

The Carver CNC has a range of 175–200 parts per hour—not within the standard.

3. To set off a definition

The total time commitment—contract duty time plus travel time—cannot exceed 40 hours per month.

4. To introduce a series less formally than with a colon

This sophisticated application allows several types of instruction sets—stacks, queues, and trees.

Parentheses

You can use parentheses before and after material that interrupts or is some kind of aside in a sentence or paragraph. Compared to dashes, parentheses have one of two effects: They deemphasize the material they set off, or they give a more formal, less dramatic tone to special asides. Parentheses are used in three ways:

1. To add information about an item.

Acronym for a
lengthy phrase A
definition

This Computer Numerically Controlled (CNC) lathe costs $20,000.

The result was long manufacturing lead times (the total time from receipt of a customer order until the product is produced).

Precise technical data   This hard drive (20GB, 5400 rpm Ultra ATA/66) can handle all of our current and future storage needs.

2. To add an aside to a sentence.

The Pulstrider has wheels, which would make it easy to move the unit from its storage site (the spare bedroom) to its use site (the living room, in front of the TV).

3. To add an aside to a paragraph.

The current program provides the user with the food's fat content percentage range by posting colored dots next to the menu item on a sign in the serving area. To determine the percentage of fat in foods, one must match the colored dot to dots on a poster hanging in the serving area. The yellow dot represents a range of 30%–60%. (The green dot is 0%–29% and the red dot is 61%–100%.) This yellow range is too large.

A Note on Parentheses, Dashes, and Commas

All three of these punctuation marks may be used to separate interrupting material from the rest of the sentence. Choose dashes or parentheses to avoid making an appositive seem like the second item in a series.

Commas are confusing   The computer has an input device, a keyboard and an output device, a monitor.

Parentheses are clearer   The computer has an input device (a keyboard) and an output device (a monitor).

Commas are confusing   The categories that have the highest dollar sales increase, sweaters, outerware, and slacks, also have the highest dollar per unit cost.

Dashes are clearer   The categories that have the highest dollar sales increase—sweaters, outerware, and slacks—also have the highest dollar per unit cost.

Ellipsis Points

Ellipsis points are three periods used to indicate that words have been deleted from a quoted passage.

According to Jones (1999), "The average customer is a tourist who . . . tends to purchase collectibles and small antiques" (p. 7).

Hyphens

Use hyphens to make the following connections:

1. The parts of a compound word when it is an adjective placed before the noun.

high-frequency system   plunger-type device   trouble-free process

Do not hyphenate the same adjectives when they are placed after the word:

The system is trouble free.

2. Words in a prepositional phrase used as an adjective.

state-of-the-art printer

3. Words that could cause confusion by being misread.

energy-producing cell   eight-hour shifts   foreign-car buyers
cement-like texture

4. Compound modifiers formed from a quantity and a unit of measurement.

a three-inch beam   an eight-mile journey

Unless the unit is expressed as a plural:

a beam 3 inches wide   a journey of 8 miles

Also use a hyphen with a number plus *-odd*.

twenty-odd

5. A single capital letter and a noun or participle.

A-frame   I-beam

6. Compound numbers from 21 through 99 when they are spelled out and fractions when they are spelled out.

Twenty-seven jobs required a pickup truck.   three-fourths

7. Complex fractions if the fraction cannot be typed in small numbers.

1-3/16 miles

Do not hyphenate if the fraction can be typed in small numbers.

1½ hp

8. Adjective plus past participle (*-ed, -en*).

red-colored table

9. Compounds made from *half-, all-,* or *cross-*.

half-finished   all-encompassing   cross-country

10. Use suspended hyphens for a series of adjectives that you would ordinarily hyphenate.

10-, 20-, and 30-foot beams

11. Do not hyphenate:

    a. *-ly* adverb-adjective combinations:

        recently altered system

    b. *-ly* adverb plus participle (*-ing, -ed*):

        highly rewarding positions   poorly motivated managers

    c. chemical terms

        hydrogen peroxide

    d. colors

        red orange logo

12. Spell as one word compounds formed by the following prefixes:

anti-	co-	infra-
non-	over-	post-
pre-	pro-	pseudo-
re-	semi-	sub-
super-	supra-	ultra-
un-	under-	

Exceptions: Use a hyphen

    a. When the second element is capitalized (*pre-Victorian*).

    b. When the second element is a figure (*pre-1900*).

    c. To prevent possible misreadings (*re-cover, un-ionized*).

Quotation Marks

Quotation marks are used at the beginning and at the end of a passage that contains the exact words of someone else.

    According to Jones (1999), "The average customer is a tourist who travels in the summer and tends to purchase collectibles and small antiques" (p. 7).

Semicolons

Use semicolons in the following ways:

1. To separate independent clauses not connected by coordinating conjunctions (*and, but, or, nor, for, yet, so*).

    Our printing presses are running 24 hours a day; we cannot stop the presses even for routine maintenance.

2. To separate independent clauses when the second one begins with a conjunctive adverb (*therefore, however, also, besides, consequently, nevertheless, furthermore*).

> Set-up time will decrease 10% and materials handling will decrease 15%; consequently, production will increase 20%.

3. To separate items in a series if the items have internal punctuation.

> Plans have been proposed for Kansas City, Missouri; Seattle, Washington; and Orlando, Florida.

Underlining (Italics)

Underlining is a line drawn under certain words. In books and laser-printed material, words that you underline when typing appear in italics. Italics are used for three purposes:

1. To indicate titles of books and newspapers.

> *Thriving on Chaos*   the San Francisco *Examiner*

2. To indicate words used as words or letters used as letters.

> That logo contains an attractive *M*.

> You used *there are* too many times in this paper.

> *Note:* You may also use quotation marks to indicate words as words.

> You used "there are" too many times in this paper.

3. To emphasize a word.

> Make sure there are no empty spaces on the contract and that all the blanks have been filled in *before* you sign.

Abbreviations, Capitalization, and Numbers

Abbreviations

Use abbreviations only for long words or combinations of words that must be used more than once in a report. For example, if words such as *Fahrenheit* or phrases such as *pounds per square inch* must be used several times in a report, abbreviate them to save space. Several rules for abbreviating follow (*Chicago*).

1. If an abbreviation might confuse your reader, use it and the complete phrase the first time.

> This paper will discuss materials planning requirements (MPR).

2. Use all capital letters (no periods, no space between letters or symbols) for acronyms.

   NASA   NAFTA   COBOL   HUD   PAC

3. Capitalize just the first letter of abbreviations for titles and companies; the abbreviation follows with a period.

   Pres.   Co.

4. Form the plural of an abbreviation by adding just *s*.

   BOMs   VCRs   CRTs

5. Omit the period after abbreviations of units of measurement. Exception: use *in.* for *inch.*

6. Use periods with Latin abbreviations.

   e.g. (for example)   i.e. (that is)   etc. (and so forth)

7. Use abbreviations (and symbols) when necessary to save space on visuals, but define difficult ones in the legend, a footnote, or the text.

8. Do not capitalize abbreviations of measurements.

   10 lb   12 m   14 g   16 cm

9. Do not abbreviate units of measurement preceded by approximations.

   several pounds per square inch   15 psi

10. Do not abbreviate short words such as *acre* or *ton*. In tables, abbreviate units of length, area, volume, and capacity.

Capitalization

The conventional rules of capitalization apply to technical writing. The trend in industry is away from overcapitalization.

1. Capitalize a title that immediately precedes a name.

   Senior Project Manager Jones

   But do not capitalize it if it is generic.

   The senior project manager reviewed the report.

2. Capitalize proper nouns and adjectives.

   Asia   American   French

3. Capitalize trade names, but not the product.

   Apple computers   Cleanall window cleaner

4. Capitalize titles of courses and departments and the titles of majors that refer to a specific degree program.

   The first statistics course I took was Statistics 1.

   I majored in Plant Engineering and have applied for several plant engineering positions.

5. Do not capitalize after a colon.

   The chair has four parts: legs, seat, arms, and back.

   I recommend the XYZ lathe: it is the best machine for the price.

Numbers

The following rules cover most situations, but when in doubt whether to use a numeral or a word, remember that the trend in report writing is toward using numerals.

1. Spell out numbers below 10; use figures for 10 and above.

   four cycles   1835 members

2. Spell out numbers that begin sentences.

   Thirty employees received safety commendations.

3. If a series contains numbers above and below 10, use numerals for all of them.

   The floor plan has 2 aisles and 14 workstations.

4. Use numerals for numbers that accompany units of measurement and time.

   1 gram   0.452 minute

   7 yards   6 kilometers

5. In compound-number adjectives, spell out the first one or the shorter one to avoid confusion.

   75 twelve-volt batteries

6. Use figures to record specific measurements.

   He took readings of 7.0, 7.1, and 7.3.

7. Combine figures and words for extremely large round numbers.

   2 million miles

8. For decimal fractions of less than 1, place a zero before the decimal point.

0.613

9. Express plurals of figures by adding just *s*.

21s   1990s

10. Place the last two letters of the ordinal after fractions used as nouns:

$^{1}/_{10\text{th}}$ of a second

But not after fractions that modify nouns:

$^{1}/_{10}$ horsepower

11. Spell out ordinals below 10.

fourth part   eighth incident

12. For 10 and above, use the number and the last two letters of the ordinal.

11th week   52nd contract

Works Cited

The Chicago Manual of Style: The Essential Guide for Writers, Editors, and Publishers. 16th ed. Chicago: University of Chicago Press, 2010. Print.

Documenting Sources

APPENDIX CONTENTS

Documenting your sources means following a citation system to indicate whose ideas you are using. Two methods are commonly used: the American Psychological Association (APA) system, and the Modern Language Association (MLA) system. All will be explained briefly. For more complete details, consult:

- *Publication Manual of the American Psychological Association* (6th ed.) http://apastyle.org
- *MLA Handbook for Writers of Research Papers* (7th ed). http://www.mlahandbook.org

How In-text Citation Works

Each method has two parts: the in-text citations and the bibliography, labeled "References" (APA) or "Works Cited" (MLA). The in-text citation works in roughly the same manner in both methods. The author places certain important items of information in the text to tell the reader which entry in the bibliography is the source of the quotation or paraphrase. These items could be the author's last name, the date of publication, or the title of an article.

In the APA method, the basic items are the author's last name and the year of publication. In the MLA method, the basic item is the author's last name and sometimes the title of the work, often in shorthand form.

In each method, the number of the page on which the quotation or paraphrase appears goes in parentheses immediately following the cited material. Because the rest of the methods vary, the rest of this appendix explains each.

Here is an example of how each method would cite the following quotation from page 73 of the article "The Seas of Arabia," by Kennedy Warne and published in the magazine *National Geographic* in March 2012.

Today human hands are reaching deep into Arabia's seas and taking more treasure than the seas can possibly replenish. Overfishing, pollution, seabed dredging, and massive coastal modification are crippling marine ecosystems by degrading water quality and exacerbating toxic algal blooms.

APA Method

The APA method requires that you use just the author's last name and include the year of publication and a page number.

According to Warne (2012), "Overfishing, pollution, seabed dredging, and massive coastal modification are crippling marine ecosystems by degrading water quality and exacerbating toxic algal blooms" (p. 73).

To find all bibliographic information on the quotation, you would refer to "Warne" in the References section.

Warne, K. (2012, March). The Seas of Arabia. *National Geographic,* 66–88.

MLA Method

The MLA method of citing the passage requires that you should include at least the author's last name with the page number.

As the author notes, "Overfishing, pollution, seabed dredging, and massive coastal modification are crippling marine ecosystems by degrading water quality and exacerbating toxic algal blooms" (Warne 73).

To find all the publication information for this quotation, you would refer to "Warne" in the Works Cited list.

Warne, Kennedy. "The Seas of Arabia." *National Geographic* Mar. 2012: 66–88.
    Print.

The "Extension" Problem

A common problem with internal documentation is indicating where the paraphrased material begins and ends. If you start a paragraph with a phrase like "According to Warne," you need to indicate which of the sentences that follow come from Warne. Or if you end a long paragraph with a parenthetical citation (Warne 73), you need to indicate which preceding sentences came from Warne. To alleviate confusion, place a marker at each end of the passage. Either use the name at the start and page numbers at the end or use a term like "one authority" at the start and the citation at the end.

According to Warne, human activity is having negative impacts on Arabia's seas. The people are taking more from the seas than is sustainable. Due to this activity there is an increase in algae problems which decreases the quality of the water (73).

> One authority explains that human activity is having negative impacts on Arabia's seas. The people are taking more from the seas than is sustainable. Due to this activity there is an increase in algae problems which decreases the quality of the water (Warne 73).

The APA Method

APA Citations

Once you understand the basic theory of the method—to use names and page numbers to refer to the References—you need to be aware of the variations possible in placing the name in the text. Each time you cite a quotation you give the page number preceded by *p.* or *pp.* Do not use *pg.* If you are paraphrasing, page numbers are not required. The following variations are all acceptable.

1. The author's name appears as part of the introduction of the quotation or paraphrase.

   > As Warne (2012) noted, "Overfishing, pollution, seabed dredging, and massive coastal modification are crippling marine ecosystems by degrading water quality and exacerbating toxic algal blooms" (p. 73).

2. The author is not named in the introduction to the quotation or paraphrase.

   > It is noted that "Overfishing, pollution, seabed dredging, and massive coastal modification are crippling marine ecosystems by degrading water quality and exacerbating toxic algal blooms" (Warne, 2012, p. 73).

3. The author has several works listed in the References. If they have different dates, no special treatment is necessary; if an author has two works dated in the same year, differentiate them in the text and in the References with a lowercase letter after each date (2012a, 2012b).

   > Warne (2012a) notes that "Overfishing, pollution, seabed dredging, and massive coastal modification are crippling marine ecosystems by degrading water quality and exacerbating toxic algal blooms" (p. 73).

4. Paraphrases are handled like quotations. Give the author's last name, the date, and the appropriate page numbers.

   > Warne (2012) states that human activity is having negative impacts on Arabia's seas.

5. When citing block quotations (more than 40 words), the period is placed *before* the page in parentheses. Do not place the quotation marks before and

after a block quotation. Starting on a new line, indent the left margin a half inch and double-space. Do not indent the right margin.

According to Warne (2012),

> Overfishing, pollution, seabed dredging, and massive coastal modification are crippling marine ecosystems by degrading water quality and exacerbating toxic algal blooms. In 2010 a group of marine scientists described the region's most strategic waterway, the Persian Gulf, as "a sea in decline," bedeviled by a storm of malign influences. "If current trends continue," they wrote, we will "lose a unique marine environment." (p. 73)

6. If no author is given for the work, treat the title as the author using an abbreviated title for the parenthetical in-text citation, and list the full title first in the References.

In an unstable investment year, "TIPS, like other kinds of Treasuries, gained as investors rushed into investments perceived as safe" ("Kiplinger 25," 2012, p. 64).

The Kiplinger 25 Update. (2012, April). *Kiplinger's Personal Finance,* 64.

APA References

The references list (titled "References") contains the complete bibliographic information on each source you use. The list is arranged alphabetically by the last name of the author or the first important word of the title. Follow these guidelines.

- Present information for all the entries in this order: Author's name. Date. Title. Publication information.
- Double-space the entire list. Entries should have a hanging indent, with the second and subsequent lines indented.
- Use only the initials of the author's first and middle names. *Note:* Many local style sheets suggest using the full first name; if this is the style at your place, follow that style.
- Place the date in parentheses immediately after the name.
- Capitalize only the first word of the title and subtitle and proper nouns.
- The inclusion of *p.* and *pp.* depends on the type of source. In general, use *p.* and *pp.* when the volume number does not precede the page numbers (or for a newspaper article).
- Place the entries in alphabetical order.

Baron, N. (2008). *Always on: Language in an online and mobile world.* New York, NY: Oxford University Press, Inc.

Bauerlein, M. (2008). *The dumbest generation.* [Kindle version]. New York, NY: Penguin Group.

Brooks, K. (2011). Death to high school English. *Salon*. Retrieved from http://www.salon.com

▶ If there are two or more works by one author, arrange them chronologically, earliest first.

Baron, N. (2000).

Baron, N. (2008).

Several common entries are shown below.

Book with One Author

Crystal, D. (2008). *Txtng: The gr8 db8*. New York, NY: Oxford University Press.

▶ Capitalize the first word after the colon.
▶ Use postal abbreviations for states.

Book with Two Authors

Reinking, J., & von der Osten, R. (2012). Strategies for successful writing: A rhetoric, reader and research guide (9th ed.). New York, NY: Longman.

Book with Editors

St. Amant, K., &Sapienza, F. (Eds.). (2011). *Culture, communication, and cyber-space: Rethinking technical communication for international online environments*. Amityville, NY: Baywood.

Essay in an Anthology

Cacho, L. (2011). Racialized hauntings or the devalued dead. In G. K. Hong & R. A. Ferguson (Eds.), *Strange affinities: The sexual and gender politics of comparative racialization*. (pp. 25–52). Durham, NC: Duke University Press.

▶ Capitalize only the first word of the essay title and subtitle (and all proper nouns).
▶ Use *p*. with inclusive page numbers.

Corporate or Institutional Author

Teachers Insurance and Annuity Association-College Retirement Equities Fund. (2012, March 1). *Summary prospectus*. New York, NY: Author.

▶ When the author is also the publisher, write *Author* for the publisher.
▶ In the text, the first citation reads this way (Teachers Insurance and Annuity Association-College Retirement Equities Fund [TIAA-CREF], 2012, March 1). Subsequent citations read (TIAA-CREF, 2012, March 1).

▶ This entry could also read

> *Summary prospectus.* (2012, March 1). New York, NY: Teachers Insurance and
> Annuity Association-College Retirement Equities Fund.

▶ Cite this entry as (*Summary prospectus*).

Work Without Date or Publisher

> Radke, J. (n.d.). *Writing for electronic sources.* Atlanta, GA: Center for Electronic
> Communication.

Brochure or Pamphlet

> *Explore the outdoors in the summer program at the child and family study center*
> [Brochure]. (2012). Menomonie, WI: Child and Family Study Center.

▶ Treat brochures like books.
▶ Place any identification number after the title.
▶ Place the word *Brochure* in brackets.
▶ This entry could also read

> Child and Family Study Center. (2012). *Explore the outdoors in the summer pro-*
> *gram at the child and family study center* [Brochure]. Menomonie, WI:
> Author.

▶ In the text, reference this entry as (Child).

Later Edition of a Book

> American Psychological Association. (2010). *Publication manual of the American*
> *psychological association* (6th ed.). Washington, DC: Author.

Article in a Journal with Continuous Pagination

> Carr, B., Haas, C., & Takayoshi, P. (2011, July). Building and maintaining con-
> texts in interactive networked writing: An examination of deixis and
> intertextuality in instant messaging. *Journal of Business and Technical*
> *Communication, 25,* 276-298.

Article in a Journal without Continuous Pagination

> Cook, D. (2010). Dr. Johnson's heart. *The Cambridge Quarterly, 39*(2), pp.
> 186–195.

▶ Put the issue number in parentheses after the volume.
▶ You could also give the month or season, if that helps identify the work:
(2011, Summer; 2012, January).

Article in a Monthly or Weekly Magazine

Kiplinger, K. (2012, April). Straight Talk on Taxes. *Kiplinger's Personal Finance, 22.*

▶ If the article has discontinuous pages, a comma indicates a break in sequence (4, 22–23).

Article from an Online Periodical

An article from an online periodical is similar to a hard copy article. Include the volume number and digital object identifier (DOI) when available. If the DOI is not listed, include the URL of the retrieval homepage:

Parry, M. (2012, March 7). Could many universities follow Borders bookstores into oblivion? *The Chronicle of Higher Education.* Retrieved from http://www.chronicle.com

▶ *Note:* The DOI, if available, would be put in the place of the retrieval URL.

Newspaper Article

Melo, F. (2012, April 12). Echos of the old roar: The renovation of Union Depot nears completion, with officials hoping it can be a transit hub again. *St. Paul Pioneer Press,* p. B1.

▶ *Note:* If the article has multiple pages, use *pp.* (pp. B1, B15).

Personal Communication

Note: The *Publication Manual of the American Psychological Association* suggests that personal communication—private letters, memos, e-mail, personal interviews and telephone conversations—should appear only in the text and not in the References. However, because these entries might be critical in research reports, a suggested form for their use in the References is given here.

1. In the text, reference personal communication material this way:

T. Henry (personal communication) suggests that . . .

2. In the References, enter it this way:

Henry, T. (2011, February 21). [Personal communication]

▶ Arrange the date so the year is first.
▶ If the person's title is pertinent, place it in brackets.

Henry, T. (2011, February 21). [Personal communication, Founder, Accidental Creative, Cincinnati, OH].

Professional or Personal Website—Homepage

Essential for citing webpages is that you give the URL where you retrieved the site. Give as much other information as possible.

> Henry, T. (n.d.). *Accidental Creative*. Retrieved from: http://www.accidentalcreative.com

> ❱ Cite this version as (Henry).

Explanation: Web owner; if available. (Date of the last update, if available). Title of article or document. Title of website. The site and the site's URL. *Note:* If the owner and the date are not available, the above entry would look like this:

> *Accidental creative.* (n.d.). Retrieved from http://www.accidentalcreative.com Cite this version as ("Accidental").

Professional or Personal Website—Internal Page

For an internal page of a website, give the URL of the document, not the homepage.

> Henry, T. (n.d.). Would you kill your "precious"? *Accidental creative.* Retrieved from http://www.accidentalcreative.com/productivity/would-you-kill-your-precious

Explanation: Web owner, if available. Date of the last updating, if available—note that internal page updates and homepage updates can be different; use the date of the page whose information you use. Title that appears on the document page. *Title that appears on that homepage.* The URL.

Online Forums and Discussion Group Messages

> Gould, D. (2012, March 2). Re: Would you kill your "precious"? [Online forum comment]. Retrieved from http://www.accidentalcreative.com/productivity/would-you-kill-your-precious#comment-454491121

> ❱ If available, use the author's last name followed by initials. If not, use the screen name.

Blog Post

A blog post, or weblog post, is cited similarly to an online forum or discussion group post. Use this form:

> Henry, T. (2012, February 29). I'm writing another book! (why!?!). [Web log post]. Retrieved from http://www.toddhenry.com/miscellaneous/im-writing-another-book-why

As with an online forum, or discussion group post, comments to blogs also use the author's full name, if available, screen name otherwise. Use this form:

McBreen, C. (2012, March 2). Re: I'm writing another book! (why!?!). [Web log comment]. Retrieved from http://www.toddhenry.com/miscellaneous/ im-writing-another-book-why

Podcast

Henry, T. (Producer). (2012). *AC podcast: Peter Bregman* [Audio podcast]. Retrieved from http://www.accidentalcreative.com/ac-podcast-peter-bregman-on-18-minutes

Video Blog Post (i.e. YouTube)

Dretzin, R. (2010, January 12). PBS frontline digital nation: Going digital at 83 [Video file]. Retrieved from http://www.youtube.com/watch?v=CoJOHu-eieI&feature=channel

Social Media Posts—Twitter and Facebook

Note: The *Publication Manual of the American Psychological Association* currently does not have a standard for citing social media posts such as Tweets and Facebook statuses. However, this type of entry may be necessary to include. A suggested form for such entries is given here.

1. In the text, reference social media posts this way:

   nprnews, (tweet) states that . . .

2. In the references section, list it this way:

   nprnews. Digital Technologies Give Dying Languages New Life n.pr/FS72Kd. 2012, March 19.2:50 PM. Tweet.

   ▶ If available, use the author's last name followed by first initial and put the username following in parentheses. Otherwise, just use the username.
   ▶ For a Facebook post, just put *Facebook* in place of *Tweet* at the end—likewise for other social media applications.

e-Book

Maushart, S. (2011). *The winter of our disconnect* [Kindle version]. Retrieved from http://www.amazon.com

The MLA Method

The following section describes variations in MLA citation and explains entries in the MLA Works Cited section.

MLA Citations

Once you understand the basic theory of the method—to use names and page numbers to refer to the Works Cited—you need to be aware of the possible variations of placing the name in the text. In this method, unlike APA, each time you refer to a quotation or paraphrase, you give the page number only; do not use *p.* or *pg.*

1. The author's name appears as part of the introduction to the quotation or paraphrase.

   > As Warne notes, "Overfishing, pollution, seabed dredging, and massive coastal modification are crippling marine ecosystems by degrading water quality and exacerbating toxic algal blooms" (73).

2. Author is not named in introduction to quotation.

   > What seems quite evident is that "Overfishing, pollution, seabed dredging, and massive coastal modification are crippling marine ecosystems by degrading water quality and exacerbating toxic algal blooms" (Warne 73).

3. Author has several sources in the Works Cited.

   > Warne points out that "Overfishing, pollution, seabed dredging, and massive coastal modification are crippling marine ecosystems by degrading water quality and exacerbating toxic algal blooms" (*Seas*, 73).

4. Paraphrases are usually handled like quotations. Give the author's last name and the appropriate page numbers.

   > Warne states that human activity is having negative impacts on Arabia's seas (73).

5. In block quotations (more than four typed lines), place the period before the page parentheses. Do not place quotation marks before and after the block quotation. Indent the left margin 10 spaces and double-space. Do not indent the right margin.

   > According to Warne,
   >> Overfishing, pollution, seabed dredging, and massive coastal modification are crippling marine ecosystems by degrading water quality and exacerbating toxic algal blooms. In 2010 a group of marine scientists described the region's most strategic waterway, the Persian Gulf, as "a sea in decline," bedeviled by a storm of malign influences. "If current trends continue," they wrote, we will "lose a unique marine environment." (73)

6. If no author is given for the work, treat the author as the title because the title is listed first in the Works Cited list.

   > In an unstable investment year, "TIPS, like other kinds of Treasuries, gained as investors rushed into investments perceived as safe" ("Kiplinger" 64).

MLA Works Cited List

The Works Cited list contains the complete bibliographic information on each source you use. The list is arranged alphabetically by the last name of the author or, if no author is named, by the first important word of the title.

Follow these guidelines:

▶ Begin the list on a new page
▶ Present information for all the entries in this order: Author's name. Title. Publication information (including date).
▶ Capitalize the first letter of every important word in the title.
▶ Enclose article titles in quotation marks.
▶ Double-space an entry if it has two or more lines.
▶ Indent the second and succeeding lines one half inch.
▶ If the author appears in the Works Cited list two or more times, type three hyphens and a period instead of repeating the name for the second and succeeding entries. Alphabetize the entries by the first word of the title.

Several common entries appear below. For more detailed instructions, use the *MLA Handbook,* 7th ed., (New York: MLA, 2009).

Book with One Author

Crystal, David. *Txtng: The Gr8 DB8*. New York: Oxford UP, 2008. Print.

▶ Only the name of the publishing company needs to appear: You may drop "Co." or "Inc."

Book with Two Authors

Reinking, James A., and Robert von der Osten. *Strategies for Successful Writing: A Rhetoric, Reader and Research Guide* 9th ed. New York: Longman, 2012. Print.

A long title may be shortened in the text, in this case to *Strategies*.

Book with Editors

St. Amant, Kirk, and Filipp Sapienza, eds. *Culture, Communication, and Cyberspace: Rethinking Technical Communication for International Online Environments*. Amityville: Baywood, 2011. Print.

Essay in an Anthology

Cacho, Lisa M. "Racialized Hauntings or the Devalued Dead." *Strange Affinities: The Sexual and Gender Politics of Comparative Racialization*. Ed. Grace K. Hong and Roderick A. Ferguson. Durham: Duke UP, 2011. 25–52. Print.

▶ In the text, both the article title and the book title may be shortened; for example the article title could be "Racialized Hauntings" and the book title could be "Strange Affinities."

Corporate or Institutional Author

Teachers Insurance and Annuity Association-College Retirement Equities Fund. *Summary Prospectus.* New York: TIAA-CREF, 2012. Print.

▶ In the text, the first citation reads this way (Teachers Insurance and Annuity Association-College Retirement Equities Fund [TIAA-CREF]). Subsequent citations read (TIAA-CREF).

▶ This entry could also read

Summary Prospectus. New York: Teachers Insurance Annuity Association-College Retirement Equity Fund, 2012. Print.

▶ Cite this entry as (*Summary*).

Work without Date or Publisher

Radke, Jean. *Writing for Electronic Sources.* Atlanta: Center for Electronic Communication, n.d.

▶ Use *n.p.* for no publisher or no place.

Brochure or Pamphlet

Explore the Outdoors in the Summer Program at the Child and Family Study Center. Menomonie: Child and Family Study Center, 2012. Print.

Later Edition of a Book

American Psychological Association. *Publication Manual of the American Psychological Association.* 6th ed. Washington, DC: APA, 2010. Print.

Article in a Journal with Continuous Pagination

Carr, Brandon J., Christina Haas, and Pamela Takayoshi. "Building and Maintaining Contexts in Interactive Networked Writing: An Examination of Deixis and Intertextuality in Instant Messaging." *Journal of Business and Technical Communication* 25 (2011): 276–98. Print.

Article in a Journal Without Continuous Pagination

Cook, Daniel. "Dr. Johnson's Heart." *The Cambridge Quarterly* 39.2 (2010): 186–95. Print.

Article in a Monthly or Weekly Magazine

Kiplinger, Knight A. "Straight Talk on Taxes." *Kiplinger's Personal Finance* Apr. 2012: 22. Print.

▶ If the article has discontinuous pages, give the first page only, followed by a plus sign: 22+.

Newspaper Article

Melo, Frederick. "Echos of the old roar: The renovation of Union Depot nears completion, with officials hoping it can be a transit hub again." *St. Paul Pioneer Press* 12 Apr. 2012, B1. Print.

◗ Identify the edition, section, and page number: A reader should be able to find the article on the page.
◗ Omit the definite article (*the*) in the title of the newspaper in the text of the article: If the newspaper is a city newspaper and the city is not given in the title, supply it in brackets after the title (e.g. *Guardian* [London]).

Personal or Telephone Interview

◗ In the text, interviews are cited like any other source: (Henry).
◗ In the Works Cited list, enter it this way:

Henry, Todd. Personal interview. 21 Feb. 2012.

◗ If the person's title or workplace are important, add them after the name:

Henry, Todd, Founder, Accidental Creative, Cincinnati, OH. Telephone interview. 22 Feb. 2012.

Personal Letter or E-mail

Henry, Todd. Letter to author. 23 Feb. 2012.

◗ For E-mail messages include the subject (if available) and the recipient.

Henry, Todd. "Re: Latest Blog Post." Message to the author. 24 Feb. 2012. E-mail.

Professional or Personal Website

Essential for citing webpages is that you give the date of the retrieval on which you viewed the site. MLA no longer requires the URL; however, if you feel it is necessary to include, place it after the date inside angle brackets like this:

<http://www.accidentalcreative.com>

Otherwise, cite websites like this:

Henry, Todd. *Accidental Creative*. N.p., n.d.Web. 11 Apr. 2012.

Explanation: Web owner, if available. *Title of the Website*. Publisher or sponsor (if available). Date of last update (if available). Date of retrieval.

Note: Use this same format for blog posts.

Professional or Personal Webpage

For an internal page of a website, include the title of the page and the title of the overall website.

> Henry, Todd. "Would You Kill Your 'Precious'?" *Accidental Creative*. N.p., n.d. Web. 11 Apr. 2012.

Explanation: Author (if available). "Title of Internal Page." *Title of the Entire Website* (from the homepage). Publisher or Sponsor (if available). Date of publication (if available), Web. Date of retrieval.

Note: Use this same format for online forums, discussion group messages, blog post comments, podcasts accessed from the Web and video blog posts (i.e., YouTube).

Social Media Posts—Twitter and Facebook

> nprnews. "Digital Technologies Give Dying Languages New Life n.pr/FS72Kd." 19 Mar. 2012, 2:50 PM. Tweet.

▶ If available, use the author's name (last name, first name) with the username following in parentheses. Otherwise, just use the username.

e-Books

> Maushart, Susan. *The Winter of Our Disconnect.* 2011. Amazon.com. Kindle file.

Explanation: Author. *Title.* Publication date. File location. Medium of file.

Article Available from an Online Source

Many libraries and companies use online services like EBSCO host to find full-text articles. In your text, cite the full-text articles by using the author's last name. Include page numbers when available. An entry in the Works Cited section would look like this:

> Yu, Han. "Intercultural Competence in Technical Communication: A Working Definition and Review of Assessment Methods." *Technical Communication Quarterly* 21.2 (2012): 168–86. *EBSCOhost.* Web. 11 Apr. 2012.

Explanation: Author. "Title of Article." *Title of the Hard-Copy Periodical* date of original publication: page numbers, if available. *Title of the database.* Web. Date of access.

Article Available from an Online Periodical

Treat an article from an on-line periodical like a hard copy article. Note that you must add the medium (Web) and date of retrieval.

> Wortham, Jenna. "A Billion-Dollar Turning Point for Mobile Apps." *New York Times.* The New York Times Company, 10 Apr. 2012. Web. 11 Apr. 2012.

Explanation: Author. "Title of Article." *Title of On-Line Periodical* date of original publication. Web. Date of retrieval.

Examples

The following examples present are two sample papers, one in each of the formats described here.

TEXTING AND LITERACY

Demographics of a Texter

Who is texting? Is it only the "younger" generation? According to a Pew Research Center Publications article, "Teens, Cell Phones and Texting: Text Messaging Becomes a Centerpiece Communication," "Cell-phone texting has become the preferred channel of basic communication between teens and their friends . . . [s] ome 75% of 12–17 year olds now own cell phones. Fully 72% of all teens or 88% of teen cell phone users—are text-messagers." The study also shows that "One in three teens sends more than 100 text messages a day . . . [and] half of teens send 50 or more text messages a day" (Lenhart, 2010). However, teens are not the only group sending text messages. David Crystal's (2008) book, *Txtng: The Gr8 Db8,* describes a study in the UK saying, "80 percent of under-25s texted rather than called. On the other hand, so did 14 percent of people over 55." This clearly shows that texting is not just a passing teenage fad. Table 4 outlines the texting trends in the United States reported in September 2011 by the Pew Research Center.

TABLE 4
2011 Texting Trends

Age Range	Mean Texts/Day	Median Texts/Day
18–24	50	109.5
25–34	20	41.8
35–44	10	25.9
45–54	6	14
55–64	2	9.8
65+	2	4.7

Crystal (2008) goes on to discuss the global trend in texting replacing traditional phone calls. In the article "Don't Call Me, I Won't Call You," published in the March 18, 2011 edition of the *New York Times*, Pamela Paul writes how recently "full-fledged adults have seemingly given up the telephone—landline, mobile, voice mail and all. According to Nielsen Media, even on cellphones, voice spending has been trending downward, with text spending expected to surpass it within three years." Texting, however, is not just something for personal lives. Naomi Baron (2008) notes in her book *Always On* how texting has made its way into the workplace as well. Baron (2008) elaborates on the use of texting with virtual business teams, as well as within the office to maintain contact when one coworker may be at a lunch meeting or conference. Seeing the prevalence of texting across many genres of society it is apparent that it is affecting numerous facets of life—on a global scale.

Literacy Redefined

Literacy is no longer merely being able to read and write as traditionally done with paper and books. It has evolved to encompass a much larger scope of material forcing readers to maintain various different literacies to accommodate all of the technology media available to communicate with. Alan Porter (2010) talks about the newly defined literacy in his piece "Preparing for the Next Generation." Porter (2010) relates a first-hand account of watching his daughter doing research for a paper, using a text book, realizing, "she was 'browsing' just as if she were online" (p. 20). Tony Self (2009) reiterates this idea of a shift in reading preference in his article "What if Readers Can't Read?" Self (2009) discusses the present day reader's inability to focus on lengthy texts noting that "Dr. Bruce Friedman, Professor of Pathology at the University of Michigan, found that he has almost lost the ability to read and absorb a longish article on the web or in print" (p.13). Self (2009) also shows the dramatic shift in literacy through noting, "since 2006, New Zealand high school students have been permitted to use 'text speak' in national exams" (p. 12). This dramatic shift in literacy curriculum shows the heavy impact of redefined literacy on culture and society.

References

Baron, N. (2008). *Always on: Language in an online and mobile world* [Kindle version]. Retrieved from http://www.amazon.com

Crystal, D. (2008). *Txtng: the gr8 db8* [Kindle version]. Retrieved from http://www.amazon.com

Lenhart, A. (2010, April 20). Teens, Cell Phones and Texting: Text Messaging Becomes a Centerpiece Communication. *(Pew Internet & American Life Project).* Retrieved from Pew Research Center website: http://pewresearch.org/pubs/1572/teens-cell-phones-text-messages

Paul, P. (2011, March 18). Don't call me, I won't call you. *The New York Times.* Retrieved from http://www.nytimes.com

Porter, A. (2010, June). Preparing for the next generation. *Intercom*, 18–21.

Self, T. (2009, February). What if readers can't read? *Intercom*, 11–14.

Example B.2

Excerpt in
MLA Format

TEXTING AND LITERACY

Demographics of a Texter

Who is texting? Is it only the "younger" generation? According to a Pew Research Center Publications article, "Teens, Cell Phones and Texting: Text Messaging Becomes a Centerpiece Communication," "Cell-phone texting has become the preferred channel of basic communication between teens and their friends . . . [s]ome 75% of 12–17 year olds now own cell phones. Fully 72% of all teens or 88% of teen cell phone users—are text-messagers." The study also shows that "One in three teens sends more than 100 text messages a day . . . [and] half of teens send 50 or more text messages a day" (Lenhart). However, teens are not the only group sending text messages. David Crystal's book, *Txtng: The Gr8 Db8,* describes a study in the UK saying, "80 percent of under-25s texted rather than called. On the other hand, so did 14 percent of people over 55." This clearly shows that texting is not just a passing teenage fad. Table 4 outlines the texting trends in the United States reported in September 2011 by the Pew Research Center.

TABLE 4
2011 Texting Trends

Age Range	Mean Texts/Day	Median Texts/Day
18–24	50	109.5
25–34	20	41.8
35–44	10	25.9
45–54	6	14
55–64	2	9.8
65+	2	4.7

Crystal goes on to discuss the global trend in texting replacing traditional phone calls. In the article "Don't Call Me, I Won't Call You," published in the March 18, 2011 edition of the *New York Times*, Pamela Paul writes how recently "full-fledged adults have seemingly given up the telephone—landline, mobile, voice mail and all. According to Nielsen Media, even on cellphones, voice spending has been trending downward, with text spending expected to surpass it within three years." Texting, however, is not just something for personal lives. Naomi Baron notes in her book *Always On* how texting has made its way into the workplace as well. Baron elaborates on the use of texting with virtual business teams, as well as within the office to maintain contact when one co-worker may be at a lunch meeting or conference. Seeing the prevalence of texting across many genres of society it is apparent that it is affecting numerous facets of life—on a global scale.

Literacy Redefined

Literacy is no longer merely being able to read and write as traditionally done with paper and books. It has evolved to encompass a much larger scope of material forcing readers to maintain various different literacies to accommodate all of the technology media available to communicate with. Alan Porter talks about the newly defined literacy in his piece "Preparing for the Next Generation." Porter relates a first-hand account of watching his daughter doing research for a paper, using a text book, realizing, "she was 'browsing' just as if she were online" (20). Tony Self reiterates this idea of a shift in reading preference in his article "What if Readers Can't Read?" Self discusses the present day reader's inability to focus on lengthy texts noting that "Dr. Bruce Friedman, Professor of Pathology at the University of Michigan, found that he has almost lost the ability to read and absorb a longish article on the web or in print" (13). Self also shows the dramatic shift in literacy through noting, "since 2006, New Zealand high school students have been permitted to use 'text speak' in national exams" (12). This dramatic shift in literacy curriculum shows the heavy impact of redefined literacy on culture and society.

Work Cited

Baron, Naomi. *Always On: Language in an Online and Mobile World.* 2008. Amazon.com. Kindle file.

Crystal, David. *Txtng: The Gr8 Db8.* 2008. Amazon.com. Kindle file.

Lenhart, Amanda. "Teens, Cell Phones and Texting: Text Messaging Becomes a Centerpiece Communication." (Pew Internet & American Life Project). *Pew Research Center.* Pew Research Center. 20 Apr. 2010. Web. 10 Nov. 2011.

Paul, Pamela. "Don't Call Me, I Won't Call You." *New York Times.* The New York Times Company. 18 Mar. 2011. Web. 9 Nov. 2011.

Porter, Alan. "Preparing for the Next Generation." *Intercom.* June 2010:18–21. Print.

Self, Tony. "What If Readers Can't Read?" *Intercom.* Feb. 2009: 11–14. Print.

Exercises

1. Edit the following sentences to place an APA citation correctly and/or to place an MLA citation correctly.

   a. Kate Baggott discusses youth and texting citing associate professor Eric Paulson noting, "they can text 'IMHO' on their cell phones, write 'my opinion is' in a school essay, and read 'it is my belief that your scar hurts when Lord Voldemort is near you' without getting discombobulated" (p. 1, 2006).

b. In 2011, Ben Yagoda discusses in his piece *The Elements of Clunk*, published in the online edition of the Chronicle of Higher Education, how college students are not using abbreviated cryptic language in their writing, but rather they "make it longer and more prosaic. They give a new sound to prose. I call it clunk."

c. Todd Henry notes, on his website *Accidental Creative*, that the definition of an "Accidental Creative" is a "person who structures their life so as to experience frequent creative insights."

d. "Dr. Bruce Friedman, Professor of Pathology at the University of Michigan, found that he has almost lost the ability to read and absorb a longish article on the web or in print" according to Tony Self, a noted communication scholar, on p.13 in 2009.

2. Turn these sets of data into an APA References list and an MLA Works Cited list.

  a. **Author:** VladSavov
      Title: OMG, FYI, and LOL enter Oxford English dictionary, foreshadow the apocalypse
      Website Title: *AOL Tech*
      Publication Date: 24 Mar. 2011
      Retrieval Date: 11 Apr. 2012
      Sponsor/Publisher: AOL Inc.
      URL: <http:// www.engadget.com>

  b. **Author:** Mark Bauerlein
      Title: The Dumbest Generation
      Publication Date: 2008
      Publisher: Penguin Group
      Publisher Location: NewYork, NY

3. Rewrite the following excerpt using APA format:

  The ancient Greeks laid the foundations of present-day rhetoric. Since it was so infused in daily life, rules were established for the practice of good rhetoric. Aristotle contributes immensely to this study with his book *The Art of Rhetoric*. In his book, *An Introduction to Classical Rhetoric: Essential Readings*, James Williams outlines Aristotle's "artificial proofs [that] 'make a man a master of rhetorical argument' . . . *ethos* (character), *logos* (reason or the speech itself), and *pathos* (emotion)" (228). Of these three, *ethos* is likely the most influential regarding trust and credibility. Williams ties *ethos* with the idea of "ethical virtue" showing that "[the] moral character necessary to utilize *ethos* in a speech derives from living within the well-ordered community, from which one can learn justice and proper conduct, and from performing ethical actions as part of that community" (229). Trust was built simply on the presence and reputation of the rhetorician. This presence is established through a "knowledgeable, charismatic, savvy, precise, persuasive, polished, nonconfrontational, reflective, lyrical, and respectful" rhetorician (Aftat). Quite possibly, the audience could have determined the credibility of the speaker before a single word was uttered from their mouth. In *The Art of Rhetoric,* Aristotle says "[f]or the orator to produce conviction, three qualities are necessary . . . independently of demonstrations . . . [t]hese qualities are good sense, virtue, and goodwill" (240). Aristotle attempted to infuse his ideas of proper rhetoric into Greek society, seeing it as an important element in cultural tradition and growth.

Works Cited

Aftat, Mokhtar. "Rhetoric and Modern Life." *NCTE Ning.* NCTE, 12 Apr. 2010. Web. 28 Apr. 2010.

Williams, James. *An Introduction to Classical Rhetoric: Essential Readings.* Chichester, United Kingdom: Wiley-Blackwell, 2009. Print.

4. Rewrite the following excerpt using MLA format:

Since texting seems to be prevalent among school-aged children it is natural that educators would seek to find a correlation between texting habits and school performance. In a study done by Beverly Plester, Clare Wood and Victoria Bell (2008) they explore the texting habits of 11–12 year olds and the effect they have on the student's written language skill. As described in the article *Txt msg n school literacy: does texting and knowledge of text abbreviations adversely affect children's literacy attainment?* Plester et al. (2008) relate the study in which students were asked to translate between standard English and text messages. The study determined that children with high textism aptitude scored high on verbal reasoning and concluded that "good writing attainment was associated with greater use of textisms" (Plester et al., 2008, p. 1). In the end the research concluded that texting does not correlate to decreased written language skills for 11–12 year olds. While it is encouraging to know that texting is not detrimental to the writing of preteens, the level of writing done at this age is not something to be classified as scholarly writing demonstrating an acute grasp of the English language such as that possibly expected by a college student.

References

Plester, B., Wood, C., & Bell, V. (2008). Txt msg n school literacy: does texting and knowledge of text abbreviations adversely affect children's literacy attainment? [Abstract]. *Literacy, 42,* 137–144. doi: 10.1111/j.1741-4369.2008.00489.x

Writing Assignment

Find three sources to create a one-page grant proposal for something your company is seeking funding for. Synthesize the main points from each source to make a strong, cohesive proposal. Use either APA or MLA for your documenting.

Works Cited

American Psychological Association. *Publication Manual of the American Psychological Association*. 6th ed. Washington, DC: APA, 2010. Print.

Modern Language Association of America. *MLA Handbook for Writers of Research Papers*. 7th ed. New York: MLA, 2009. Print.

Index